Lecture Notes in Physics

Springer-Verlag Berlin Heidelberg GmbH

The Editorial Policy for Proceedings

The series Lecture Notes in Physics reports new developments in physical research and teaching – quickly, informally, and at a high level. The proceedings to be considered for publication in this series should be limited to only a few areas of research, and these should be closely related to each other. The contributions should be of a high standard and should avoid lengthy redraftings of papers already published or about to be published elsewhere. As a whole, the proceedings should aim for a balanced presentation of the theme of the conference including a description of the techniques used and enough motivation for a broad readership. It should not be assumed that the published proceedings must reflect the conference in its entirety. (A listing or abstracts of papers presented at the meeting but not included in the proceedings could be added as an appendix.)
When applying for publication in the series Lecture Notes in Physics the volume's editor(s) should submit sufficient material to enable the series editors and their referees to make a fairly accurate evaluation (e.g. a complete list of speakers and titles of papers to be presented and abstracts). If, based on this information, the proceedings are (tentatively) accepted, the volume's editor(s), whose name(s) will appear on the title pages, should select the papers suitable for publication and have them refereed (as for a journal) when appropriate. As a rule discussions will not be accepted. The series editors and Springer-Verlag will normally not interfere with the detailed editing except in fairly obvious cases or on technical matters.
Final acceptance is expressed by the series editor in charge, in consultation with Springer-Verlag only after receiving the complete manuscript. It might help to send a copy of the authors' manuscripts in advance to the editor in charge to discuss possible revisions with him. As a general rule, the series editor will confirm his tentative acceptance if the final manuscript corresponds to the original concept discussed, if the quality of the contribution meets the requirements of the series, and if the final size of the manuscript does not greatly exceed the number of pages originally agreed upon. The manuscript should be forwarded to Springer-Verlag shortly after the meeting. In cases of extreme delay (more than six months after the conference) the series editors will check once more the timeliness of the papers. Therefore, the volume's editor(s) should establish strict deadlines, or collect the articles during the conference and have them revised on the spot. If a delay is unavoidable, one should encourage the authors to update their contributions if appropriate. The editors of proceedings are strongly advised to inform contributors about these points at an early stage.
The final manuscript should contain a table of contents and an informative introduction accessible also to readers not particularly familiar with the topic of the conference. The contributions should be in English. The volume's editor(s) should check the contributions for the correct use of language. At Springer-Verlag only the prefaces will be checked by a copy-editor for language and style. Grave linguistic or technical shortcomings may lead to the rejection of contributions by the series editors. A conference report should not exceed a total of 500 pages. Keeping the size within this bound should be achieved by a stricter selection of articles and not by imposing an upper limit to the length of the individual papers. Editors receive jointly 30 complimentary copies of their book. They are entitled to purchase further copies of their book at a reduced rate. As a rule no reprints of individual contributions can be supplied. No royalty is paid on Lecture Notes in Physics volumes. Commitment to publish is made by letter of interest rather than by signing a formal contract. Springer-Verlag secures the copyright for each volume.

The Production Process

The books are hardbound, and the publisher will select quality paper appropriate to the needs of the author(s). Publication time is about ten weeks. More than twenty years of experience guarantee authors the best possible service. To reach the goal of rapid publication at a low price the technique of photographic reproduction from a camera-ready manuscript was chosen. This process shifts the main responsibility for the technical quality considerably from the publisher to the authors. We therefore urge all authors and editors of proceedings to observe very carefully the essentials for the preparation of camera-ready manuscripts, which we will supply on request. This applies especially to the quality of figures and halftones submitted for publication. In addition, it might be useful to look at some of the volumes already published. As a special service, we offer free of charge LaTeX and TeX macro packages to format the text according to Springer-Verlag's quality requirements. We strongly recommend that you make use of this offer, since the result will be a book of considerably improved technical quality. To avoid mistakes and time-consuming correspondence during the production period the conference editors should request special instructions from the publisher well before the beginning of the conference. Manuscripts not meeting the technical standard of the series will have to be returned for improvement.
For further information please contact Springer-Verlag, Physics Editorial Department II, Tiergartenstrasse 17, D-69121 Heidelberg, Germany

Series homepage – http://www.springer.de/phys/books/lnpp

Peter Breitenlohner Dieter Maison (Eds.)

Quantum Field Theory

Proceedings of the Ringberg Workshop Held at Tegernsee, Germany, 21-24 June 1998 On the Occasion of Wolfhart Zimmermann's 70th Birthday

Springer

Editors

Peter Breitenlohner
Dieter Maison
Werner-Heisenberg-Institut
Max-Planck-Institut für Physik
Föhringer Ring 6
80805 München, Germany

Library of Congress Cataloging-in-Publication Data applied for.

Die Deutsche Bibliothek - CIP-Einheitsaufnahme

Quantum field theory : proceedings of the Ringberg Workshop, held at Tegernsee, Germany, 21 - 24 June 1998 on the Occasion of Wolfhart Zimmermann's 70th Birthday / Peter Breitenlohner ; Dieter Maison (ed.). - Berlin ; Heidelberg ; New York ; Barcelona ; Hong Kong ; London ; Milan ; Paris ; Singapore ; Tokyo : Springer, 2000
(Lecture notes in physics ; Vol. 558)

ISSN 0075-8450

ISBN 978-3-662-14270-7 ISBN 978-3-540-44482-4 (eBook)
DOI 10.1007/978-3-540-44482-4

Springer-Verlag Berlin Heidelberg New York
a member of BertelsmannSpringer Science+Business Media GmbH

Typesetting: Camera-ready by the authors/editors
Cover design: *design & production*, Heidelberg

Printed on acid-free paper
SPIN: 10780725 55/3141/du - 5 4 3 2 1 0

Preface

On the occasion of Wolfhart Zimmermann's 70th birthday a conference on Quantum Field Theory was held on June 21–24, 1998 at the Ringberg Castle situated in the splendid scenery of Upper Bavaria.

Twelve invited speakers gave talks on topics related in one or the other way to the scientific work of Wolfhart Zimmermann. All participants agreed that the conference was a great success and we would like to thank the speakers for their interesting and well-prepared talks covering a wide range of interesting actual quantum field theoretical problems.

With great regret we had received the short-term cancellation of Harry Lehmann's participation due to his worsening health conditions. Harry Lehmann, like Wolfhart Zimmermann, one of the pioneers of modern QFT had planned to give a historical talk on the "Göttinger Feldverein", which we are sure would have been spiced with many personal anecdotes. Harry Lehmann died a few months later on Nov. 22, 1998.

In addition to the talks held at the conference we decided to include in these proceedings a small selection of Wolfhart Zimmermann's scientific papers, which proved fundamental to various important developments of modern QFT as there are the treatment of bound states in QFT, Renormalized Perturbation Theory, Composite Operators and Operator Product Expansions. We resisted the temptation to include the famous LSZ papers from the fifties, whose content has already entered in most of the standard text books on QFT.

We would like to thank the Max Planck Society for funding the conference and A. Hörmann and his crew at the Ringberg Castle for their warm hospitality and perfect organization.

Munich,
July 2000

Peter Breitenlohner
Dieter Maison

Contents

Anomalies

William A. Bardeen

Fermi National Accelerator Laboratory,
P.O. Box 500, Batavia, IL 60510, USA

Abstract. I discuss the role of anomalies in the modern development of quantum field theory and the implications for physics.

1 Introduction

Symmetries play an essential role in our understanding of elementary particle physics. Global symmetries in the form of conserved charges label the physical states and reflect the existence of conserved local currents. Local symmetries in the form of gauge field theories are used to describe practically all aspects of elementary particle physics phenomena and imply the existence of vector gauge fields coupled to conserved local currents.

In electromagnetism, the photons are the quanta of the electromagnetic gauge field. In the theory of electroweak interactions, the massive W and Z particles are the quanta of the electroweak gauge fields in addition to the massless photon. The strong dynamics of the quarks and gluons are controlled by the color interactions of the quantum chromodynamic gauge fields. Local Lorentz symmetries are used to describe the gravitational interactions.

In some cases the symmetries are not realized explicitly although these invisible symmetries still involve exact symmetries at the fundamental level. In quantum chromodynamics, the color confinement phenomena results from an exact local color gauge symmetry. However color confinement implies that there are no asymptotic states with color, such as the fundamental quarks and gluons, and only color singlet particles can be directly observed as isolated states.

Symmetries can also be dynamically broken without destroying the exact underlying symmetry. Spontaneous magnetization occurs when the spins in a material tend to align in a particular direction breaking the explicit rotational symmetry. This spontaneous breaking of the rotational spin symmetry implies the existence of spin waves which govern the long range fluctuations of the spins. Chiral symmetries reflect the independent rotations of the left and right handed components of fermions which is an exact symmetry of a gauge field theory of massless Dirac fermions as in the case of quantum electrodynamics with massless electrons. PCAC and the dynamics of massless pions are thought to reflect the dynamical breaking of the approximate chiral symmetries of the strong interactions. At the fundamental level these global chiral symmetries are due to local gauge dynamics of the color interactions becoming exact in the limit where the light quarks are massless.

P. Breitenlohner and D. Maison (Eds.): Proceedings 1998, LNP 558, pp. 3–14, 2000.

Local gauge symmetries can also be spontaneously broken. Superconductivity results from the dynamical breaking of the electromagnetic gauge symmetry. This dynamical breaking implies the existence of supercurrents and the Meissner effect which is related to the generation of a dynamical magnetic mass for the photon, the gauge quanta of the electromagnetic gauge field. In a similar manner, the electroweak interactions described by an exact local gauge symmetry which is dynamically broken generating masses for all of the presently observed particles including the massive gauge bosons, the W and Z particles, which mediate the observed electroweak forces.

At present the Standard Model is used to explain all of the observed phenomena of elementary particle physics. The Standard Model is based on exact local gauge symmetries and the dynamics generated by the local gauge fields coupled to the corresponding local conserved currents. The Standard Model currently invokes the local gauge symmetries,

$$SU(3)_{\text{color}} \otimes (SU(2) \otimes U(1))_{\text{ew}} \otimes \text{Gravity} ,$$

of the strong, electroweak and gravitation interactions. Of course there are many speculations about additional dynamical structure including supersymmetry, technicolor grand unification and strings.

2 Anomalies — Clashing Symmetries

Symmetries provide the fundamental framework for our present formulation of theoretical particle physics. However, anomalies arise when apparent classical symmetries come in conflict. This clashing of symmetries has an important impact on both the dynamics and the symmetry structure of the theories we use to describe elementary particles. In the following sections, I will discuss the origins of anomalies, the structure of anomalies and some of the implications of anomalies for physics.

The original anomaly puzzle arose in attempts to apply the newly formulated quantum field theory to the two photon decay of the neutral pion [1,2]. J. Steinberger computed the decay from the one loop, triangle diagram for a virtual proton with point couplings to the neutral pion and photons (Fig. 1). The decay

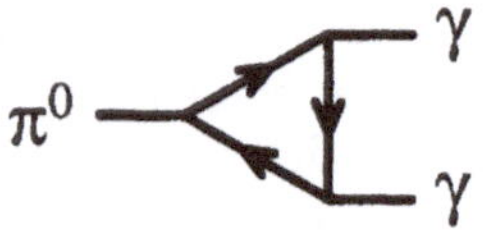

Fig. 1. One loop diagram for the two photon decay of the neutral pion.

amplitude seemed to depend strongly upon whether pseudoscalar or pseudovector couplings were used for the interaction of the pion with the proton. However, these interactions seemed to be equivalent if one integrated by parts and used the field equations. This contradiction between the naive application of the quantum field equations and the direct calculation of the triangle diagrams became known as the anomaly. Although the pseudoscalar coupling eventually proved to give the correct experimental result, it is possible that this early attempt to apply quantum field theory to problems beyond QED convinced Steinberger to focus his future career on experimental physics instead of struggling with unreliable theories.

A formal resolution of the problem of the equivalence of pseudovector and pseudoscalar couplings was provided J. Schwinger [3] using proper time methods. He showed that a careful definition of singular operator products was required before the equations of motion could be used to study the anomaly using the equations of motion of the quantum field theory.

A more complete understanding of the anomaly and its physical impact came from the study of the anomalous divergence equations for the axial-vector current [4,5]. Axial-vector currents had become an important focus of research because of their role in understanding hadronic chiral symmetry or PCAC. The partial conservation of the axial-vector current followed from this chiral symmetry and implied particular couplings for the pions at low energy. Massless pions are identified as the Goldstone bosons of dynamical chiral symmetry breaking. Adler showed that the anomaly required the existence of specific operator corrections to the fermion axial-vector divergence equation.

$$\partial^\mu\{\bar{\psi}\gamma_\mu\gamma_5\psi\} = 2m\{\bar{\psi}\gamma_5\psi\} + \frac{\alpha}{4\pi}F^{\mu\nu}\cdot {}^*F_{\mu\nu} .$$

This result for a free fermion can be generalized to the axial-vector current for hadronic chiral symmetry. The anomaly modifies the divergence equation and predicts the decay amplitude for the Goldstone pion,

$$\begin{aligned} \partial^\mu J_{5\mu} &= \mathrm{tr}_{\mathrm{fermions}}\{T^3Q^2\}\frac{\alpha}{4\pi}F^{\mu\nu}\cdot {}^*F_{\mu\nu} , \\ A_{\pi^0\to\gamma\gamma} &= \mathrm{tr}_{\mathrm{fermions}}\{T^3Q^2\}\frac{1}{f_\pi} , \end{aligned}$$

where the anomaly coefficient is determined by the fundamental fermion structure of the theory. The anomalous divergence equation implies that the axial vector current can not be conserved in the presence of electromagnetism even in the symmetric limit where the pions are massless. From this perspective, the chiral symmetry associated with axial-vector current clashes with local gauge symmetries of electromagnetism.

Because the magnitude of the pion decay amplitude is directly related to the strength of the anomaly, it is a sensitive measure of the fundamental fermion structure of a dynamical theory of hadrons. The measured values of the anomalous pion decay amplitude and the e+e- annihilation cross-section could be combined with current algebra and operator product expansion methods to provide the first convincing evidence for the dynamical color triplet quark picture [6]. Of course, the observed pion decay rate was also consistent with the original Steinberger calculation if pseudoscalar pion-nucleon couplings were used to compute the proton loop amplitude.

3 The Nonabelian Anomaly

Anomalies have a more complex structure than the abelian anomaly observed in the anomalous divergence of the neutral axial-vector current. Nonabelian anomalies can be studied using generalized fermion loops for nonabelian currents where the fermions have arbitrary nonabelian couplings to vector, axial-vector, scalar and pseudoscalar densities.

$$L = \bar{\psi}\{\gamma^\mu V_\mu + \gamma^\mu \gamma_5 A_\mu - \Sigma - i\gamma_5 \Pi\}\psi = \bar{\psi}\{\Gamma\}\psi \, .$$

Explicit perturbative computations of general fermion loops for arbitrary external fields can be made where the short distance singularities are controlled by a well-defined cutoff or regularization procedure. This vacuum functional, or fermion loop effective potential (Fig. 2), can be used to define consistent

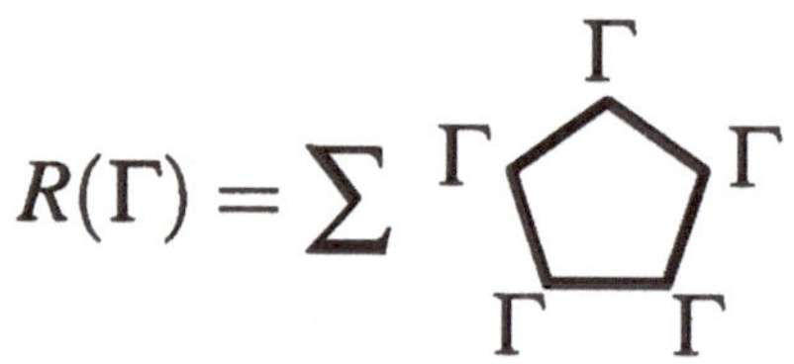

Fig. 2. Vacuum functional for fermions with arbitrary nonabelian couplings.

matrix elements of the nonabelian current and other operators. The covariant derivative of these currents can then be studied for possible anomalous terms. This study corresponds to an explicit check of the gauge covariance of the effective potential. Anomalous terms reflect the explicit breaking of the nonabelian gauge symmetries. A general regularization procedure will normally break many of these symmetries. Local counter-terms can then be added to the effective potential to restore the classical gauge symmetries. When this is not possible, the fermion loops are said to contain anomalies. By explicit calculation [7], all anomalous terms can be made to cancel except those involving certain external vector and axial-vector fields. For a particular choice of counter-terms, the gauge variation of the general fermion loop effective action be reduced to an especially simple form,

$$\begin{aligned} D(\Lambda_+, \Gamma) &= R(\Gamma i\Lambda_+ - i\Lambda_- \Gamma - \gamma \cdot \partial \Lambda_+ \Gamma) \\ &= \frac{1}{6}\frac{\pi^2}{(2\pi)^4} i \int dz \epsilon_{\mu\nu\sigma\tau} \mathrm{tr}\gamma_5 \{2i\Lambda_+ \partial^\mu V_+^\nu \partial^\sigma V_+^\tau - \partial^\mu V_+^\nu V_+^\sigma V_+^\tau\} \, , \end{aligned}$$

where Λ_+ is a left-handed gauge transformation and V_+ is the left-handed external gauge field. Right handed gauge transformations yield a corresponding result. The generalized anomalous divergence equation for nonabelian currents yields

$$D^\mu J^a_{+\mu}(x) = \frac{1}{6}\frac{\pi^2}{(2\pi)^4} \epsilon_{\mu\nu\sigma\tau} \mathrm{tr}\lambda^a_+ \{2\partial^\mu V_+^\nu \partial^\sigma V_+^\tau - \partial^\mu (V_+^\nu V_+^\sigma V_+^\tau)\} \, ,$$

where D^μ is the appropriate covariant derivative. Since the anomalous divergence only involves other external gauge fields, the breaking of the nonabelian symmetries can be viewed as a clash between the symmetries associated with the current and the symmetries associated with the external gauge fields.

The form of the nonabelian anomaly is not arbitrary but is constrained by consistency conditions which must be satisfied by any proper formulation of the quantum theory [8]. The Wess-Zumino consistency conditions provide a powerful constraint on the algebraic structure of the anomaly and a simple test for the consistency of any specific calculation of anomalous terms.

The general result for the fermion loop anomaly obtained above has been confirmed by many different methods. A particularly elegant derivation of the anomaly invokes the path integral formulation of quantum field theory [9]. Fermion loops are generated by the functional integral,

$$\int D\psi\, D\bar{\psi}\, \exp\left\{ i \int dx \{\bar{\psi} i\gamma \cdot D\psi\} \right\} ,$$

where the classical fermion action is presumed to be covariant under generalized gauge transformations, but the fermionic measure may not preserve this covariance. Even here great care must be used in giving precise meaning to these formal expressions. In this formalism, anomalies are directly related to the noninvariance of the fermionic measure and not to problems associated with defining composite operators. Of course, this approach gives the same result as the direct calculation of the fermion loop diagrams, but it adds an important perspective to our understanding of anomalies.

4 Nonrenormalization Theorem

A remarkable feature of anomalies concerns their behavior under renormalization. A careful study of higher order radiative corrections shows that these corrections do not modify the fermion loop anomaly computed above. Since anomalies reflect unavoidable gauge symmetry breaking, they are determined solely by the structure of the small fermion loops and their symmetries [10]. The nonrenormalization theorem was originally checked by explicit two loop computations and confirmed by general regularization arguments to all orders and extended to arbitrary renormalizable quantum field theories in four dimensions [10,11]. The nonrenormalization theorem was also proven using renormalization group methods [12].

The nonrenormalization theorem is extremely important as it establishes the fundamental significance of the anomaly. The anomaly is not simply an artifact of a particular method of calculation or order of perturbation theory. As stated in our discussion of the evidence for color triplet quarks, the anomaly directly measures properties related to the fundamental fermion structure of the underlying quantum field theory. This feature has great significance in the many applications of anomalies to physics.

5 Classical Applications

Anomalies have many different implications for quantum field theory. The consistency of gauge field theory requires the absence of anomalies associated with the dynamical currents (D) which implies that the fermion loop anomalies must cancel between different kinds of fermions in the theory. If the anomalous current divergence involves dynamical gauge fields, then the global symmetries (G) associated with the anomalous current are explicitly broken by dynamics of the gauge fields. Even if no dynamical currents are involved, anomalies can have important implications for the global current algebra associated with the external symmetries of a quantum field theory.

5.1 Anomaly Cancellation $\langle DDD \rangle$

Anomalies reflect an intrinsic breaking of local gauge symmetries which can not be compensated by simply adding local counter-terms in higher order calculations. Since gauge field theories are consistent only if the local gauge symmetries are preserved by the quantum theory, the presence of anomalies implies that certain gauge models simply do not exist at the quantum level. Hence, anomalies associated with the dynamical gauge currents must cancel if the dynamical gauge symmetries are to be preserved. The fermion loop anomalies depend only on the charge structure of the dynamical fermions, and their cancellation constrains the fermion matter content of many gauge field theories. The nonrenormalization theorem then guarantees that this cancellation will be preserved to all orders. From the form of the nonabelian anomaly, it can be shown that models with vectorlike gauge couplings, such as QED or QCD, do not have dynamical anomalies. Only theories where the fermions have chiral gauge couplings can have nontrivial anomalies.

The Standard Model of the electroweak interactions provides an interesting example of a chiral gauge theory where anomalies do occur but are canceled between the various quark and lepton contributions [13,14]. The anomalies for a single generation of quarks and leptons are listed in Table 1. It is a remarkable feature of the Standard Model that a theory involving only quarks or only leptons would not be consistent, but the combined theory of quarks and leptons is free of all dynamical anomalies. Anomaly cancellation is a central element

Table 1. Contributions of leptons and quarks to nonabelian anomalies.

Standard Model	Leptons	Quarks	Sum
$SU(2)^2 \otimes U(1)$	$-\frac{1}{2}$	$3 \cdot \frac{1}{6}$	0
$U(1)^3$	$1 - \frac{1}{4}$	$\frac{1}{36} - \frac{8}{9} + \frac{1}{9}$	0

in building models beyond the Standard Model including grand unification, extended technicolor or any other theory which adds new fermions or additional gauge interactions.

5.2 Global Symmetry Breaking $\langle GDD \rangle$

In gauge field theories, the anomalous divergence equations imply that various global symmetries can be broken by anomalies. In the original calculation of the axial current anomaly, the chiral symmetry of the neutral pion current was broken by the coupling to the electromagnetic gauge fields which modified the low energy theorem for the coupling of pions to photons.

Global symmetries can be broken more dramatically by the presence of nontrivial gauge dynamics. The $U(1)$ problem of QCD is a classic example. The original formulation of quark model seemed to have too much symmetry as there were nine conserved chiral currents in the limit where the light quarks are massless. Weinberg had argued that there should be an extra Goldstone boson, an η', nearly degenerate with the pion. Instead, the physical η' has a mass of order 1 GeV. In quantum chromodynamics, the singlet axial-vector current has an anomaly involving the QCD gauge fields. An explicit calculation by 't Hooft [15] showed that instanton effects could break the $U(1)$ symmetries and generate a mass for the η' [16].

In a similar vein, instanton effects can be shown to generate explicit breaking of the baryon number symmetry in the Standard Model [15]. This may be somewhat surprising as the baryon number current is a vector current and not normally associated with anomalies. However, the Standard Model requires that the $SU(2) \times U(1)$ gauge symmetries be exactly preserved. Since these currents have chiral structure, the anomaly must be shifted away from the dynamical currents, and it reappears as an anomaly in the baryon number current. Hence, the anomaly predicts the proton will decay in the normal Standard Model although the explicit calculation shows that the vacuum decay rate is so highly suppressed that a proton has yet to decay via this mechanism in the entire lifetime of the universe.

Another implication of the QCD anomaly concerns the strong CP problem. Naively, all CP violating phases in the quark and lepton masses matrices can be rotated away leaving only the weak CP phases of the CKM matrix. However, the anomaly induced $U(1)$ breaking of QCD implies that the $U(1)$ phase cannot be freely rotated and a strong CP violation remains. Since there are precise limits on the size of any strong CP violation, alternative models beyond the standard model were considered where a new Peccei-Quinn symmetry [17] would allow the strong CP phase to be rotated away. However, Wilczek and Weinberg [18] argued that this new symmetry would imply the existence a new pseudo-Goldstone boson, the axion. Detailed predictions about the mass and couplings of the axion could be made using the anomalous current algebra reflecting the strong breaking of the $U(1)$ symmetry in QCD [19]. Extensive tests of these predictions show that axions associated with the scale the electroweak interactions are now ruled out [20] and only much higher scales are consistent with the axion picture.

The resolution of the strong CP problem remains an outstanding puzzle of the Standard Model.

5.3 Global Current Algebra $\langle GGG \rangle$

Anomalies also modify the current algebra relations associated with purely global symmetries. This is clear from the anomalous divergence equation where the external gauge fields in the anomalous divergence are associated with global symmetry currents and not the dynamical gauge fields. These anomalies reflect the clash of symmetries generated by the quantum effects of the fermion loops. In many applications of current algebra one combines the constraints of current algebra with low energy theorems associated with the infrared dynamics of the system. Wess and Zumino used their consistency conditions to derive an effective action for the Goldstone pions consistent with the anomalous couplings to the electromagnetic field [8]. Witten showed that this could be extended to derive anomalous terms in the purely strong strong dynamics of pseudoscalar mesons [21].

The anomaly has both ultraviolet and infrared implications. The anomaly associated with the global symmetries of a given theory provides a set of consistency conditions which must be satisfied by any infrared realization of the theory. These consistency conditions place severe constraints on the massless spectrum of fermions and Goldstone bosons even when the dynamics is highly nonperturbative [22].

6 Topology and Geometry

Anomalies have important relations to the topology and geometry of gauge fields. Atiyah and Singer [23] showed that index theorems and the spectral properties of the Dirac operator relate the anomaly to the topological structure of gauge fields. The eigenvalues of the Dirac operator,

$$\gamma \cdot D = \gamma^\mu D_\mu = \gamma^\mu \partial_\mu - iT^a \gamma^\mu A^a_\mu \ ,$$

depend upon the deformations of the background gauge fields and reflect their topological structure. The anomalous divergence of the axial vector current,

$$\partial^\mu J_{5\mu} = \frac{N_f}{8\pi^2} \mathrm{tr}\{G^{\mu\nu}(A) \cdot {}^*G_{\mu\nu}(A)\} \ ,$$

is directly related to the topological index of the gauge field

$$\nu = \frac{1}{16\pi^2} \int dz \mathrm{tr}\{G^{\mu\nu}(A(z)) \cdot {}^*G_{\mu\nu}(A(z))\} \ ,$$

which takes on integer values.

Differential geometry has been used to analyze the structure of anomalies in arbitrary dimensions of space-time [24]. The descent equations can be used to connect various aspects of the anomaly structure. As in the case of the Wess-Zumino consistency conditions, the descent equations strongly constrain the anomalous structure allowed for any theory.

The anomaly also implications for topological objects which occur in gauge field theories. Instantons, sphalerons and similar objects are related to fermion number changing processes which are determined by the anomaly structure of the underlying theory [22]. Anomalies are related to the mechanisms of charge fractionalization and induced charge on topological defects such as dyons, skyrmeons and polyacetylene. Anomalies also have an important impact on the physics of magnetic monopoles, cosmic strings, domain walls, vacuum bubbles, D-branes, etc. In many cases where the physics is highly nonperturbative, the anomaly structure provides the only precise information on the behavior of complex systems.

7 Gravitational Anomalies

Anomalies also occur for systems interacting with gravitational fields. In precise analogy with the axial-vector current anomaly in a background electromagnetic field, the fermion loop processes generate a gravitational anomaly in the divergence of the axial-vector current [26],

$$\partial^\mu J_{5\mu} = \frac{1}{768\pi^2}\epsilon_{\mu\nu\sigma\tau}R^{\mu\nu\alpha\beta}R^{\sigma\tau}_{\alpha\beta} ,$$

where the anomaly is related to a topological index of the gravitational field. Since the Standard Model contains chiral $U(1)$ currents, the potential for gravitational anomalies exists. Such an anomaly would imply a clash between the Standard Model gauge symmetries and the general covariance of the background gravitational field. We would expect the gravitational anomalies to cancel if we wish to preserve our normal picture of gravity. In the Standard model, the individual fermions do have anomalous contributions, but the sum over all fermionic contributions cancels (Table 2). Contrary to the case of the gauge anomalies, the cancellation occurs separately for quarks and leptons.

Table 2. Contributions of leptons and quarks to the gravitational anomaly.

Standard Model	Leptons	Quarks	Sum
$\mathbb{R}^2 \otimes U(1)$	$@(-\frac{1}{2})+1$	$3(\frac{1}{3})+3(-\frac{2}{3})+6(\frac{1}{6})$	0

Pure gravitational anomalies can also exist in 2, 6 and 10 dimensions [26]. As in gauge theories, it is important to determine the precise form of the consistent gravitational anomaly as distinguished from the covariant anomalies associated with various currents or densities. In theories with fermions, the vierbein field must be introduced to define the spin using the tangent space symmetries. In this case, the local gravitational symmetries can be viewed from the perspectives of local Lorentz symmetry or general covariance. By using the veirbein field, the gravitational anomalies can be transformed from one perspective to the other by adding the analogue of Wess-Zumino counter-terms to the gravitational action [27].

8 Supersymmetry

Supersymmetry adds additional complexity to the anomaly picture. Here there is potential for the gauge symmetries or global symmetries to clash with supersymmetry. Indeed, there was initially considerable confusion between the nonrenormalization theorem associated with the axial-vector currents, the renormalization of the supersymmetric b-function and the nonrenormalization theorems associated with the holomorphy of the superpotential [27–30]. Anomalies also have an important impact on the nonperturbative structure of the superpotential, holomorphy and duality [31].

9 Superstrings

The modern superstring era began in 1984 with the observation by Green and Schwarz [32] that the anomalies which affected earlier formulations of string theory could be made to cancel. The apparent loop anomalies were found to cancel against anomalous couplings of the graviational sector. Consistent superstring theories were found to exist in 10 dimensions (four visible dimensions and six compact dimensions) for particular gauge groups. The most interesting early string model was the heterotic string [33]. The low energy spectrum of the theory is determined by anomalies in terms of index theorems and the topological structure of the compact six dimensional manifolds. In this way the anomalies could be used to predict the generation structure of the chiral fermions [34]. More recently, theoretical efforts have focused on superstring duality, M-theory and D-branes [34]. Even here anomalies and related phenomena continue provide important insights into the structure and applications of string theory.

10 Conclusions

Anomalies started out as a troublesome ambiguity about how to apply the new ideas of quantum field theory to interesting physical problems. The resolution of this ambiguity led to a more fundamental understanding of quantum field theories and their symmetries. The discovery and analysis of the complete nonabelian anomaly showed that the anomaly was much more complex than the simple form of the anomalous divergence of the axial-vector current. Anomalies could be viewed as the fundamental clash between the classical symmetries which can occur in a quantum system. The nonrenormalization theorems showed that the anomalies reflected the fundamental structure of the quantum field theory and were not just an artifact of a particular computation in some order in perturbation theory. As nonabelian gauge theories began to take over the theoretical foundations of particle physics, the anomaly played an important role in determining the structure of the gauge models and the symmetry structure of the resulting theories. Anomalies cancellation was a required condition for model building, the global symmetry structure is modified by the presence of anomalies, and the anomaly also changed the global current algebras. In many cases, the anomaly provides the only nonperturbative information about specific gauge field theories, as reflected by the constraints of the 't Hooft anomaly matching conditions and by many other applications.

Connections to fundamental mathematical structures have led to a deeper understanding of anomalies and their implications Differential geometry provided an elegant mechanism for the analysis of anomaly structure and pointed to generalizations of the anomaly picture. Index theorems, spectral flow and related techniques revealed the deep connection between anomalies and the topological structure of gauge fields. The interplay between the mathematics and the physics has led to a much richer view of both fields.

Anomalies played an important role in the rebirth of string theory. They continue to have an important impact on recent developments of string theory, M-theory and D-branes. String theories have revealed a much richer symmetry structure that goes far beyond the symmetries of normal gauge field theory, and anomalies may help provide a path to a more complete understanding of the symmetries and the dynamics.

Many people have played important roles in understanding the mathematical structure of anomalies and in developing the vast array of applications in both physics and mathematics. In my original derivation of the nonabelian anomaly, I knew the result had fundamental significance but had little idea how pervasive anomalies would become in the future.

Acknowledgments

I would like to thank Professor Wolfhart Zimmermann for his kind hospitality over the many years that I have been fortunate to visit the Max-Planck-Institute for Physics in Munich. I would also like to thank Peter Breitenlohner, Dieter Maison and Julius Wess for the invitation to participate in this Workshop which honors the career of Professor Zimmermann.

References

1. J. Steinberger, Phys. Rev. 76, 1180(1949).
2. H. Fukuda and Y. Miyamoto, Prog. Theor. Phys. 4, 347(1949).
3. J. Schwinger, Phys. Rev. 82, 664(1951).
4. S.L. Adler, Phys. Rev. 177, 2426(1969).
5. J.S. Bell and R. Jackiw, Nuovo Cim. 60A, 47(1969).

6. W. Bardeen, H. Fritzsch and M. Gell-Mann, in Scale and Conformal Symmetry in Hadron Physics, R. Gatto, ed., John Wiley and Sons (1973), pg. 139.
7. W. Bardeen, Phys. Rev. 184, 1848(1969).
8. J. Wess and B. Zumino, Phys. Lett. 37B, 95 (1971).
9. K. Fujikawa, Phys. Rev. Lett. D42, 1195 (1979); Phys. Rev. D21, 2848 (1980).
10. S.L. Adler and W. Bardeen, Phys. Rev. 182, 1517(1969).
11. W. Bardeen, Proc. of the Marsielles Conf. on Renormalization, C.P. Korthals-Altes, ed., Marsielles(1972);
Proc. of XVI Int. Conf. on H.E.P., J.D. Jackson and A. Roberts, eds., Fermilab (1972), Vol. 2, pg. 295.
12. A. Zee, Phys. Rev. Lett. 29, 1198 (1972).
13. D. Gross and R. Jackiw, Phys. Rev. D6, 477 (1972).
14. C. Bouchiat, J. Iliopoulos and P. Meyer, Phys. Lett. 38B, 519 (1972).
15. G. 't Hooft, Phys. Rev. Lett. 37, 8 (1976); Phys. Rev. D14, 3432 (1976); see also J. Kogut and L. Susskind, Phys. Rev. D11, 1477(1975).
16. E. Witten, Nucl. Phys. B156, 269 (1976).
17. R. Peccei and H. Quinn, Phys. Rev. Lett. 38, 1440 (1977); Phys. Rev. D16, 1791 (1977).
18. S. Weinberg, Phys. Rev. Lett. D40, 223 (1978);
F. Wilczek, Phys. Rev. Lett. 40, 279 (1978).
19. W. Bardeen and H. Tye, Phys. Lett. 74B, 229(1978);
W. Bardeen, H. Tye and J. Vermaseren, Phys. Lett. 76B, 580 (1978).
20. W. Bardeen, R.D. Peccei and T. Yanagida, Nucl. Phys. B279, 401 (1987);
W. Bardeen, Proc. of the XVI Int. Symp. on Multiparticle Dynamics, M. Markytan et al, eds., World Scientific (1987).
21. E. Witten, Nucl. Phys. B223, 422 (1983); also mentioned in the preprint version of the original nonabelian anomaly paper [7].
22. G. 't Hooft, Cargèse Lectures (1979).
23. M.F. Atiyah and I.M. Singer, Ann. Math. 87, 485 (1968); Ann. Math 97, 119, 139 (1972);
M.F. Atiyah and G.B. Segal, Ann. Math. 87, 531 (1968).
24. B. Zumino, Les Houches Lectures (1983);
R. Stora, Cargèse Lectures (1983);
L. Baulieu, Nucl. Phys. B241, 557 (1984).
25. T. Kimura, Prog. Theor. Phys. 42, 1191 (1969);
R. Delbourgo and A. Salam, Phys. Lett. 40B, 381 (1972).
26. L. Alvarez-Gaumé and E. Witten, Nucl. Phys. B234, 269 (1984);
O. Alvarez, I.M. Singer and B. Zumino, Commun. Math. 94, 409 (1984).
27. J. Iliopoulos, J. Wess and B. Zumino, Phys. Lett. 49B, 52 (1974); Nucl. Phys. B76, 310 (1974).
28. Novikov et al, Nucl. Phys. B229, 381 (1983).
29. M. Shifman and A. Vainshtein, Nucl. Phys. B277, 456 (1986).
30. O. Piguet and K. Sibold, J. Mod. Phys. A1, 913 (1986); Helv. Phys. Acta, 61, 32 (1988).
31. N. Seiberg and E. Witten, Nucl. Phys. B426, 19 (1994); Nucl. Phys. B431, 484 (1994).
32. J. Schwarz and M. Green, Phys. Lett. 149, 117 (1984).
33. D. Gross, J. Harvey, M. Green and R. Rohm, Nucl. Phys. B256, 253 (1985).
34. N. Seiberg and E. Witten, Nucl. Phys. B431, 484 (1994);
E. Witten, Nucl. Phys. B443, 85 (1995).

The Algebraic Method in Renormalization Theory

Carlo Becchi, Stefano Giusto, and Camillo Imbimbo

Dipartimento di Fisica, Università di Genova,
Istituto Nazionale di Fisica Nucleare, Sezione di Genova,
via Dodecaneso 33, 16146 Genova, Italy

Abstract. We give short historical account of the origin of the algebraic quantization method from Zimmermann's construction of normal products. We also give a sketchy description of a recent application of the same method.

In his report at the Aix-en-Provence Conference in 1973 Symanzik mentioned among the most relevant achievements of renormalization theory, the "Normal Operator Products": "There exist (in renormalization theory, at least) operators which are finite, local, and transform as the naive operator product would". In his speech Symanzik referred mainly to the role of these operators in the construction of the short distance operator product expansions [1], and hence he considered only minimally subtracted operators. It had however been clear for a couple of years that the normal products provided a new, general and rigorous tool for the study of renormalization theory and in particular of symmetry properties in field theory.

Indeed, after the pioneering Zimmermann's work [2], in a sequence of papers published between 1971 and 1973 Lowenstein, Schroer, Gomes and Lam showed that the use of normal products allows a rigorous and simple formulation of the Majorana-Schwinger Quantum Action Principle [4,5]. They discussed in particular the derivation of single current Ward Identities in models with broken symmetries, the origin of a generalized class of anomalies, the structure of renormalized energy-momentum tensor [6]. With the same level of rigor and simplicity Lowenstein was able to derive a class of parametric equations generalizing Callan-Symanzik equations and, in two remarkable papers with Schroer, gave the first rigorous proof of the gauge independence of the S-matrix in massive QED [7] and of the absence of radiative corrections to the axial anomaly in the same framework [8].

The renormalized version of the Majorana-Schwinger Quantum Action Principle describes the first-order effect of an "infinitesimal" quantum field transformation on a generic T-ordered vacuum correlator. In general this introduces into the correlators new operators that, in the framework of Zimmermann's scheme, are characterized by different, possibly anisotropic, subtraction degrees [9]. These over-subtracted normal products do not transform "as they would" under field transformations. They can be reduced to linear combinations of minimally subtracted normal products that, in perturbation theory, generate a basis for the local point-like operators. This reduction yields a generalized class of anomalies.

P. Breitenlohner and D. Maison (Eds.): Proceedings 1998, LNP 558, pp. 15–25, 2000.

Lam, in his detailed study of the Quantum Action Principle, extending Lowenstein's work, put into evidence the relation between the over- and anisotropic subtractions and the diagrammatic and forest structure of the renormalized amplitudes, in particular in the case of non-linear field transformations. The need of some refinement of Lam's analysis was later shown by Breitenlhoner and Maison [10]. The extension of the analysis to models involving massless particles required a further generalization of the method of normal operator involving infra-red subtractions [11].

From a general point of view the normal product method is rigorous and simple. However its direct applicability requires a detailed analysis of the diagrams contributing to the correlators and of their subtractions. This is therefore limited to a particular but important class of models which includes QED and the simplest models with broken linear symmetries.

Taking into account the complexity of the diagrammatic and subtraction structure of the Feynman amplitudes, it is clear that a detailed and direct study of the renormalization corrections to non-abelian gauge theories and to the Ward identities corresponding to the non-abelian current algebra relations [12] is out reach even for the top experts. There are however few general properties of the quantum corrections that are independent of the particular diagrammatic and subtraction structure of the correlators and direct consequence of power counting.

Indeed, independently of any detail of the theory, the quantum corrections appear as linear combinations of (integrated) point-like normal products of the external and quantized fields. These are Lorentz scalars and have dimensions limited by that of the variation of the classical action. One has therefore a finite number of independent contributions.

To keep a sufficient level of generality it is convenient to adopt the functional framework in which the point-like operators are defined through functional derivatives with respect to corresponding external fields. In this framework the Schwinger terms, that we consider together with the contact terms required by a covariant T-ordering, appear as normal products of external and quantized fields (with subtraction indices depending on both kinds of field legs). The current algebra Ward identities describe the invariance of the vacuum functional under non-abelian gauge transformations of the external vector fields coupled to the currents and the chiral anomalies correspond to genuine quantum breakings to these identities: they are 4-dimensional normal products of external gauge fields.

For what concerns the construction of the quantum theory, we assume a subtraction scheme fixed a priori. Consider for example the minimal dimensional scheme or the Lowenstein-Zimmermann's zero-momentum subtraction method. Once the subtraction scheme is chosen, the quantum theory is identified by a Lagrangian — that we shall call " Effective Lagrangian" — whose coefficients are formal power series in $\hbar$ and are identified with the series of the finite counterterms implementing the wanted renormalization conditions. In particular a theory with prescribed invariance properties is renormalizable if there exist deformations of the Effective Lagrangian for which the quantum corrections to the wanted Ward Identities vanish to all orders of perturbation theory. Assuming this attitude, the study of the quantized theory should be extended from that of the classically invariant theories with $\hbar$-dependent coefficients to a wide class of deformations. Power counting and Lorentz invariance are the only classical properties that are generally preserved at the quantum perturbative level. Therefore by deformations of the classical theory we shall mean the Effective Lagrangians corresponding to generic Lorentz invariant and power-counting renormalizable actions with the wanted quantum and external field content.

Let the generic deformations of the classical theory be identified by a finite number of $\hbar$-dependent parameters: c^i for $i = 1, .., n$. The vanishing conditions for the quantum corrections correspond to a finite number (say N) of algebraic relations, ensuring the vanishing of the coefficients of the N independent breaking operator (Schwinger) terms. In perturbation theory these algebraic relations appear as formal power series in the perturbative parameter $\hbar$; that is, they can be written in the form:

$$B_a(\hbar, c) = 0 \qquad \text{for} \quad a = 1, .., N \tag{1}$$

with B_a, and of course c, formal power series in $\hbar$. The classical theory is identified by the system:

$$B_a(0, c_0) = 0. \tag{2}$$

At least a subset of the solutions of this system lie on a non-singular (q-dimensional, if q is the number of the parameters of the classical theory) submanifold M_c of R^n . The tangent space to this manifold at the point c_0 is defined by:

$$B_{a,i}(0, c_0)\, v^i = 0 \tag{3}$$

for every tangent vector v. Here the index i labels the c^i-partial derivative.

In perturbation theory the condition (1) is replaced by its linearized version that ensures recursively the existence of $\hbar$ formal series solutions to (1). This linearized condition requires that the system

$$B_a(\hbar, c) + B_{a,i}(0, c_0)\, \delta c^i = O(\hbar B) \tag{4}$$

has solutions $\delta c = O(B)$ for every point c_0 of M_c. Indeed, assuming that $B_a(\hbar, c)$ be of order m in $\hbar$ (m, owing to (2), must be greater than zero), one has from (4) that $B_a(\hbar, c + \delta c)$ is of order greater than m. This is expressed in closed form by

$$B_a(\hbar, c + \delta c) = O(\hbar B_a(\hbar, c)), \tag{5}$$

which is equivalent to the vanishing of B_a.

Thus, according to (4), one should prove that $B_a(\hbar, c)$ belongs to the image of the matrix $B_{a,i}(0, c_0)$ in R^n if $c_0 \in M_c$, up to higher order corrections. Notice that $B_{a,i}(0, c_0)\, \delta c^i$ represents the variation of the classical action. Therefore (4) means that the genuine quantum corrections should be compensable by deformations of the action.

Let us illustrate the above equations by a "trivial" example. Suppose we are given a massive scalar field theory carrying an orthogonal representation of a simple group. We want to analyze the renormalizability of the theory under the condition that the Green functions of the theory be invariant under the action of the above group. Let the infinitesimal action of the group on the field be given by:

$$\delta_\omega \phi^i = \omega^\alpha T^i_{\alpha j} \phi^j . \tag{6}$$

Let $\Gamma[\phi]$ be the one-particle-irreducible Green functional of the theory. The renormalized invariant theory should satisfy:

$$\delta_\omega \Gamma \equiv \int dx\, \omega^\alpha T^i_{\alpha j} \phi^j(x) \frac{\delta}{\delta \phi^j(x)} \Gamma = 0. \tag{7}$$

For a generic choice of the Effective Lagrangian, according to the normal product version of the Quantum Action Principle, this Ward Identity is written in the broken form:

$$\delta_\omega \Gamma = \int dx\, N_4 \left[\omega_\alpha b^\alpha_a \Omega^a(x)\right] \Gamma = \int dx\, \left[\omega_\alpha b^\alpha_a \Omega^a(x) + O(\hbar b)\right] \tag{8}$$

where the symbol $N_4[X]$ in front of Γ means the insertion of the corresponding operator into the one-particle irreducible Green functions and $\{\Omega^a\}$ for $a = 1, .., N$ is a basis of 4-dimensional point-like operators. Notice that in (8) the operators Ω^a play two different roles; indeed in the first equation they appear as normal product operators while in the second one they are functionals. This puts into evidence that in the tree approximation the insertion into Γ of a normal product generates the corresponding elementary vertex.

Now we can identify the coefficients $B_a(\hbar, c)$ introduced above with $\omega_\alpha b^\alpha_a$ and hence Eq. (1) corresponds to the system $\omega_\alpha b^\alpha_a(\hbar, c) = 0$, that we write in the form of a functional equation:

$$\int dx\, \omega_\alpha b^\alpha_a(\hbar, c)\, \Omega^a(x) = 0 . \tag{9}$$

In this functional form we can also represent the symmetry breaking induced in the classical approximation by an infinitesimal deformation of the action $\delta\hat{\Gamma}$ corresponding to the variation δc of the parameters. Introducing the complete system $\{\gamma_i\}$ of independent 4-dimensional integrated Lorentz scalar functionals — that is a basis for the Effective Lagrangians — we can write $\delta\hat{\Gamma} = \hat{\gamma}_i\, \delta c^i$ and the corresponding symmetry breaking is written:

$$\int dx\, B_{a,i}(0, c_0)\, \Omega^a(x) \delta c^i = \delta_\omega \hat{\gamma}_i . \tag{10}$$

Now we can translate into our functional formalism the recursive version of the renormalizability condition (4):

$$\int dx\, \omega_\alpha b^\alpha_a(\hbar, c)\, \Omega^a(x) + \delta_\omega \hat{\gamma}_i\, \delta c^i = O(\hbar b) . \tag{11}$$

Taking into account that the basis of functionals $\{\gamma_i\}$ is complete, we can write this equation in the simpler form:

$$\int dx\, \omega_\alpha b_a^\alpha (\hbar, c)\, \Omega^a(x) = \delta_\omega \hat{X} + O(\hbar b)\ . \tag{12}$$

where $\hat{X}$ is any 4-dimensional integrated Lorentz scalar functional of order b. This is the final form of the perturbative renormalizability condition for our symmetric theory.

In the present situation the diagrammatic analysis is sufficient to prove that (12) has solutions $\hat{X}$. Indeed in this example the infinitesimal field transformations are linear and homogeneous and the normal products transform as the corresponding classical functionals; therefore the left-hand side of (12) can be written as the action of δ_ω on the Effective Action.

However, for example, the same diagrammatic analysis appears particularly complex in the study of current algebra Ward identities in a linear sigma model [12]. Adopting in this specific case the same functional framework, one is led to a situation formally analogous to our example in which, however, the number and structure of the elements of the Ω^a basis is exceedingly rich. This is why the idea emerged to reduce this number by using a consistency condition introduced by Wess and Zumino [13] to characterize the non-abelian axial anomaly.

In the case of our example the consistency condition appears as a consequence of the Lie algebra commutation relations:

$$[\delta_{\omega'}, \delta_\omega] = \delta_{[\omega,\omega']}\ . \tag{13}$$

Indeed, if we compute $[\delta_{\omega'}, \delta_\omega]\,\Gamma$ by means of Eq. (8) we obtain:

$$\begin{aligned} &\delta_{\omega'} \int dx\, \omega_\alpha\ b_a^\alpha \Omega^a(x) - \delta_\omega \int dx\, \omega'_\alpha\ b_a^\alpha \Omega^a(x) = \\ &\int dx\ [\omega, \omega']_\alpha\ b_a^\alpha \Omega^a(x) + O(\hbar b)\ . \end{aligned} \tag{14}$$

This is the rigid version of Wess-Zumino consistency condition. It is easy to show by purely algebraic means that, in the case of semi-simple symmetry groups, Eq. (14) guarantees the existence of solutions to Eq. (12) and hence the renormalizability of the symmetric theory. This is the historical origin of the algebraic renormalization method.

The second important situation requiring more powerful algebraic tools appears in the study of non-abelian gauge theories in which the invariance properties of the theory are controlled by the Slavnov-Taylor identity. The essential novelties of this theory, in the Faddeev-Popov quantization scheme, are the non-linearity and nilpotency of the field transformations leaving the classical action invariant [14].

The non-linearity of the field transformations implies that they are affected by quantum corrections in much the same way as the couplings appearing in the Lagrangian. In other words the infinitesimally transformed fields are composite operators that must be written as linear combinations of normal products and suitably renormalized. In our functional scheme this implies the introduction for (almost) every field of a corresponding external field. We shall use for these external fields the name of anti-fields that has been introduced quite recently without any substantial change with respect to the original framework. The anti-fields obey the opposite statistics of the corresponding fields: those corresponding to bosonic fields are elements of a Grassmannian algebra. They anticommute with the fermionic fields whose anti-fields are c-number valued. Fields and anti-fields carry a ghost-number conserved charge (possibly anomalous, as it is the case, for example, in string theory). The sum of the field and corresponding anti-field charge is minus one.

In the case in which there is an anti-field for every field the Slavnov Taylor identity is written according:

$$\int dx \sum_i \frac{\delta}{\delta\phi^i}\Gamma \, \frac{\delta}{\delta\phi^*_i}\Gamma = 0, \tag{15}$$

where the sum runs over all the quantum fields ϕ^i and the corresponding anti-fields ϕ^*_i. There are many important situations — in particular the standard case of a gauge theory with linear gauge fixing function (for example the Feynman gauge choice) — in which the prescription of supplementary conditions on the form of the gauge fixing function allows the reduction of the effective number of pairs field-anti-field.

In a generic gauge theory with generic subtraction prescriptions the Slavnov-Taylor identity is modified by quantum corrections:

$$\int dx \sum_i \frac{\delta}{\delta\phi^i}\Gamma \, \frac{\delta}{\delta\phi^*_i}\Gamma = \int dx \, N_4\left[\Delta(x)\right]\Gamma = \int dx \left[\Delta(x) + O\left(\hbar\Delta\right)\right], \tag{16}$$

where, to keep contact with (1), we decompose Δ on a suitable basis of normal products with Faddeev-Popov charge one:

$$\Delta(x) = \sum_a B_a\left(\hbar, c\right)\Omega^a(x). \tag{17}$$

Let Γ_0 be the classical action corresponding to a generic point on M_c and

$$\hat{\Gamma} = \Gamma_0 + \hat{\gamma} \tag{18}$$

be its local generic deformation. Γ_0 satisfies Eq. (15). The functional $\hat{\gamma}$ corresponds to a tangent vector to M_c if:

$$\mathcal{D}_{\Gamma_0}\hat{\gamma} \equiv \int dx \sum_i \left[\frac{\delta}{\delta\phi^i}\Gamma_0 \, \frac{\delta}{\delta\phi^*_i}\hat{\gamma} + \frac{\delta}{\delta\phi^*_i}\Gamma_0 \, \frac{\delta}{\delta\phi^i}\hat{\gamma}\right] = 0. \tag{19}$$

This is the functional version of Eq. (3). The functional differential operator $\mathcal{D}_{\Gamma_0}$ is nilpotent since Γ_0 satisfies Eq. (15). $\mathcal{D}_{\Gamma_0}$ acts on the space of the local deformations of Γ_0 giving possible breaking terms. Given a basis of local deformations and the basis $\{\Omega^a(x)\}$ for the possible breakings, the action of $\mathcal{D}_{\Gamma_0}$ corresponds to the matrix $B_{a,i}(0, c_0)$ in (3).

Therefore the renormalizability condition Eq. (4) is written in the functional language according:

$$\int dx\, \Delta(x) + \mathcal{D}_{\Gamma_0}\hat{\gamma}_q = O(\hbar\Delta) \tag{20}$$

that should be solved in terms of $\hat{\gamma}_q$, the deformation of the classical action corresponding to δc in Eq. (4). Eq. (20) means that the Slavnov-Taylor identity is renormalizable if the breaking $\int dx\, \Delta(x)$ belongs to the image of $\mathcal{D}_{\Gamma_0}$ in the breaking space.

Also in this case the existence of $\hat{\gamma}_q$ solutions of (20) is controlled by a consistency condition following from the identity:

$$\mathcal{D}_\Gamma \int dx \sum_i \frac{\delta}{\delta\phi^i}\Gamma\, \frac{\delta}{\delta\phi_i^*}\Gamma \equiv 0 \tag{21}$$

that implies:

$$\mathcal{D}_{\Gamma_0} \int dx\, \Delta(x) = O(\hbar\Delta)\,. \tag{22}$$

For the theory to be renormalizable it is sufficient that the image of the "coboundary operator" $\mathcal{D}_{\Gamma_0}$ in the space of breakings coincides with its kernel. Otherwise the coboundary operator has a non-trivial cohomology in this space that corresponds to the potential anomalies of the theory. The study of the actual presence of these anomalies goes beyond the algebraic method [8].

Beyond the analysis of the quantum corrections, the operator $\mathcal{D}_{\Gamma_0}$ plays an essential role in the identification of the physical content of a perturbative gauge theory. Indeed, let us assume that a given theory has been fully renormalized. In other worlds, we are given the functional Γ as formal power series in $\hbar$ satisfying Eq. (15) to all orders. Γ is a function of all the parameters associated with the possible deformations of our theory and a functional of external fields coupled to all the relevant point-like operators.

Were we able to define an asymptotic space H_{as} we could find a nilpotent Fermionic charge Q [15] corresponding to $\mathcal{D}_\Gamma$ and commuting with the S-matrix. This is the direct consequence of the Slavnov-Taylor identity. One could consequently define the cohomology H_Q of Q in the asymptotic space, that is the quotient of the kernel versus the image of Q in H_{as}. At least in the perturbative framework, the original scalar product induces a Hilbert space structure into H_Q and the S-matrix defined in H_{as} induces an unitary S-matrix into H_Q. Therefore H_Q can be identified with the physical asymptotic space [14].

To any external bosonic field α commuting with $\mathcal{D}_\Gamma$

$$\mathcal{D}_\Gamma \frac{\delta}{\delta\alpha}\Gamma = 0, \tag{23}$$

there corresponds an operator O_α in H_{as} such that:

$$[Q, O_\alpha] = 0. \tag{24}$$

Furthermore among the solutions of (23) one finds those satisfying:

$$\frac{\delta}{\delta\alpha}\Gamma = \mathcal{D}_\Gamma \frac{\delta}{\delta\beta}\Gamma. \tag{25}$$

The corresponding relation in the operator formalism is

$$O_\alpha = [Q, O_\beta], \tag{26}$$

and O_α has vanishing matrix elements in the physical space.

This shows that the non-trivial point-like observables correspond to external fields α such that $\frac{\delta}{\delta\alpha}\Gamma$ belongs to a cohomology class of $\mathcal{D}_\Gamma$ and the corresponding operators belong to a cohomology class of Q.

In perturbation theory, having recourse to the implicit function theorem, one sees that $\mathcal{D}_\Gamma$ is replaced by $\mathcal{D}_{\Gamma_0}$ and Γ by Γ_0.

The above considerations about the role of $\mathcal{D}_{\Gamma_0}$ justify a deeper analysis of its structure. It is apparent from Eq. (19) that $\mathcal{D}_{\Gamma_0}$ is the sum of two terms; the second one, that we call s, is a field derivative; therefore it can be considered as the generator of a field transformation that in general is anti-field dependent. The first term is the generator of a field dependent anti-field variation. We shall limit ourselves to the case of mass-shell-closed gauge algebras which means that Γ_0 is linear in the anti-fields:

$$\Gamma_0 \equiv \Gamma_{inv} + \Gamma_s = \Gamma_{inv} + \int dx \sum_i \phi_i^*(x) s\phi^i(x). \tag{27}$$

In this situation s is anti-field independent and nilpotent, while in the general case it is nilpotent only modulo the field equations.

One can thus consider the cohomology of s and ask what is its relation with that of $\mathcal{D}_{\Gamma_0}$.

To answer this question if we consider a generic $\mathcal{D}_{\Gamma_0}$-closed functional X. It satisfies:

$$\mathcal{D}_{\Gamma_0} X \equiv sX + \int dx \sum_i \frac{\delta}{\delta\, \phi^i}\Gamma_0 \frac{\delta}{\delta\, \phi_i^*} X = 0. \tag{28}$$

The second term in the right-hand side is proportional to the field derivatives of the classical action Γ_0: this means that X is s-closed on the classical mass-shell. Henneaux and his collaborators [16] have performed a very general analysis of the solutions to the $\mathcal{D}_{\Gamma_0}$-closeness condition in the case of standard, semi-simple, gauge theories. They find that in this case the $\mathcal{D}_{\Gamma_0}$-cohomology coincides with that of s on the functional space constrained by the field equations. Therefore it corresponds to gauge invariant, anti-field independent, operators.

There is however a wide class of models in which the cohomology of $\mathcal{D}_{\Gamma_0}$ is substantially different from that of s. Since a general analysis has not yet been completed, we conclude this report mentioning a very simple example of a model of this class.

We consider a particularly simple topological σ-model [17,18] whose target space is a n-dimensional complex torus T. Given the complex structure J of T

we choose a system of complex coordinates $V, \bar{V}$ adapted to J. The basic fields in the holomorphic sector are the world-sheet scalars $V(x)$, the world-sheet one-forms $P(x)$ and the two-forms $F(x)$ carrying ghost number -1 and -2 respectively and taking values in holomorphic tangent of T. The anti-holomorphic sector contains only world-sheet scalar fields: the coordinates $\bar{V}$ and the anti-holomorphic tangent vectors Σ, Δ and H with ghost number +1, +1 and +2 respectively. Concerning the statistical properties: $V, \bar{V}, H$ and F are bosonic while Σ, Δ and the components of P are fermionic fields.

The symmetry of the model emerges from a particular heterotic twisting of a $N = 2$ supersymmetry which gives rise to the so-called B-model [17]. In the B-model the nilpotent functional differential operator s, introduced above acts according to:

$$\begin{aligned} sP &= dV \qquad & sF &= dP \\ s\bar{V} &= \Sigma \qquad & s\Delta &= H. \end{aligned} \tag{29}$$

The model being topological, its classical action reduces to a gauge fixing term that is trivial in the s-cohomology:

$$\Gamma_{inv} = s\Psi, \tag{30}$$

where Ψ is an integrated two-form with ghost number -1 constrained by the condition that the kinetic term in the Lagrangian be non-degenerate. For example:

$$\Psi = \int_{\Sigma} g_{i\bar{\jmath}} \left[F^i \Delta^{\bar{\jmath}} + *P^i dV^{\bar{\jmath}}\right], \tag{31}$$

where $*P$ is the form Hodge dual to P and d is the exterior differential on the world-sheet Σ; $g_{i\bar{\jmath}}$ is the target space Kähler metric.

The anti-field dependent part of Γ_0 is

$$\Gamma_s = \int_{\Sigma} \left[P^* dV + F^* dP + \bar{V}^* \Sigma + \Delta^* H\right]. \tag{32}$$

Notice that, in the present model, the anti-fields are world-sheet forms and the sum of the form degrees of every field-anti-field pair is two.

The anti-field-independent part of Γ_0 is obtained from Γ_s by translating the anti-fields according to:

$$\phi^* \longrightarrow \phi^* + \frac{\delta\Psi}{\delta\phi}. \tag{33}$$

Now we consider the structure of the relevant cohomology. As discussed above one is interested in the cohomology of $\mathcal{D}_{\Gamma_0}$. However it is not difficult to verify that this can be obtained from that of $\mathcal{D}_{\Gamma_s}$ by the translation (33). Therefore, for the purpose of classifying the "observables", we can choose the classical functional (27) with vanishing Γ_{inv}, i.e.:

$$\Gamma_0 = \Gamma_s. \tag{34}$$

The action of $\mathcal{D}_{\Gamma_s}$ on the fields coincides with that of s, while on the anti-fields one has:

$$\begin{array}{ll} \mathcal{D}_{\Gamma_s} V^* = dP^* & \mathcal{D}_{\Gamma_s} P^* = dF^* \\ \mathcal{D}_{\Gamma_s} \Sigma^* = -\bar{V}^* & \mathcal{D}_{\Gamma_s} H^* = \Delta^* \end{array} \tag{35}$$

The functional space upon which $\mathcal{D}_{\Gamma_s}$ acts corresponds to the polynomial functionals in the variables $dV, P, F, d\bar{V}, \Sigma, \Delta, H$ and the anti-fields, whose coefficients are target space tensors. These operators must be globally defined on the target space.

Under this condition, it is apparent that the zero-form cohomology consists of the polynomials in Σ — since this is the image of $\bar{V}$ which is not globally defined — and F^*. Further cohomology elements are associated to one and two-form valued operators which satisfy the so-called descent equations:

$$\mathcal{D}_{\Gamma_s} \Omega^{(1)} = d\Omega^{(0)} \qquad \mathcal{D}_{\Gamma_s} \Omega^{(2)} = d\Omega^{(1)} \tag{36}$$

The corresponding cohomology elements are obtained by integrating $\Omega^{(1)}$ and $\Omega^{(2)}$ over non-trivial one and two-cycles of Σ.

The zero-form cohomology is generated by:

$$\Omega^{(0)} = \mu^i_{\bar{\jmath}} F^*_i \Sigma^{\bar{\jmath}}, \tag{37}$$

where the coefficients $\mu^i_{\bar{\jmath}}$ are constant tensors. The corresponding one-form is:

$$\Omega^{(1)} = \mu^i_{\bar{\jmath}} \left[P^*_i \Sigma^{\bar{\jmath}} - F^*_i dV^{\bar{\jmath}} \right], \tag{38}$$

and the two-form is:

$$\Omega^{(2)} = \mu^i_{\bar{\jmath}} \left[V^*_i \Sigma^{\bar{\jmath}} - P^*_i dV^{\bar{\jmath}} \right]. \tag{39}$$

The construction of the corresponding operators obtained after the anti-field translation (33) is left to the reader. It turns out that integrating over the world-sheet the operator generated by $\Omega^{(2)}$, one finds the variation of Γ_0 under a deformation of the complex structure of the target space parametrized by $\mu^i_{\bar{\jmath}}$.

This remark is just the starting point of a field theory construction of special geometry [19] that will be developed elsewhere. It is however important to notice here that the presence of anti-field dependent elements of the $\mathcal{D}_{\Gamma_s}$-cohomology implies that the corresponding elements of the $\mathcal{D}_{\Gamma_0}$-cohomology depend on the parameters appearing in the gauge-fixing fermion Ψ. This can transform "unphysical" into "physical" parameters inducing an "holomorphic anomaly" [20,18].

We think that this sketchy example gives an idea of the extension of the range of the algebraic quantization method that has originated from Zimmermann's construction of the method of normal products.

References

1. W. Zimmermann, Annals of Phys. 77 (1973) 570.
2. W. Zimmermann, Commun. Math. Phys. 6 (1967) 161; 10 (1968) 325; Brandeis Lectures 1970, Vol. 1.
3. J.H. Lowenstein, Commun. Math. Phys. 24 (1971) 1.
4. J.H. Lowenstein, Phys. Rev. D4 (1971) 2281;
M. Gomes and J.H. Lowenstein, Phys.Rev. D7 (1973) 550; Nucl.Phys. B 45 (1972) 252.
5. Y.M.P. Lam, Phys. Rev. D6 (1972) 2145; D7 (1973) 2943.
6. B. Schroer, Lettere al Nuovo Cimento, 2 (1971) 867.
7. J.H. Lowenstein and B. Schroer, Phys.Rev. D6 (1972) 1553.
8. J.H. Lowenstein and B. Schroer, Phys.Rev. D7 (1973) 1929.
9. W. Zimmermann, Annals of Phys. 77 (1973) 536.
10. P. Breitenlohner and D. Maison, Commun. Math. Phys. 52 (1977) 11.
11. W. Zimmermann and J.H. Lowenstein, Commun. Math. Phys. 44 (1975) 73;
J.H. Lowenstein, Commun. Math. Phys. 47 (1976) 53.
12. C. Becchi, Commun. Math. Phys. 39 (1975) 329.
13. J. Wess and B. Zumino, Phys. Letters B37 (1971) 95.
14. C. Becchi, A. Rouet and R. Stora, Phys. Letters 52B (1974) 344; Commun. Math. Phys. 42 (1975) 127; Ann. Phys. 98 (1976) 287.
15. T. Kugo and I. Ojima, Prog. Theor. Phys. Suppl. 66 (1979) 1.
16. M.Henneaux, Commun. Math. Phys. 140 (1991) 1;
G. Barnich, F. Brandt and M. Henneaux, Commun. Math. Phys. 174 (1995) 57 and 93.
17. E. Witten, "Mirror Manifolds and Topological Field Theory", in *Essays on Mirror Manifolds*, ed. S. Yau, International Press (Hong Kong 1992).
18. C. Becchi, S. Giusto and C. Imbimbo, "The Holomorphic Anomaly of Topological Strings", hep-th/9801100.
19. P. Candelas and X. de la Ossa, Nucl. Phys. B355 (1991) 455.
20. M. Bershadsky, S. Cecotti, H. Ooguri and C. Vafa, Commun. Math. Phys. 165 (1994) 311.

Modular Groups in Quantum Field Theory

Hans-Jürgen Borchers

Institut für Theoretische Physik, Universität Göttingen,
Bunsenstraße 9, D-37073 Göttingen, Germany

1 Introduction

Quantum Field theory appears in several different settings:

1. Lagrangean quantum field theory together with perturbation theory.
2. L.S.Z.–theory, which is useful for scattering problems [1].
3. Wightman's quantum field theory [2] and its derivative, the Euclidean field theory.
4. The theory of local observables in the sense of Araki, Haag and Kastler [3].

It is believed that these different branches of quantum field theory describe essentially the same physics. But there is only very little known about the rigorous equivalence of the above–mentioned theories. Wolfhart Zimmermann and myself [4] were the first to look at the passage from Wightman's theory to the theory of local observables. We found only sufficient conditions. Meanwhile, there exists a large number of them, but necessary and sufficient conditions are still missing. The situation is not better for the reverse direction or the other equivalence problems.

If one thinks that two theories are equivalent then one should at least try to transcribe the great achievements of one of the theories to the other. However, in many cases it is not so simple as it seems to be at first inspection. I will try to discuss this problem by looking at two examples. One is the PCT–thoerem and the other is the tensor product problem.

2 The PCT–theorem

This theorem tells us that the product of time reversal, space reflection, and charge conjugation is always a symmetry. Reading the paper of Pauli [5] on this subject one gets the impression that a precurser of the PCT–theorem has been discovered by Schwinger [6]. But it was a mysterious transformation containing the interchange of operators. The first development of the PCT–theorem in the frame of Lagrangean field theory is due to Lüders [7]. This result has triggered the clarification of the connection between spin and statistics and the role of the positive energy. (See W. Pauli [5] and also G. Lüders and B. Zumino [8].)

1957 R. Jost [9] gave a proof of the PCT–theorem in the frame of Wightman's field theory. The beauty of this proof is the clarification of the role of the different conditions one has to impose. These are

P. Breitenlohner and D. Maison (Eds.): Proceedings 1998, LNP 558, pp. 26–42, 2000.

1. Covariance of the theory under the (connected part of the) Poincaré group.
2. Positivity of the energy.
3. There are only fields, which transform with respect to finite dimensional representations of the Lorentz group. (Transformation of the index space.)
4. Locality, which means that for spacelike distances the Bose fields commute with all other fields and the Fermi fields anti–commute with eachother.
5. The Minkowski space has even dimensions.
6. To every field in the theory appears its conjugate complex partner.

From the spectrum condition it follows that the Wightman functions have an analytic continuation into the forward tube T_n^+

$$T_n^+ = \{z_1, \ldots, z_n \in \mathbb{C}^4; \Im m\,(z_i - z_{i+1}) \in V^+\}\,.$$

Using locality, Poincaré covariance of the theory, and the appearence of only finite dimensional representation of the Lorentz group, Hall and Wightman [10] could show that the analytically continued Wightman functions can be considered as functions on the complex Lorentz group. If the index space transforms under infinite dimensional representation of the Lorentz group then the Hall Wightman theorem fails because of lack of analyticity. Examples are given by Streater [11] and by Oksak and Todorov [12]. The Hall Wightman theorem was the starting point of Jost's investigation. If the Minkowski space has even dimensions then the complex Lorentz group contains the element $-\mathbb{1}$

$$-\mathbb{1} = \begin{pmatrix} -1 & 0 & 0 & 0 \\ 0 & -1 & 0 & 0 \\ 0 & 0 & -1 & 0 \\ 0 & 0 & 0 & -1 \end{pmatrix}\,.$$

This transformation is the product of time reversal and space reflection. But there is the time translation e^{iEt} with the positive energy operator. In order to keep the energy positive one has to change $\imath$ into $-\imath$. Therefore, the time reversal has to be an antiunitary operator. If Θ is an antiunitary total reflection one obtains for a scalar field

$$\Theta\Phi(x)\Theta = \Phi^*(-x)\,.$$

The passage to the conjugate complex is closely related to the charge conjugation. Therefore, one has to look at the product of C and PT. One remark more to the role of locality: The passage to the conjugate complex interchanges the order of an operatorproduct. At totally spacelike points the original order can be restored. Putting things together one gets the PCT–theorem for scalar fields. The general case needs in addition the handling of finite dimensional matrices which appear with fields of higher spin.

For a long time it was impossible to show the PCT–theorem in the theory of local observables because one did not know the meaning of condition 3 and 6 in the setting of local observables. These are:

3. There are only fields which transform with respect to finite dimensional representations of the Lorentz group. (Transformation of the index space.)
6. To every field in the theory appears its conjugate complex partner.

That it took such a long time to understand the two conditions in the setting of local observables is due to the lack of proper mathematics. This new technique is the Tomita–Takesaki theory. At the 1967 Baton Rouge conference Tomita [13] distributed a preprint describing his theory of modular Hilbert algebras. As one knows by now this is the biggest progress in the theory of operator algebras since von Neumann. Tomita himself did not publish his result and it was Takesaki [14] who presented this theory in form of a lecture note. This theory is concerned with the following:

Let $\mathcal{H}$ be a Hilbert space and $\mathcal{M}$ be a von Neumann algebra acting on this space with commutant $\mathcal{M}'$. A vector Ω is cyclic and separating for $\mathcal{M}$ if $\mathcal{M}\Omega$ and $\mathcal{M}'\Omega$ are dense in $\mathcal{H}$. If these conditions are fulfilled then a modular operator Δ and a modular conjugation J is associated to the pair $(\mathcal{M}, \Omega)$ such that
(i) Δ is self-adjoint, positive and invertible

$$\Delta\Omega = \Omega, \qquad J\Omega = \Omega\,.$$

(ii) The operator J is a conjugation, i.e. J is antilinear, $J^* = J$, $J^2 = 1$, and J commutes with Δ^{it}. This implies the relation

$$\operatorname{Ad} J\Delta = \Delta^{-1}\,.$$

(iii) The unitary group Δ^{it} defines a group of automorphisms of $\mathcal{M}$

$$\operatorname{Ad}\Delta^{it}\mathcal{M} = \mathcal{M}, \qquad \forall t \in \mathbb{R}\,.$$

(iv) For every $A \in \mathcal{M}$ the vector $A\Omega$ belongs to the domain of $\Delta^{\frac{1}{2}}$.
(v) J maps $\mathcal{M}$ onto its commutant

$$\operatorname{Ad} J\mathcal{M} = \mathcal{M}'\,.$$

Because of the Reeh–Schlieder theorem [15] the vacuum vector is cyclic and separating for every algebra $\mathcal{M}(G)$, where G is any domain which has a spacelike complement with interior points.

In order to get a better understanding for these new symmetries it is important to find examples where these groups can explicitly be computed. So far one knows the following examples:

(a) G is a spacelike wedge and the local algebras are generated by Wightman fields, which transform covariantly with a finite dimensional representation of the Lorentz group [16,17].
(b) G is a forward light cone and $\mathcal{M}(V^+)$ is generated by a massless, non-interacting field [18].
(c) G is a double cone and $\mathcal{M}(D)$ is generated by conformally covariant fields [19].

(d) G is a spacelike wedge and the local algebras are generated by generalized free fields of a certain type, which break Lorentz covariance [20].
(e) G is the forward light cone or the wedge for a quantum field in a thermal equilibrium state of two dimensional models that factorize in light-cone coordinates [21].

In (a) the modular group is the group of Lorentz boosts that leave the wedge invariant, and the conjugation is the PCT operator (combined with a rotation). The precise connection of the modular group and the Lorentz boosts of the wedge is

$$\Delta_W^{it} = \Lambda_W(t) := \begin{pmatrix} \cosh 2\pi t & -\sinh 2\pi t & 0 & 0 \\ -\sinh 2\pi t & \cosh 2\pi t & 0 & 0 \\ 0 & 0 & 1 & 0 \\ 0 & 0 & 0 & 1 \end{pmatrix} .$$

At the same time Bisognano and Wichmann showed that in their situation wedge duality holds, this is the relation

$$\mathcal{M}(W) = \mathcal{M}(W') ,$$

where W' denotes the opposite wedge.

From this result one can learn the following: Let $\Phi_i(x)$ be Wightman fields. Then one has

$$U(\Lambda_W(t))\Phi_i(x)U^*(\Lambda_W(t)) = D_i^k(\Lambda_W(t))\Phi_k(\Lambda_W(t)x) .$$

If $D_i^k(\Lambda_W(t))$ has an analytic continuation into the complex Lorentz group then the spectrum condition implies that for $x_i \in W$ the expression

$$\begin{aligned} &U(\Lambda_W(t))\Phi_{i_1}(x_1)\ldots\Phi_{i_n}(x_n)\Omega \\ &\quad = D_{i_1}^{k_1}(\Lambda_W(t))\Phi_{k_1}(\Lambda_W(t)x_1)\ldots D_{i_n}^{k_n}(\Lambda_W(t))\Phi_{k_n}(\Lambda_W(t)x_n)\Omega , \end{aligned}$$

has an analytic continuation into the strip $-\frac{1}{2} < \Im m\, t < 0$. At the lower boundary one finds $\Lambda_W(t-\frac{i}{2})x \in W'$.

This gives the hint how to solve problem (3) for local observables. The result [22] needs some explanation:
Let K_0 be a double cone in the characteristic two–plane of the wedge with center at the origin and K be the cylindrical set with the same cylindrical direction as that of W, such that the intersection with the characteristic two–plane is K_0. Let $A \in \mathcal{M}(K)$ and denote by $A(K,x)$ the translated operator $T(x)AT(-x)$, where $T(x)$ is the given representation of the translations. With this notation one introduces the following set:

Let $\mathcal{A}_r$ be the set of operators $A(K,0)$ with the properties:
(i) The operator $A(K,x)$ with $K+x \subset W_r$ is such that $U(\Lambda(t))A(K,x)\Omega$ has a bounded analytic continuation into the strip $S(-\frac{1}{2},0)$ with continuous boundary–values and

(*ii*) $A^*(K,x)$ *with* $K+x \subset W_l$ *is such that* $U(\Lambda(t))A^*(K,x)\Omega$ *has a bounded analytic continuation into the strip* $S(0,\frac{1}{2})$ *with continuous boundary–values.*

The set $\mathcal{A}_r^*$ *will be denoted by* $\mathcal{A}_l$. *It has the corresponding property with respect to* $\mathcal{M}(W_l)$.

Recall the main result about wedge duality:

Theorem:

Consider a Lorentz covariant theory of local observables in the vacuum–sector. This theory fulfils wedge–duality exactly if

$$\{A(K,x); A(K,0) \in \mathcal{A}_r,\ K+x \subset W_r\}$$
$$\{A(K,x); A(K,0) \in \mathcal{A}_l,\ K+x \subset W_l\}$$

are $*$*–strong dense in* $\mathcal{M}(W_r)$ *and* $\mathcal{M}(W_l)$, *respectively.*

In the above result it has not been mentioned what one knows about the structure of the elements $U(\Lambda(\frac{-i}{2}))A(K,x)\Omega$, where $K+x$ is in the right wedge. This question has been answered in

Theorem 3.6 of [22]:

(*i*) *For every* $A(K,0) \in \mathcal{A}_r$ *and every* x *with* $K+x \subset W_r$ *there exists an element* $\hat{A}(K,0) \in \mathcal{A}_l$, *such that the following relation holds:*

$$U(\Lambda(-\frac{i}{2}))A(K,x)\Omega = \hat{A}(K,P_W x)\Omega\ ,$$

where P_W *is the reflection in the characteristic two–plane, which does not change the perpendicular directions.*

(*ii*) *For every* y *with* $K+y \in W_l$ *and* $A(K,0) \in \mathcal{A}_l$ *there exists an element* $\tilde{A}(K,0) \in \mathcal{A}_r$ *fulfilling the relation*

$$U(\Lambda(\frac{i}{2}))A(K,y)\Omega = \tilde{A}(K,P_W y)\Omega\ .$$

The above statement

$$U(\Lambda(-\frac{i}{2}))A(K,x)\Omega = \hat{A}(K,P_W x)\Omega\ ,$$

says nothing about what is happening if with $A(K)$ also $A(K)^*$ belongs to $\mathcal{A}_r$. In Wightman's theory one has for $x \in W$:

$$\widehat{\Phi^*(x)}\Omega = \{\hat{\Phi}(P_W(x))\}^*\Omega\ .$$

The reality condition to be introduced is of the same form.

Reality condition:

We say a Poincaré covariant theory of local observables in the vacuum sector with the property of wedge duality fulfils the reality condition if (*i*) *every* $A(K,0) \in \mathcal{A}_r \cap \mathcal{A}_l$ *and every* x *such that* $K + x \subset W_r$ *fulfils the relation*

$$\widehat{A^*}(K, P_W x) = \{\hat{A}(K, P_W x)\}^* .$$

(*ii*) Ω *is cyclic for the set*

$$\{A(K,x); A(K,0) \in \mathcal{A}_r \cap \mathcal{A}_l, \text{ and } K + x \subset W_r\} .$$

With the wedge duality and this condition it is possible to give a proof of the CPT–theorem. Before formulating the result it is useful to understand the reason for this. Together with the modular group of the algebra $\mathcal{M}(W)$ there is a conjugation J_W which maps $\mathcal{M}(W)$ onto its commutant

$$J_W \mathcal{M}(W) J_W = \mathcal{M}(W)' .$$

If the theory fulfils wedge duality one has $\mathcal{M}(W)' = \mathcal{M}(W')$. A good candidate for the CPT–operator is

$$\Theta = J_W U(R_W(\pi)) ,$$

provided the origin is contained in the edge of the wedge. $R_W(\alpha)$ denotes the rotation in the two–plane perpendicular to the characteristic two–plane of the wedge. This can only be correct if J_W acts local. From a result of [23] one knows only that J_W maps the cylinder $K+x$ onto the cylinder $K+P_W(x)$. If one wants to have local action for J_W then the modular group Δ_W^{it} must also act local.

If Δ_W^{it} acts local then the result of Bisognano and Wichmann suggests that Δ_W^{it} and $U(\Lambda_W(t))$ coincide. This is guaranteed by the reality condition [24].

Theorem:

In a representation of a Poincaré covariant theory of local observables in the vacuum sector the modular group associated with the algebra of any wedge coincides with the corresponding Lorentz boosts iff the theory fulfils wedge duality and the above reality condition with respect to the Lorentz transformations.

Using this one gets the CPT–theorem by a result of Brunetti, Guido and Longo [25] and Guido and Longo [26].

Theorem:

Assume the theory of local observables is such that for every wedge W *the modular group acts as the corresponding Lorentz boosts, i.e.*

$$\Delta_W^{\mathrm{it}} \mathcal{M}(D) \Delta_W^{-\mathrm{it}} = \mathcal{M}(\Lambda_W(t) D) ,$$

where D *is any double cone, then:*

(1) *The theory is Poincaré covariant, i.e. there exists a continuous unitary representation* $U(g)$ *of the Poincaré group with*

$$U(\Lambda_W(t)) = \Delta_W^{\mathrm{it}} .$$

(2) *The theory fulfils the CPT–theorem and one has for every wedge with 0 in the edge of the wedge*

$$\Theta = J_W U(R_W(\pi)) .$$

We collect the results obtained previously and get:

Theorem:

If a Poincaré covariant theory of local observables fulfils wedge duality and the reality condition then this theory is CPT–covariant.

Remark:

Brunetti, Guido and Longo and also Guido and Longo have used group theoretical methods for their proof. That the theory is Poincaré covariant if the modular groups fulfil the Bisognano Wichmann property can be shown also directly [27]. The construction of the CPT–symmetry using only the duality and reality condition is still missing. The problem is the following: For every wedge the expression $U(\Lambda_W(t)R_W(\varphi))A(D+x)\Omega$ can be analytically continued to the group element -1. If one starts from two different wedges such that $A(D+x)$ belongs to the corresponding algebras one has to show that the two different analytic continuations do not land on different sheets. This part is still missing.

3 Tensor product decomposition

As an example which is solvable in the theory of local observables is the tensor product problem. In Wightman's field theory there are two operations which are related to the tensor product. These are the s– and the p–products. If $\Phi_1(x)$ and $\Phi_2(x)$ are Wightman fields, then the s–product corresponds to

$$\Phi_1(x) s \Phi_2(x) = \Phi_1(x) \otimes \mathbb{1}_2 + \mathbb{1}_1 \otimes \Phi_2(x) ,$$

while the p–product corresponds formally to

$$\Phi_1(x) p \Phi_2(x) = \Phi_1(x) \otimes \Phi_2(x) .$$

As operators the product on the righthand side is not well defined. But since one can multiply Wightman functions in the complex one can give the above expression a well defined meaning.

For the s–product there exists a reduction theory due to Hegerfeldt [28], but his investigation has not solved all the problems associated with the s–product. In particular a characterization of indecomposible fields is missing. There exists no decomposition theory for the p–product.

In order to understand the problems involved with the tensor product decomposition in the theory of local observables let us start with two theories $\{\mathcal{M}_i(O), U_i(\Lambda, x), \mathcal{H}_i, \Omega_i\}$, $i = 1, 2$. One can define a new theory on $\mathcal{H}_1 \overline{\otimes} \mathcal{H}_2$ by

$$\mathcal{M}(O) = \mathcal{M}_1(O) \overline{\otimes} \mathcal{M}_2(O), \quad U(\Lambda, x) = U_1(\Lambda, x) \otimes U_2(\Lambda, x) , \quad \Omega = \Omega_1 \otimes \Omega_2 .$$

The new theory $\{\mathcal{M}(O), U(\Lambda,x), \mathcal{H}, \Omega\}$ fulfils again all axioms of local quantum field theory. In order to uncover the direct product structure one has to look at the sub–theory

$$\{\mathcal{M}_1(O) \otimes \mathbb{1}, U(\Lambda,x), \mathcal{H}, \Omega\}\ , \quad \text{and} \quad \{\mathbb{1} \otimes \mathcal{M}_2(O), U(\Lambda,x), \mathcal{H}, \Omega\}\ ,$$

which fulfils the assumptions of the theory of local observables, except the cyclicity assumption for the vacuum vector. If one denotes the algebras $\mathcal{M}_1(O)\otimes\mathbb{1}$ by $\mathcal{N}_1(O)$ and $\mathbb{1}\otimes\mathcal{M}_2(O)$ by $\mathcal{N}_2(O)$ then one has to answer the following questions:

1) In order to obtain a tensor product it is necessary that all the $\mathcal{N}_1(O)$ commute with all the $\mathcal{N}_2(O)$. Does this imply that the algebra $\mathcal{N}_1(O) \vee \mathcal{N}_2(O)$ can be written as tensor product?
2) It is a very strong assumption to start with sub–quantum–field–theories $\mathcal{N}_1(O)$ and $\mathcal{N}_2(O)$. Therefore, one would like to start with only one domain, for instance a wedge W_0, and assume that $\mathcal{M}(W_0)$ can be written as a tensor product
$$\mathcal{N}(W_0) \cong \mathcal{N}_1(W_0)\overline{\otimes}\mathcal{N}_2(W_0)\ .$$
Is it possible to construct a tensor product decomposition for all other wedges and for all double cones?
3) In order to obtain a tensor product decomposition of the whole theory it is not only necessary to have a decomposition for every wedge and every double cone but the family of decompositions has to fulfil some coherence property. If one wants to get a tensor product decomposition of the whole field theory, then all the algebras $\mathcal{N}_1(W)$ and $\mathcal{N}_1(D)$ must act on the same Hilbert space $\mathcal{H}_1$ and the $\mathcal{N}_2(W)$ and $\mathcal{N}_2(D)$ must act on the Hilbert space $\mathcal{H}_2$.

The technique by which one can handle the questions is the Tomita–Takesaki theory. Besides the usual axioms of local quantum field theory in the vacuum sector one has to assume that the theory fulfils the Bisognano–Wichmann property. This is the assumption used also in the derivation of the PCT–theorem, namely, for every wedge W the modular group Δ_W^{it} of $\mathcal{M}(W)$ coincides with the corresponding Lorentz boosts $U(\Lambda_W(t))$.

Remark:

(1) *As a consequence of the Bisognano–Wichmann property one concludes that the theory fulfils the wedge duality, i.e, for every wedge the relation*
$$\mathcal{M}(W)' = \mathcal{M}(W')$$
holds, where W' denotes the opposite wedge of W. For the proof see [25] *or* [27].

(2) *If one identifies the algebra of the double cone D with*
$$\mathcal{M}(D) = \cap\{\mathcal{M}(W);\ D \subset W\}\ .$$
then the general duality property
$$\mathcal{M}(D)' = \mathcal{M}(D')$$
holds, where D' denotes the (interior) of the spacelike complement of D.

The existence of a cyclic vacuum vector together with the locality condition implies that the global algebra $\mathcal{M} := \mathcal{M}(\mathbb{R}^d)$ is of type I. More precisely, the commutant of $\mathcal{M}$ is an abelian algebra [29,30] which must be the same as the center of $\mathcal{M}$.

If the algebra $\mathcal{M}$ is of the form

$$\mathcal{M} = \mathcal{M}_1 \overline{\otimes} \mathcal{M}_2 \quad \text{on} \quad \mathcal{H} = \mathcal{H}_1 \overline{\otimes} \mathcal{H}_2 \quad \text{with} \quad \Omega = \Omega_1 \otimes \Omega_2 \,,$$

then also the modular group splits, i.e.

$$\Delta^{it} = \Delta_1^{it} \otimes \Delta_2^{it} \,.$$

If this is the case, then $\mathcal{M}_1 \otimes \mathbb{1}$ is a subalgebra of $\mathcal{M}$, which is mapped by σ^t onto itself.

$$\sigma^t(\mathcal{M}_1 \otimes \mathbb{1}) = \mathcal{M}_1 \otimes \mathbb{1} \,.$$

Subalgebras, which are mapped by σ^t onto itself, are "modular covariant subalgebras".

The treatment of the modular covariant subalgebras and the tensor product decomposition of von Neumann algebras is taken from an article of Takesaki [31]. Modular covariant subalgebras have the following well known and easy to verify properties. (See [31–33] and [34].):

Lemma:
Let $\mathcal{N}$ be a modular covariant subalgebra of $\mathcal{M}$. Let $\mathcal{H}_{\mathcal{N}}$ be the closure of $\mathcal{N}\Omega$ and denote by $E_{\mathcal{N}}$ the projection onto $\mathcal{H}_{\mathcal{N}}$. By $\widehat{\mathcal{N}}$ one denotes the restriction of $\mathcal{N}$ to $\mathcal{H}_{\mathcal{N}}$. Then:

1. *$E_{\mathcal{N}}$ commutes with Δ^{it} and J. The restriction of Δ and J to $\mathcal{H}_{\mathcal{N}}$ will be denoted by $\widehat{\Delta}$ and $\widehat{J}$.*
2. *$\widehat{\Delta}$ and $\widehat{J}$ are the modular group and modular conjugation of $(\widehat{\mathcal{N}}, \Omega)$.*
3. *The commutant of $\widehat{\mathcal{N}}$ in $\mathcal{H}_{\mathcal{N}}$ coincides with $\widehat{J}\widehat{\mathcal{N}}\widehat{J}$.*
4. *The map $\mathcal{N} \longrightarrow \widehat{\mathcal{N}}$ is an isomorphism of von Neumann algebras.*
5. *$A \in \mathcal{M}$ and $[A, E_{\mathcal{N}}] = 0$ implies $A \in \mathcal{N}$.*
6. *$A \in \mathcal{M}$ then*
$$E_{\mathcal{N}} A E_{\mathcal{N}} \in \widehat{\mathcal{N}} \,.$$

The results of the last lemma have been strengthened.

Theorem (Takesaki [31]):
With the previous assumptions and notations one obtains:

1) *There exists a normal faithful conditional expectation $\mathcal{E}$ from $\mathcal{M}$ onto $\mathcal{N}$.*
2) *$\mathcal{E}$ commutes with the modular action:*
$$\mathcal{E}(\operatorname{Ad} \Delta^{it} A) = \operatorname{Ad} \Delta^{it} \mathcal{E}(A), \;\; A \in \mathcal{M} \,.$$
3) *There exists also a conditional expectation $\mathcal{E}'$ from $\mathcal{M}'$ to $J\mathcal{E}(\mathcal{M})J$ defined by*
$$\mathcal{E}'(A') = J\mathcal{E}(JA'J)J, \qquad A' \in \mathcal{M}' \,.$$

4) *Let E be a projection with $E\Omega = \Omega$. If there is a von Neumann algebra $\mathcal{N} \subset \mathcal{M}$ with $E \in \mathcal{N}'$ and the central support of E in $\mathcal{N}'$ is $\mathbb{1}$, and if in addition one has $E\mathcal{M}E = \mathcal{N}E$, then $\mathcal{N}$ is a modular covariant subalgebra of $\mathcal{M}$.*

Notice that the conditions of 4) imply that there exists a conditional expectation from $\mathcal{M}$ onto $\mathcal{N}$.

Let $\mathcal{N}^c = \mathcal{N}' \cap \mathcal{M}$ be the relative commutant of $\mathcal{N}$ in $\mathcal{M}$. Since $\mathcal{N}$ is a modular covariant subalgebra of $\mathcal{M}$ the same is true for $\mathcal{N}^c$. Hence exists a second conditional expectation $\mathcal{E}^c$ with

$$\mathcal{E}^c : \mathcal{M} \longrightarrow \mathcal{N}^c .$$

The existence of the two conditional expectations $\mathcal{E}$ and $\mathcal{E}^c$ has some important consequences. These have been discovered by Takesaki [31].

Theorem:

Let $\mathcal{M}$ be a von Neumann algebra with cyclic and separating vector Ω. Assume the modular covariant subalgebra $\mathcal{N}$ of $\mathcal{M}$ is a von Neumann subfactor. Let $\mathcal{N}^c$ be the relative commutant of $\mathcal{N}$ in $\mathcal{M}$ and let $\mathcal{R} = \mathcal{N} \vee \mathcal{N}^c$ be the von Neumann algebra generated by $\mathcal{N}$ and $\mathcal{N}^c$. Then the map

$$\pi : \sum A_i \otimes B_i \in \mathcal{N} \otimes \mathcal{N}^c \longrightarrow \sum A_i B_i \in \mathcal{R} \subset \mathcal{M}$$

extends to an isomorphism of $\mathcal{N} \bar{\otimes} \mathcal{N}^c$ onto $\mathcal{R} = \mathcal{N} \vee \mathcal{N}^c$. Moreover, the vacuumstate $(\Omega, . \Omega)$ is a product state on $\mathcal{R}$, i.e. $A \in \mathcal{N}$ and $B \in \mathcal{N}^c$ implies

$$(\Omega, AB\Omega) = (\Omega, A\Omega)(\Omega, B\Omega) .$$

In order to apply the concepts just described one starts with a wedge W and assumes that the algebra $\mathcal{M}(W)$ has a modular covariant subalgebra $\mathcal{N}(W)$. Let $\mathcal{E}_W$ be the associated conditional expectation and E_W the projection onto $[\mathcal{N}(W)\Omega]$. If one now changes the wedge to $\Lambda W + x$ then, of course, $U(\Lambda, x)\mathcal{N}(W)U(\Lambda, x)^*$ is a modular covariant subalgebra of $\mathcal{M}(\Lambda W + x)$. But in order to obtain a decomposition of the global field theory the projections E_W and $E_{\Lambda W + x}$ have to coincide. We do not only have to transport the conditional expectations to different wedges, but one also needs conditional expectations for the algebras $\mathcal{M}(D)$ associated with double cones. In order to be able to construct such conditional expectations the algebras must be closely related to the algebras of wedges. Therefore, one sets

$$\mathcal{M}(D) = \cap\{\mathcal{M}(\Lambda W + x);\ D \subset \Lambda W + x\} .$$

Now the coherence property can be defined.

Definition:

Assume that the modular covariant subalgebras $\mathcal{N}(D) \subset \mathcal{M}(D)$ and $\mathcal{N}(W) \subset \mathcal{M}(W)$ are associated with every double cone D and ever wedge W. Then this family is called coherent if the projections E_D and E_W coincide for all double cones D and for all wedges W.

One can show the following result:

Theorem:
Assume one is dealing with a quantum field theory fulfilling the usual axioms and the Bisognano–Wichmann property. Assume also that for one wedge W_0 the algebra $\mathcal{M}(W_0)$ has a modular covariant subalgebra $\mathcal{N}(W_0)$. Then there exists a coherent family of modular covariant subalgebras $\{\mathcal{N}(W), \mathcal{N}(D)\}$ of $\{\mathcal{M}(W), \mathcal{M}(D)\}$ such that $\mathcal{N}(W_0)$ is the given subalgebra of $\mathcal{M}(W_0)$.

The key to the demonstration is the following result taken from [24] Thm. 4.11, which needs some explanation of the notation. Let $\mathcal{M}$ be a von Neumann algebra with cyclic and separating vector Ω. Then a one–parametric group $V(t)$ of unitaries is called half–sided translation for $\mathcal{M}$ if

i $V(t)\Omega = \Omega$ for all $t \in \mathbb{R}$,
ii $V(t) = e^{iHt}$ with $H \geq 0$,
iii $V(t)\mathcal{M}V^*(t) \subset \mathcal{M}$ for $t \geq 0$ (or for $t \leq 0$).

Proposition:
Let $\mathcal{M}$ be a von Neumann algebra on $\mathcal{H}$ with cyclic and separating vector Ω and let $V(t)$ be a half–sided translation for $\mathcal{M}$. Assume $\mathcal{N}$ is a modular covariant subalgebra of $\mathcal{M}$ and $\mathcal{E}$ the associated conditional expectation. Then the map $\mathrm{Ad}\,V(t)$ commutes with $\mathcal{E}$ for $t \geq 0$ if $V(t)$ is a + half–sided translation and for $t \leq 0$ if $V(t)$ is a − half–sided translation. This implies

$$[V(t), E_{\mathcal{N}}] = 0 \qquad \text{for all} \qquad t \in \mathbb{R}\,.$$

In the boundary of a wedge W there are two different lightlike vectors ℓ_1 and ℓ_2. The translations in the direction of these lightlike vectors are half–sided translations for the algebra $\mathcal{M}(W)$, so that one can transport the modular covariant subalgebra $\mathcal{N}(W_0)$ of $\mathcal{M}(W_0)$ in a coherent way to all translated wedges. Therefore, the projection E_{W_0} onto $[\mathcal{N}(W_0)\Omega]$ commutes with all translations in the two–plane spanned by ℓ_1, ℓ_2, which is called the characteristic two–plane of the wedge. From this one can derive

Lemma:
Let the dimension of the Minkowski space be larger than 2. Let $\mathcal{N}(W)$ be a modular covariant subalgebra of $\mathcal{M}(W)$. Then E_W commutes not only with the translations in the characteristic two–plane but E_W commutes with all translations.

With this result one can transport the modular covariant subalgebras from one wedge to all translated wedges in a coherent way. In order to get to Lorentz transformed wedges one has to use the Bisognano–Wichmann property. This implies that there exist other half–sided translations for $\mathcal{M}(W)$ besides the translations in the two lightlike directions. These translations do not have an easy interpretation and need some explanation.

If $\mathcal{N}$ is a von Neumann subalgebra of $\mathcal{M}$ with the same cyclic and separating vector Ω, then one says that $\mathcal{N}$ fulfils the condition of half-sided modular inclusion with respect to $\mathcal{M}$ if

$$\mathrm{Ad}\,\Delta^{it}_{\mathcal{M}}\mathcal{N} \subset \mathcal{N} \quad \text{for} \quad t \leq 0 \quad (\text{or} \quad t \geq 0)$$

holds. This situation implies by a result of H.-W. Wiesbrock [35,36] that there exists a half-sided translation $U(t)$ such that one has

$$\mathcal{N} = \mathrm{Ad}\,U(1)\mathcal{M} \qquad \text{or} \quad (\mathcal{N} = \mathrm{Ad}\,U(-1)\mathcal{M})\,.$$

For the following it is essential that the theory enjoys the Bisognano–Wichmann property. Let $W(\ell, \ell_1)$,$W(\ell, \ell_2)$ be two wedges with the same first vector then the algebra

$$\mathcal{M}(W(\ell, \ell_1) \cap W(\ell, \ell_2))$$

fulfils the condition of half-sided modular inclusion with respect to both algebras $\mathcal{M}(W(\ell, \ell_1))$ and $\mathcal{M}(W(\ell, \ell_2))$ [27]. Using this result one can transport the modular covariant subalgebras from one wedge to all others in such a manner that the coherence property is fulfilled. Starting from $W(\ell_1^0, \ell_2^0)$ and choosing $W(\ell_1^0, \ell_2^1), \ell_2^0 \neq \ell_2^1$ then one obtains a modular covariant subalgebra of $\mathcal{M}(W(\ell_1^0, \ell_2^0) \cap W(\ell_1^0, \ell_2^1))$ by means of the half-sided translation which maps $\mathcal{M}(W(\ell_1^0, \ell_2^0))$ onto $\mathcal{M}(W(\ell_1^0, \ell_2^0) \cap W(\ell_1^0, \ell_2^1))$. Since this transportation is done by a half-sided translation the coherence property is fulfilled. There exists also a half-sided translation which maps $\mathcal{M}(W(\ell_1^0, \ell_2^1))$ onto $\mathcal{M}(W(\ell_1^0, \ell_2^0) \cap W(\ell_1^0, \ell_2^1))$ which is at the same time a half-sided translation for $\mathcal{M}(W(\ell_1^0, \ell_2^0) \cap W(\ell_1^0, \ell_2^1))$. Reversing the last construction one is able to find a modular covariant subalgebra of $\mathcal{M}(W(\ell_1^0, \ell_2^1))$, such that the coherence condition is satisfied.

$$\mathcal{N}(W(\ell_1^0, \ell_2^0) \longrightarrow \mathcal{N}(W(\ell_1^0, \ell_2^0) \cap W(\ell_1^0, \ell_2^1)) \longrightarrow \mathcal{N}(W(\ell_1^0, \ell_2^1))\,.$$

Repeating this procedure with the second and eventually with a third lightlike vector, one can get to every other wedge. By this procedure one has to use alternatively the first and the second lightlike vector defining the wedge.

Using also the translations one can construct to every wedge W a modular covariant subalgebra $\mathcal{N}(W)$ of $\mathcal{M}(W)$, such that E_W is independent of W.

It remains to construct a modular covariant subalgebra for every double cone.

Lemma:
Let $\mathcal{N}(W)$ be a coherent family of modular covariant subalgebras of $\mathcal{M}(W)$. Define for any double cone

$$\mathcal{N}(D) = \cap\{\mathcal{N}(W);\ D \subset W\}\,.$$

Then $\mathcal{N}(D)$ is a modular covariant subalgebra of

$$\mathcal{M}(D) = \cap\{\mathcal{M}(W);\ D \subset W\}\,.$$

Moreover, one has

$$[\mathcal{N}(D)\Omega] = [\mathcal{N}(W)\Omega]\,.$$

These are the steps to show the theorem. Since the projection E commutes with the Poincaré group one obtains:

Theorem:

Let $\{\mathcal{M}(D), U(\Lambda, x), \mathcal{H}, \Omega\}$ be a theory of local observables fulfilling the usual axioms and the Bisognano–Wichmann property. Let $\mathcal{N}(W_0)$ be a modular covariant subalgebra of $\mathcal{M}(W_0)$. Then one obtains a coherent sub-theory

$$\{\mathcal{N}(D), U(\Lambda, x), \mathcal{H}, \Omega\} ,$$

which fulfils the axioms of local quantum field theory except the cyclicity of the vacuum vector. Let E be the projection onto $[\mathcal{N}(W_0)\Omega]$, which commutes with all $\mathcal{N}(D)$ and the representation of the Poincaré group, then

$$\{\widehat{\mathcal{N}}(D), \widehat{U}(\Lambda, x), E\mathcal{H}, \Omega\}$$

defines a local quantum field theory with cyclic vacuum. $\widehat{\mathcal{N}}$ denotes the restriction of $\mathcal{N}$ to $E\mathcal{H}$. In particular one has for every wedge

$$\widehat{\mathcal{N}}(W) = \vee\{\widehat{\mathcal{N}}(D); D \subset W\} .$$

In order to apply Takesaki's result on tensor products the modular covariant subalgebra $\mathcal{N}(W)$ of $\mathcal{M}(W)$ must be a factor. This can be shown under the assumption that $\mathcal{M}(W)$ itself is a factor. This is known to be the case if the global algebra is a factor. Since the factor property for $\mathcal{M}(D)$ is not known one is not able to show that $\mathcal{N}(D)$ is a factor. Hence Takesaki's result can not be used. Here one has to use a characterization of tensor products due to Ge and Kadison [37].

For the factor property of $\mathcal{N}(W)$ the following result can be used ([24] Lemma II.3):
Let $U(t)$ be a half-sided translation for the von Neumann algebra $\mathcal{M}$. Denote by E_0 the projection onto the $U(t)$ invariant vectors and by F_1 the projection onto the eigenvectors of $\Delta_{\mathcal{M}}$ to the eigenvalue 1. Then one has

$$F_1 \leq E_0 .$$

From this one concludes:

Proposition:

Let $\{\mathcal{M}(D), U(\Lambda, x), \mathcal{H}, \Omega\}$ be a theory of local observables. Assume the global algebra is a factor and hence $\mathcal{M}(W)$ is a factor. Then every modular covariant subalgebra of $\mathcal{M}(W)$ is a factor.

Now one is in the following situation: Starting with one wedge W_0 and a modular covariant subalgebra $\mathcal{N}(W_0)$ of $\mathcal{M}(W_0)$ there are the modular covariant subalgebras

$$\mathcal{N}(W_0), \quad \mathcal{N}^c(W_0), \quad \mathcal{N}^p(W_0) := \mathcal{N}(W_0) \vee \mathcal{N}^c(W_0) .$$

Associated with these algebras there are three field theories

$$\{\mathcal{N}(D), U(\Lambda, x), \mathcal{H}, \Omega\}, \ \{\mathcal{N}^c(D), U(\Lambda, x), \mathcal{H}, \Omega\}, \ \{\mathcal{N}(D)^p, U(\Lambda, x), \mathcal{H}, \Omega\}.$$

For every wedge Takesaki's theorem implies

$$\mathcal{N}^p(W) \cong \mathcal{N}(W) \overline{\otimes} \mathcal{N}^c(W) .$$

For the double cones it is not known whether or not the algebras $\mathcal{N}(D)$ are factors. Therefore, one has to apply different techniques in order to conclude that $\mathcal{N}^p(D)$ is isomorphic to a tensor product. Here one can use a result of L. Ge and R. Kadison [37]. They characterize subalgebras of a tensor product $\mathcal{R}\overline{\otimes}\mathcal{S}$ which themselves are tensor products $\mathcal{R}_1\overline{\otimes}\mathcal{S}_1$, with $\mathcal{R}_1 \subset \mathcal{R}$ and $\mathcal{S}_1 \subset \mathcal{S}$. Since every double cone is contained in a wedge their result can be applied and one finds

$$\mathcal{N}^p(D) \cong \mathcal{N}(D) \overline{\otimes} \mathcal{N}^c(D) .$$

Collecting the results one obtains:

Theorem:

Let $\{\mathcal{M}(O), U(\Lambda, x), \mathcal{H}, \Omega\}$ be a theory of local observables fulfilling the usual axioms and the Bisognano–Wichmann property. Let W_0 be a wedge and assume that $\mathcal{N}(W_0)$ is a modular covariant subalgebra of $\mathcal{M}(W_0)$. Let $\mathcal{N}^c(W_0)$ be the relative commutant of $\mathcal{N}(W_0)$ in $\mathcal{M}(W_0)$ and $\mathcal{N}^p(W_0) = \mathcal{N}(W_0) \vee \mathcal{N}^c(W_0)$. Then:

(1) *There exists on $\mathcal{H}$ a sub–theory of local observables*

$$\{\mathcal{N}^p(D), \mathcal{N}^p(W), U(\Lambda, x)\}$$

covariant under the existing unitary group $U(\Lambda, x)$. Moreover, $\{\mathcal{N}^p(D), \mathcal{N}^p(W)\}$ are modular covariant subalgebras of $\{\mathcal{M}(D), \mathcal{M}(W)\}$ such that $\mathcal{N}^p(W)$ has a trivial relative commutant in $\mathcal{M}(W)$. In addition, for W_0 the algebra $\mathcal{N}^p(W)$ coincides with the given $\mathcal{N}^p(W_0)$. If E^p denotes the projection onto $[\mathcal{N}^p(W_0)\Omega]$ then E^p commutes with $\mathcal{N}^p(D), \mathcal{N}^p(W)$ and the group representation $U(\Lambda, x)$. Moreover, Ω is cyclic for $\mathcal{N}^p(D)$ in $E^p\mathcal{H}$. If one denotes the restriction of $\mathcal{N}^p(D)$ and $U(\Lambda, x)$ by $\widehat{\mathcal{N}}^p(D)$ and $\widehat{U}(\Lambda, x)$ respectively then

$$\{\widehat{\mathcal{N}}^p(D), \widehat{U}(\Lambda, x), E^p\mathcal{H}, \Omega\}$$

defines a theory of local observables satisfying the usual axioms and the Bisognano–Wichmann property.

(2) *There exist two coherent families $\{\mathcal{N}(D), \mathcal{N}(W)\}$ and $\{\mathcal{N}^c(D), \mathcal{N}^c(W)\}$ of modular covariant subalgebras of $\{\mathcal{M}(D), \mathcal{M}(W)\}$ extending $\mathcal{N}(W_0)$ and $\mathcal{N}^c(W_0)$, respectively. If E and E^c are the projections onto $[\mathcal{N}(W_0)]$ and $[\mathcal{N}^c(W_0)]$ then these projections commute with $U(\Lambda, x)$, and E with $\mathcal{N}(D)$ and E^c with $\mathcal{N}^c(D)$. With this one obtains:*

$$\{\widehat{\mathcal{N}}^p(D), \widehat{U}(\Lambda, x), E^p\mathcal{H}, \Omega\} \cong$$
$$\{\widehat{\mathcal{N}}^0(D)\overline{\otimes}\widehat{\mathcal{N}}^c(D), \widehat{U}^0(\Lambda, x) \otimes \widehat{U}^c(\Lambda, x), E\mathcal{H}\overline{\otimes}E^c\mathcal{H}, \Omega^0 \otimes \Omega^c\} .$$

In this formula $\widehat{X}^0$ denotes the restriction to $E\mathcal{H}$, $\widehat{X}^c$ the restriction to $E^c\mathcal{H}$, and $\widehat{X}^p$ the restriction to $E^p\mathcal{H}$.

It remains to discuss the situation where the relative commutant is trivial, i.e. $\mathcal{N}^c(W) = \mathbb{C}\mathbb{1}$. This situation appears in the hidden charge problem, which is the following: If we start with a theory of local observables $\{\mathcal{N}(O), U(\Lambda, x), \mathcal{H}, \Omega\}$, such that the theory has charged sectors, which are connected by localized Bose fields, then we can add these Bose fields and obtain a field algebra $\{\mathcal{F}(O), \widehat{U}(\Lambda, x), \widehat{\mathcal{H}}, \Omega\}$, which also fulfils the assumptions of the theory of local observables. Knowing only the latter theory one would like to discover the local net $\{\mathcal{N}(O), U(\Lambda, x), \mathcal{H}, \Omega\}$ and the structure of the charged fields.

Unfortunately the proof is missing that the case of the trivial relative commutant is only connected with the hidden charge problem. There are strong indications that this is true. Let us assume that this is the case. Then, in general, one is in the situation characterized by the diagram Fig. 1.

$$
\begin{array}{ccccc}
\{\mathcal{N}, \mathcal{N}^c\} & \xrightarrow{\text{B.f.}} & \{\mathcal{N}^{cc}, \mathcal{N}^c\} & & \\
\Big\downarrow \text{t.p.} & & \Big\downarrow \text{t.p.} & & \\
\mathcal{N}\overline{\otimes}\mathcal{N}^c & \xrightarrow{\text{B.f.}} & \mathcal{N}^{cc}\overline{\otimes}\mathcal{N}^c & \xrightarrow{\text{B.f.}} & \mathcal{M}
\end{array}
$$

Fig. 1. General situation for the hidden charge case: $\mathcal{N}$ is a modular covariant subalgebra of $\mathcal{M}$, and $\mathcal{N}^c = \mathcal{M} \cap \mathcal{N}'$.
t.p. stands for the construction of the tensor product;
B.f. stands for the construction of the Bose field.

Looking at the diagram it would be interesting to understand why one does not obtain the whole algebra by going first from $\mathcal{N}$ to $\mathcal{N}^{cc}$ and forming the tensor product $\mathcal{N}^{cc}\overline{\otimes}\mathcal{N}^c$ afterwards. It can happen that both theories $\mathcal{N}^{cc}$ and $\mathcal{N}^c$ have superselection sectors connected to the vacuum by Fermi fields. Forming the tensor product of the two Fermi fields one obtains a field which commutes for spacelike distances, i.e. a Bose field. In field theories on low dimensional Minkowski space the same phenomenon can be achieved with help of anyonic fields. Examples are due to K.-H. Rehren [38].

Since by philosophy of the theory of local observables one should start from the observable algebra and reconstruct the whole theory from it. Therefore, one has to start from elementary theories. A theory is indecomposible with respect to the tensor product and the hidden charge problem iff the algebra of any wedge contains no non–trivial modular covariant subalgebra. Since every non–trivial operator of $\mathcal{M}(W)$ and the modular group generate all of $\mathcal{M}(W)$ it follows that the vacuum Hilbert space is separable.

References

1. H. Lehmann, K. Symanzik and W. Zimmermann: *Zur Formulierung quantisierter Feldtheorien*, Nuovo Cimento **1**, 425, (1955).
2. A.S. Wightman: *Quantum field theory in terms of vacuum expectation values*, Phys. Rev. **101**, 860 (1956).
3. R. Haag: *Local Quantum Physics*, Springer Verlag, Berlin-Heidelberg-New York (1992).
4. H.-J. Borchers and W. Zimmermann: *On the Self–Adjointness of Field Operators.* Nuovo Cimento **31**, 1047-1059 (1964).
5. W. Pauli: *Exclusion Principle, Lorentz Group and Reflection of Space-Time and Charge*, Niels Bohr and the Development of Physics, W. Pauli, l. Rosenfeld, and V. Weisskopf ed.,Pergamon Press, London (1955).
6. J. Schwinger: *The Theory of Quantized Fields I*, Phys. Rev. **82**, 914-927, (1951).
7. G. Lüders: *On the equivalence of invariance under time reversal and under particle-antiparticle conjugation for relativistic field theories*, Danske Vidensk. Selskab, Mat.-fysiske Meddelelser **28,no 5**, 1-17, (1954).
8. G. Lüders and B. Zumino: *On the connection between spin and statistics*, Phys. Rev. **110**, 1450-1453, (1958).
9. R. Jost: *Eine Bemerkung zum CTP Theorem*, Helv. Phys. Acta **30**, 409-416, (1957).
10. D. Hall and A.S. Wightman: *A theorem on invariant analytic functions with applications to relativistic quantum field theory*, Danske Vidensk. Selskab, Mat.-fysiske Meddelelser **31,no 5**, 1-41, (1957).
11. R. Streater: *Local Fields with the Wrong Connection Between Spin and Statistics*, Commun. Math. Phys. **5**, 88-96 (1967).
12. A.I. Oksak and I.T. Todorov: *Invalidity of the TCP–Theorem for Infinite-Component Fields*, Commun. Math. Phys. **11**, 125, (1968).
13. M. Tomita: *Quasi-standard von Neumann algebras*, Preprint (1967).
14. M. Takesaki: *Tomita's Theory of Modular Hilbert Algebras and its Applications*, Lecture Notes in Mathematics, Vol. **128** Springer-Verlag Berlin, Heidelberg, New York (1970).
15. H. Reeh and S. Schlieder: *Eine Bemerkung zur Unitäräquivalenz von Lorentzinvarianten Feldern*, Nuovo Cimento **22**, 1051 (1961).
16. J. Bisognano and E.H. Wichmann: *On the duality condition for a Hermitean scalar field*, J. Math. Phys. **16**, 985-1007 (1975).
17. J. Bisognano and E.H. Wichmann: *On the duality condition for quantum fields*, J. Math. Phys. **17**, 303-321 (1976).
18. D. Buchholz: *On the Structure of Local Quantum Fields with Non–trivial Interactions*, In: Proceedings of the International Conference on Operator Algebras, Ideals and their Applications in Theoretical Physics, Leipzig 1977, Teubner–Texte zur Mathematik (1978) p. 146-153.
19. P.D. Hislop and R. Longo: *Modular structure of the local algebra associated with a free massless scalar field theory*, Commun. Math. Phys. **84**, 71-85 (1982).
20. J. Yngvason: *A Note on Essential Duality*, Lett.Math.Phys. **31**, 127-141, (1994).
21. H.-J. Borchers and J. Yngvason: *Modular Groups of Quantum Fields in Thermal States*, Preprint (1998).
22. H.-J. Borchers: *When Does Lorentz Invariance Imply Wedge Duality*, Lett. Math. Phys. **35**, 39-60 (1995).
23. H.-J. Borchers: *The CPT-Theorem in Two-dimensional Theories of Local Observables*, Commun. Math. Phys. **143**, 315-332 (1992).

24. H.-J. Borchers: *On Poincaré transformations and the modular group of the algebra associated with a wedge*, to be published in Lett. Math. Phys.
25. R. Brunetti, D. Guido and R. Longo: *Group cohomology, modular theory and space-time symmetries*, Rev. Math. Phys. **7**, 57-71 (1995).
26. D. Guido and R. Longo: *An Algebraic Spin and Statistics Theorem*, Commun. Math. Phys. **172**, 517-533 (1995).
27. H.-J. Borchers: *Half-sided Modular Inclusion and the Construction of the Poincaré Group*, Commun. Math. Phys. **179**, 703-723 (1996)
28. G.C. Hegerfeldt: *Prime Field Decompositions and Infinitely Divisible States on Borchers' Tensor Algebra*, Commun. Math. Phys. **45**, 137-157 (1975).
29. H.-J. Borchers: *On the structure of the algebra of field operators* , Nuovo Cimento **24**, 214-236 (1962).
30. S. Doplicher, R.V. Kadison, D. Kastler. and D.W. Robinson: *Asymptotically abelian systems* , Commun. Math. Phys. **6**, 101-120 (1967).
31. M. Takesaki: *Conditional Expectations in von Neumann Algebras*, Jour. Func. Anal. **9**, 306-321 (1972).
32. H. Kosaki: *Extension of Jones' Theory on Index to Arbitrary Factors*, J. Funct. Anal. **66**, 123-140 (1986).
33. L. Kadison and D. Kastler: *Cohomological aspects and relative separability of finite Jones index factors*, Nachr. Akad. d. Wissensch. Göttingen, (1992) p. 95-105.
34. H.-J. Borchers: *Half-sided Modular Inclusions and Structure Analysis in Quantum Field Theory*, in Operator Algebras and Quantum Field Theory, S.Doplicher, R. Longo, J. Roberts, L. Zsido ed. International Press (1997).
35. H.-W. Wiesbrock: *Symmetries and Half-Sided Modular Inclusions of von Neumann Algebras*, Lett. Math. Phys. **28**, 107-114 (1993).
36. H.-W. Wiesbrock: *Half-Sided Modular Inclusions of von Neumann Algebras*, Erratum, Commun. Math. Phys. **184**, 683-685 (1997)
37. L. Ge, R. Kadison: *On tensor products of von Neumann algebras*, Invent. math. **123**, 453-466 (1966).
38. K.-H. Rehren: *Bounded Bose fields*, Lett. Math. Phys. **40**, 299-306 (1997).

Current Trends in Axiomatic Quantum Field Theory

Detlev Buchholz

Institut für Theoretische Physik, Universität Göttingen,
Bunsenstraße 9, D-37073 Göttingen, Germany

Abstract. In this article a non–technical survey is given of the present status of Axiomatic Quantum Field Theory and interesting future directions of this approach are outlined. The topics covered are the universal structure of the local algebras of observables, their relation to the underlying fields and the significance of their relative positions. Moreover, the physical interpretation of the theory is discussed with emphasis on problems appearing in gauge theories, such as the revision of the particle concept, the determination of symmetries and statistics from the superselection structure, the analysis of the short distance properties and the specific features of relativistic thermal states. Some problems appearing in quantum field theory on curved spacetimes are also briefly mentioned.

1 Introduction

Axiomatic Quantum Field Theory originated from a growing desire in the mid–fifties to have a consistent mathematical framework for the treatment and interpretation of relativistic quantum field theories. There have been several profound solutions of this problem, putting emphasis on different aspects of the theory. The Ringberg Symposium on Quantum Field Theory has been organized in honor of one of the founding fathers of this subject, Wolfhart Zimmermann. It is therefore a pleasure to give an account of the present status of the axiomatic approach on this special occasion.

It should perhaps be mentioned that the term "axiomatic" is no longer popular amongst people working in this field since its mathematical connotations have led to misunderstandings. Actually, Wolfhart Zimmermann never liked it and called this approach Abstract Quantum Field Theory. Because of the modern developments of the subject, the presently favored name is Algebraic Quantum Field Theory. So the invariable abbreviation AQFT seems appropriate in this survey.

The early successes of AQFT are well known and have been described in several excellent monographs [1–3]. They led, on the one hand, to an understanding of the general mathematical structure of the correlation functions of relativistic quantum fields and laid the foundations for the rigorous perturbative and non–perturbative construction of field theoretic models. On the other hand they provided the rules for the physical interpretation of the theory, the most important result being collision theory and the reduction formulas.

It was an at first sight perhaps unexpected bonus that the precise formulation of the foundations of the theory payed off also in other respects. For it led

P. Breitenlohner and D. Maison (Eds.): Proceedings 1998, LNP 558, pp. 43–64, 2000.

to the discovery of deep and general features of relativistic quantum field theories, such as the PCT–theorem, the relation between spin and statistics, dispersion relations, bounds on the high energy behavior of scattering amplitudes, the Goldstone theorem etc. These results showed that the general principles of relativistic quantum field theory determine a very rigid mathematical framework which comprises surprisingly detailed information about the physical systems fitting into it.

These interesting developments changed their direction in the seventies for two reasons. First, it became clear that, quite generally, the linear and nonlinear properties of the correlation functions of quantum fields following from the principles of AQFT admit an equivalent Euclidean description of the theory in terms of classical random fields. Thereby, the construction of relativistic field theories was greatly simplified since one could work in a commutative setting [4]. Most results in constructive quantum field theory were obtained by using this powerful approach which also became an important tool in concrete applications.

The simplification of the constructive problems outweighed the conceptual disadvantage that the Euclidean theory does not have a direct physical interpretation. To extract this information from the Euclidean formalism is frequently a highly non–trivial task and has been the source of mistakes. So for the interpretation of the theory the framework of AQFT is still indispensable.

The second impact came from physics. It was the insight that gauge theories play a fundamental role in *all* interactions. It was already clear at that time that quantum electrodynamics did not fit into the conventional setting of AQFT. For the local, covariant gauge fields require the introduction of unphysical states and indefinite metric. The idea that one could determine from such fields in an *a priori* manner the physical Hilbert space had finally to be given up. Features such as the phenomenon of confinement in non–Abelian gauge theories made it very clear that the specification of the physical states is in general a subtle dynamical problem.

A way out of these problems had already been discovered in the sixties, although its perspectives were perhaps not fully recognized in the beginning. Namely it became gradually clear from the structural analysis in AQFT that the local observables of a theory carry all relevant physical information. In particular, the (charged) physical states and their interactions can be recovered from them. The situation is analogous to group theory, where the set of unitary representations can be determined from the abstract structure of the group.

From this more fundamental point of view the gauge fields appear to be nothing but a device for the construction of the local (gauge invariant) observables of the theory in some faithful representation, usually the vacuum representation. The determination of the physical states and their analysis is then regarded as a problem in representation theory.

It also became clear that one does not need to know from the outset the specific physical significance of the local observables for the interpretation of the theory. All what matters is the information about their space–time localization properties. From these data one can determine the particle structure, collision cross sections, the charges appearing in the theory and, finally, identify individual observables of physical interest, such as the charge and energy–densities.

These insights led to the modern formulation of AQFT in terms of families (nets) of algebras of observables which are assigned to the bounded space–time regions [5]. Field operators, even observable ones, no longer appear explicitly in this formalism. They are regarded as a kind of coordinatization of the local algebras without intrinsic meaning; which fields one uses for the description of a specific theory is more a matter of convenience than of principle.

This more abstract point of view received, in the course of time, its full justification. First, the framework proved to be flexible enough to incorporate non–pointlike localized observables, such as the Wilson loops, which became relevant in gauge theory. Second, it anticipated to some extent the phenomenon of quantum equivalence, i.e., the fact that certain very differently looking theories, such as the Thirring model and the Sine–Gordon theory or the recently explored supersymmetric Yang–Mills theories, describe the same physics. The basic insight that fields do not have an intrinsic meaning, in contrast to the system of local algebras which they generate, found a striking confirmation in these examples. Third, the algebraic approach proved natural for the discussion of quantum field theories in curved spacetime and the new types of problems appearing there [6]. There is also evidence that it covers prototypes of string theory [7].

So the general framework of AQFT has, for many decades, proved to be consistent with the progress in our theoretical understanding of relativistic quantum physics. It is the appropriate setting for the discussion of the pertinent mathematical structures, the elaboration of methods for their physical interpretation, the solution of conceptual problems and the classification of theories on the basis of physical criteria. Some major achievements and intriguing open problems in this approach are outlined in the remainder of this article.

2 Fields and algebras

As mentioned in the Introduction, the principles of relativistic quantum field theory can be expressed in terms of field operators and, more generally, nets of local algebras. In this section we give an account of the relation between these two approaches.

We proceed from the for our purposes reasonable idealization that spacetime is a classical manifold $\mathcal{M}$ with pseudo–Riemannian metric g. In the main part of this article we assume that $(\mathcal{M}, g)$ is four–dimensional Minkowski space with its standard Lorentzian metric and comment on curved spacetime only in the last section.

Fields: In the original formulation of AQFT one proceeds from collections of field operators $\phi(x)$ which are assigned to the space–time points $x \in \mathcal{M}$,

$$x \longmapsto \{\, \phi(x)\, \}. \tag{1}$$

(In order to simplify the discussion we assume that the fields $\phi(x)$ are observable and omit possible tensor indices.) As is well known, this assignment requires some mathematical care, it is to be understood in the sense of operator–valued distributions. With this precaution in mind the fundamental principles of relativistic quantum field theory, such as Poincaré covariance and Einstein causality (locality), can be cast into mathematically precise conditions on the field operators [1–3]. As a matter of fact, for given field content of a theory one can encode these principles into a universal algebraic structure, the Borchers–Uhlmann algebra of test functions [1].

It is evident that such a universal algebra does not contain any specific dynamical information. That information can be put in by specifying a vacuum state (expectation functional) on it. This step is the most difficult task in the construction of a theory. Once it has been accomplished, one can extract from the corresponding correlation functions, respectively their time–ordered, advanced or retarded counterparts, the desired information.

The fact that the construction of a theory can be accomplished by the specification of a vacuum state on some universal algebra has technical advantages and ultimately led to the Euclidean formulation of quantum field theory. But it also poses some problems: Given two such states, when do they describe the same theory? That this is a non–trivial problem can be seen already in free field theory. There the vacuum expectation values of the basic free field ϕ_0 and those of its Wick power ϕ_0^3, say, correspond to quite different states on the abstract algebra. Nevertheless, they describe the same physics. A less trivial example, where the identification of two at first sight very differently looking theories required much more work, can be found in [8]. So in this respect the field theoretic formalism is not intrinsic.

Algebras: In the modern algebraic formulation of AQFT one considers families of W^*–algebras[1] $\mathcal{A}(\mathcal{O})$ of bounded operators which are assigned to the open, bounded space–time regions $\mathcal{O} \subset \mathcal{M}$,

$$\mathcal{O} \longmapsto \mathcal{A}(\mathcal{O}). \qquad (2)$$

Each $\mathcal{A}(\mathcal{O})$ is regarded as the algebra generated by the observables which are localized in the region $\mathcal{O}$; it is called the *local algebra* affiliated with that region. Again, the principles of locality and Poincaré covariance can be expressed in this setting in a straightforward manner [5]. In addition there holds the property of isotony, i.e.,

$$\mathcal{A}(\mathcal{O}_1) \subset \mathcal{A}(\mathcal{O}_2) \quad \text{if} \quad \mathcal{O}_1 \subset \mathcal{O}_2. \qquad (3)$$

This condition expresses the obvious fact that the set of observables increases with the size of the localization region. Despite its at first sight almost tautological content, it is this net structure (nesting) of the local algebras which comprises the relevant physical information about a theory. To understand this fact one has to recognize that the assignment of algebras to a given collection of space–time regions will be very different in different theories.

[1] The letter W indicates that the respective algebras are closed with respect to weak limits and * says that they are stable under taking adjoints.

The algebraic version of AQFT defines a conceptually and mathematically compelling framework of local relativistic quantum physics and has proved very useful for the general structural analysis. It is rather different, however, from the field theoretic formalism which one normally uses in the construction of models. The clarification of the relation between the two settings is therefore an important issue.

From fields to algebras: The problems appearing in the transition from the field theoretic setting to the algebraic one are of a similar nature as in the transition from representations of Lie algebras to representations of Lie groups: one has to deal with regularity properties of unbounded operators. Heuristically, one would be inclined to define the local algebras by appealing to von Neumann's characterization of concrete W^*–algebras as double commutants of sets of operators,

$$\mathcal{A}(\mathcal{O}) = \{\phi(x) : x \in \mathcal{O}\}'' , \tag{4}$$

that is they ought to be the smallest weakly closed algebras of bounded operators on the underlying Hilbert space which are generated by the (smoothed–out) observable fields in the respective space–time region. Because of the subtle properties of unbounded operators it is, however, not clear from the outset that the so–defined algebras comply with the physical constraint of locality assumed in the algebraic setting.

The first courageous steps in the analysis of this problem were taken by Borchers and Zimmermann [9]. They showed that if the vacuum $|0\rangle$ is an analytic vector for the fields, i.e., if the formal power series of the exponential function of smeared fields, applied to $|0\rangle$, converge absolutely, then the passage from the fields to the local algebras can be accomplished by the above formula. Further progress on the problem was made in [10], where it was shown that fields satisfying so–called linear energy bounds also generate physically acceptable nets of local algebras in this way.

The latter result covers all interacting relativistic quantum field theories which have been rigorously constructed so far. As for the general situation, the most comprehensive results are contained in [11] and references quoted there. In that analysis certain specific positivity properties of the vacuum expectation values of fields were isolated as crucial pre–requisite for the passage from fields to algebras. In view of these profound results, it can now safely be stated that the algebraic framework is a proper generalization of the original field theoretic setting.

From algebras to fields: As already mentioned, the algebraic version of AQFT is more general than the field theoretic one since it covers also finitely localized observables, such as Wilson loops or Mandelstam strings, which are not built from observable pointlike localized fields. Nevertheless, the point–field content of a theory is of great interest since it includes distinguished observables, such as currents and the stress energy tensor.

Heuristically, the point–fields of a theory can be recovered from the local algebras by the formula

$$\{\phi(x)\} = \bigcap_{\mathcal{O} \ni x} \overline{\mathcal{A}(\mathcal{O})} . \tag{5}$$

It should be noticed here that one would not obtain the desired fields if one would simply take the intersection of the local algebras themselves, which is known to consist only of multiples of the identity. Therefore, one first has to complete the local algebras in a suitable topology which allows for the appearance of unbounded operators (respectively forms). This step is indicated by the bar.

The first profound results on this problem were obtained in [12], where it has been proposed to complete the local algebras with the help of suitable energy norms which are sensitive to the energy–momentum transfer of the observables. Using this device, it was shown that one can reconstruct from local algebras, if they are generated by sufficiently "tame" fields, the underlying field content by taking intersections as above. These results were later refined in various directions [13]. They show that the step from fields to algebras can be reversed.

From a general point of view it would, however, be desirable to clarify the status of point fields in the algebraic setting in a more intrinsic manner, i.e., without assuming their existence from the outset. An interesting proposal in this direction was recently made in [14]. There it was argued that the presence of such fields is encoded in phase space properties of the net of local algebras[2] and that the field content can be uncovered from the algebras by using notions from sheaf theory. In [15] this idea was put into a more suitable mathematical form and was also confirmed in models. The perspectives of this new approach appear to be quite interesting. It seems, for example, that one can establish in this setting the existence of Wilson–Zimmermann expansions [16] for products of field operators. Such a result would be a major step towards the ambitious goal, put forward in [14], to characterize the dynamics of nets of local algebras directly in the algebraic setting.

3 Local algebras and their inclusions

Because of their fundamental role in the algebraic approach, much work has been devoted to the clarification of the structure of the local algebras and of their inclusions. We cannot enter here into a detailed discussion of this subject and only give an account of its present status. To anticipate the perhaps most interesting perspective of the more recent results: There is evidence that the dynamical information of a relativistic quantum field theory is encoded in, and can be uncovered from the relative positions of a few (depending on the number of space–time dimensions) algebras of specific type. This insight may be the starting point for a novel constructive approach to relativistic quantum field theory.

[2] A quantitative measure of the phase space properties of local algebras is given in Sec. 7.

To begin, let us recall that the center of a W^*–algebra is the largest sub–algebra of operators commuting with all operators in the algebra. A W^*–algebra is called a factor if its center consist only of multiples of the identity, and it is said to be hyperfinite if it is generated by its finite dimensional sub–algebras. The hyperfinite factors have been completely classified, there exists an uncountable number of them.

Because of this abundance of different types of algebras it is of interest that the local algebras appearing in quantum field theory have a universal (model–independent) structure [17], they are generically isomorphic to the tensor product

$$\mathcal{A}(\mathcal{O}) \simeq \mathcal{M} \otimes \mathcal{Z}, \tag{6}$$

where $\mathcal{M}$ is the unique hyperfinite type III_1 factor according to the classification of Connes, and $\mathcal{Z}$ is an Abelian algebra. That the local algebras are hyperfinite is encoded in phase space properties of the theory, the type III_1 property is a consequence of the short distance structure, cf. [17] and references quoted there. The possible appearance of a non–trivial center $\mathcal{Z}$ in a local algebra is frequently regarded as a nuisance, but it cannot be excluded from the outset.

Under the above generic conditions also the global (quasilocal) algebra

$$\mathcal{A} = \overline{\bigcup_{\mathcal{O}} \mathcal{A}(\mathcal{O})}, \tag{7}$$

which is the C^*–inductive limit of all local algebras, is known to be universal. It is the so–called "proper sequential type I_∞ funnel". So one has very concrete information about the mathematical objects appearing in the algebraic setting. From the conceptual point of view, these results corroborate the insight that the individual local algebras as well as the global one do not comprise any specific physical information. This information is entirely contained in the "arrow" in (2), i.e., the map from space–time regions to local algebras.

In view of these results it is natural to have a closer look at the possible "relative positions" of the local algebras with respect to each other. Depending on the location of the regions it has been possible to characterize in purely algebraic terms the following geometric situations.

(a) The closure of $\mathcal{O}_1$ is contained in the interior of $\mathcal{O}_2$.

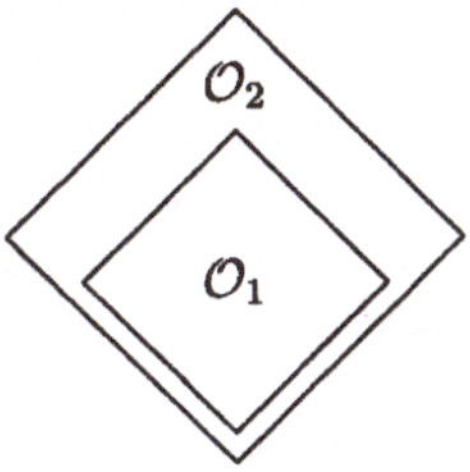

In that case there holds generically

$$\mathcal{A}(\mathcal{O}_1) \subset \mathcal{N} \subset \mathcal{A}(\mathcal{O}_2), \tag{8}$$

where $\mathcal{N}$ is a factor of type I_∞, i.e., an algebra which is isomorphic to the algebra of all bounded operators on some separable Hilbert space. This "split property" of the local algebras has been established in all quantum field theories with reasonable phase space properties, cf. [17] and references quoted there. It does *not* hold if the two regions have common boundary points [18].
(b) $\mathcal{O}_1$ and $\mathcal{O}_2$ are spacelike separated.

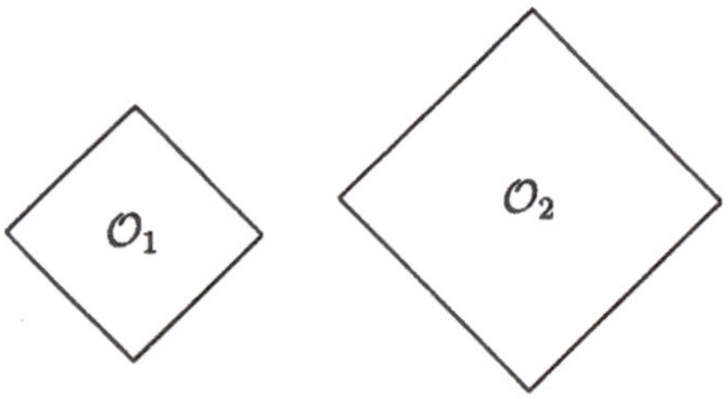

Then, under the same conditions as in (a), it follows that the W^*–algebra which is generated by the two local algebras associated with these regions is isomorphic to their tensor product,

$$\mathcal{A}(\mathcal{O}_1) \vee \mathcal{A}(\mathcal{O}_2) \simeq \mathcal{A}(\mathcal{O}_1) \otimes \mathcal{A}(\mathcal{O}_2). \tag{9}$$

Thus the local algebras satisfy a condition of causal (statistical) independence, which may be regarded as a strengthened form of the locality postulate.

(c) $\mathcal{O}_1$ and $\mathcal{O}_2$ are wedge–shaped regions, bounded by two characteristic planes, such that $\mathcal{O}_1 \subset \mathcal{O}_2$ and the edge of $\mathcal{O}_1$ is contained in a boundary plane of $\mathcal{O}_2$.

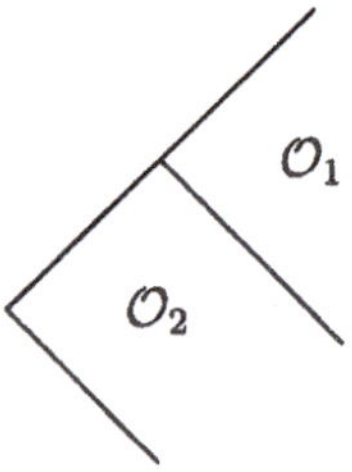

In this geometric situation the corresponding algebras give rise to so–called "half–sided modular inclusions" or, more generally, "modular intersections". It is a striking fact that one can reconstruct from a few algebras in this specific position a unitary representation of the space–time symmetry group, the PCT operator and the net of local algebras [19,20]. This observation substantiates the claim that the dynamical information of a theory is contained in the relative position of the underlying algebras. Thus the concept of modular inclusions and intersections seems to be a promising starting point for the direct construction of nets of local algebras.

The preceding results rely heavily on modular theory, which has become an indispensable tool in the algebraic approach. It is not possible to outline here the many interesting applications which are based on these techniques. Some recent results and further pertinent references can be found in [21].

4 Particle aspects

We turn now to the physical interpretation of the mathematical formalism of AQFT. Here the basic ingredient is the notion of particle. According to Wigner the states of a single particle are to be described by vectors in some irreducible representation of the Poincaré group, or its covering group. This characterization of particles has been extremely useful in the solution of both, constructive and conceptual problems.

As for the interpretation of the theory, the particle concept enters primarily in collision theory, whose first precise version was given by Lehmann, Symanzik and Zimmermann [22]. By now collision theory has been rigorously established in AQFT, both for massive and massless particles [5]. These results formed the basis for the derivation of analyticity, crossing and growth properties of the scattering amplitudes [23]. There remain, however, many open problems in this context. In view of the physical relevance of gauge theories it would, for example, be desirable to determine the general properties of scattering amplitudes of particles carrying a gauge charge. In physical gauges such particles require a description in terms of non–local fields, hence the classical structural results cannot be applied. Some remarkable progress on this difficult problem has been presented in [24].

Another longstanding question is the problem of asymptotic completeness (unitarity of the S–matrix): Even in the models which have been rigorously constructed, a complete solution of this problem is not known [25]. So the situation is quite different from quantum mechanics, where the problem of asymptotic completeness was solved in a general setting almost two decades ago.

In order to understand the origin of these difficulties, one has to realize that, in contrast to quantum mechanics, one deals in quantum field theory with systems with an infinite number of degrees of freedom. In such systems there can occur the formation of superselection sectors which require an extension of the original Hilbert space. So one first has to determine in a theory the set of all superselection sectors and particle types before a discussion of the problem of asymptotic completeness becomes meaningful. In models this step can sometimes be avoided by technical assumptions, such as restrictions on the size of coupling constants, by which the formation of superselection sectors and new particles can be excluded. But the determination of the full physical Hilbert space from the underlying local operators is an inevitable step in any general discussion of the problem. Some progress on this problem will be outlined below.

Still another important problem which deserves mentioning here is the treatment of particles carrying charges of electromagnetic type. As is well known, the states of such particles cannot consistently be described in the way proposed by Wigner, cf. [26] and references quoted there. In the discussion of scattering processes involving such particles this problem can frequently be circumvented by noting that an infinite number of soft massless particles remains unobserved. Because of this fact one can proceed to an "inclusive description", where the difficulties disappear. This trick obscures, however, the specific properties of the electrically charged particles. It seems therefore worthwhile to analyze their uncommon features which may well be accessible to experimental tests.

It is apparent that progress on these problems requires a revision of the particle concept. A proposal in this respect which is based on Dirac's idea of improper states of sharp momentum has been presented in [27]. The common mathematical treatment of improper states as vector–valued distributions would not work in the general setting since it assumes the superposition principle which is known to fail for momentum eigenstates carrying electric charge. Instead one defines the improper states as linear maps from some space (more precisely, left ideal) of localizing operators $L \in \mathcal{A}$ into the physical Hilbert space,

$$L \longmapsto L\,|\boldsymbol{p}, \iota\rangle. \tag{10}$$

Here $\boldsymbol{p}$ is the momentum of the particle and ι subsumes its charges, mass and spin. In quantum mechanics, the simplest example of a localizing operator L which transforms the "plane waves" $|\boldsymbol{p}, \iota\rangle$ into normalizable states is the projection onto a compact region of configuration space. In quantum field theory such localizing operators can be constructed out of local operators by convolution with suitable test functions which restrict the energy–momentum transfer of the operators to spacelike values. It can be shown that by acting with any such operator on an improper momentum eigenstate of a particle one obtains a Hilbert space vector. Thus, from a mathematical point of view, the improper states are weights on the algebra of observables $\mathcal{A}$.

Using this device one can, on the one hand, determine the particle content of a theory from the states $|\Phi\rangle$ in the vacuum sector by means of the formula [27]

$$\lim_{t \to \pm\infty} \int d^3\boldsymbol{x}\, \langle \Phi | \alpha_{t,\boldsymbol{x}}(L^* L) | \Phi \rangle = \sum_{\iota} \int d\mu_{\pm}(\boldsymbol{p}, \iota)\, \langle \boldsymbol{p}, \iota | L^* L | \boldsymbol{p}, \iota \rangle. \tag{11}$$

Here L is any localizing operator, $\alpha_{t,\boldsymbol{x}}$ are the automorphisms inducing the space–time translations on $\mathcal{A}$ and $\mu_{\pm}$ are measures depending on $|\Phi\rangle$. Thus if one analyzes the states $|\Phi\rangle$ at asymptotic times by spatial averages of localizing operators, they look like mixtures of improper single particle states. These mixtures are formed by the members of the incoming respectively outgoing particle configurations in the state $|\Phi\rangle$ which generically include also pairs of oppositely charged particles. By decomposing the mixtures in (11) one can therefore recover all particle types. As outlined in [28], this result also allows one to recover from the underlying local observables in the vacuum sector the pertinent physical Hilbert space of the theory.

Relation (11) does not only establish a method for the determination of the particles in a theory, but it also provides a framework for their general analysis. It is of interest that this framework also covers particles carrying charges of electromagnetic type. Whereas for a particle of Wigner type the corresponding improper states lead, after localization, to vectors in the same sector of the physical Hilbert space, this is no longer true for electrically charged particles. There one finds that the vectors $L|\boldsymbol{p}, \iota\rangle$ and $L'|\boldsymbol{p}', \iota\rangle$ are orthogonal for any choice of localizing operators L, L' if the momenta $\boldsymbol{p}, \boldsymbol{p}'$ are different. So the superposition principle fails in this case and wave packets of improper states cannot be formed. Nevertheless, the charges, mass and spin of such particles can be defined and have the values found by Wigner in his analysis. The only possible exception are massless particles whose helicity need not necessarily be restricted to (half–)integer values [27].

So there is progress in our general understanding of the particle aspects of quantum field theory. Further advancements seem to require, however, new methods such as a more detailed harmonic analysis of the space–time automorphisms [29].

5 Sectors, symmetries and statistics

One of the great achievements of AQFT is the general understanding of the structure of superselection sectors in relativistic quantum field theory, its relation to the appearance of global gauge groups and the origin and classification of statistics. Since this topic has been expounded in the monographs [5,30] we need to mention here only briefly the main results and open problems.

The superselection sectors of interest in quantum field theory correspond to specific irreducible representations of the algebra of observables $\mathcal{A}$, more precisely, to their respective unitary equivalence classes. It is an important fact that each sector has representatives which are (endo)morphisms ρ of the algebra $\mathcal{A}$ (or certain canonical extensions of it), so there holds

$$\rho(\mathcal{A}) \subset \mathcal{A}. \tag{12}$$

Using this fact one can distinguish various types of superselection sectors according to the localization properties of the associated morphisms.

Localizable Sectors: These sectors have been extensively studied by Doplicher, Haag and Roberts. Each such sector can be characterized by the property that for any open, bounded space–time region $\mathcal{O} \subset \mathcal{M}$ there is a morphism $\rho_{\mathcal{O}}$, belonging to this sector, which is localized in $\mathcal{O}$ in the sense that it acts trivially on the observables in the causal complement $\mathcal{O}'$ of $\mathcal{O}$,

$$\rho_{\mathcal{O}} \upharpoonright \mathcal{A}(\mathcal{O}') = \mathrm{id}. \tag{13}$$

Localizable sectors describe charges, such as baryon number, which do not give rise to long range effects.

Sectors in Massive Theories: It has been shown in [31] that the superselection sectors in quantum field theories describing massive particles can always be represented by morphisms $\rho_{\mathcal{C}}$ which are localized in a given spacelike cone $\mathcal{C}$,

$$\rho_{\mathcal{C}} \upharpoonright \mathcal{A}(\mathcal{C}') = \mathrm{id}. \tag{14}$$

This class of sectors includes also non–localizable charges appearing in certain massive gauge theories. The preceding result says that the long range effects of such charges can always be accommodated in extended string–like regions.

It is possible to establish for both classes of sectors under very general conditions, the most important one being a maximality property of the underlying algebras of local observables (Haag–duality), the existence of a composition law. Namely, if ι, κ, λ are labels characterizing the sectors there holds for the corresponding morphisms the relation

$$\rho_\iota \circ \rho_\kappa = \sum_\lambda c(\iota, \kappa, \lambda)\ \rho_\lambda, \tag{15}$$

where $c(\iota, \kappa, \lambda)$ are integers and the summation is to be understood in the sense of direct sums of representations. Moreover, for each ρ_ι there is a charge conjugate morphism $\overline{\rho}_\iota$ such that $\rho_\iota \circ \overline{\rho}_\iota$ contains the identity.

It is a deep result of Doplicher and Roberts that these general facts imply the existence of charged fields ψ of Bose or Fermi type which connect the states in the various superselection sectors. More precisely, each morphism ρ can be represented in the form

$$\rho(\,\cdot\,) = \sum_{m=1}^{d} \psi_m \cdot \psi_m^*\,, \tag{16}$$

where the fields ψ_m have the same localization properties as ρ and satisfy Bose or Fermi commutation relations at spacelike distances. Moreover, these fields transform as tensors under the action of some compact group G,

$$\gamma_g\,(\psi_n) = \sum_{m=1}^{d} D_{nm}(g^{-1})\,\psi_m\,, \tag{17}$$

where $\gamma.$ are the automorphisms inducing this action and $D_{nm}(\cdot)$ is some d–dimensional representation of G. The observables are exactly the fixed points under the action of G and the whole structure is uniquely determined by the underlying sectors.

In order to appreciate the strength of this result one has to notice that it does not hold in low space–time dimensions. In that case there can occur fields with braid group statistics and the superselection structure can in general not be described by the representation theory of compact groups, while more complex symmetry structures, such as "quantum groups", seem to emerge [32]. These facts have stimulated much work in the general analysis of low dimensional quantum field theories in recent years, yet we cannot comment here on these interesting developments.

With regard to physical space–time, there are several interesting problems which should be mentioned here. First, it would be of interest to understand in which way the presence of supersymmetries is encoded in the superselection structure of a theory. It seems that this problem has not yet been thoroughly studied in the general framework of AQFT.

Second, there is the longstanding problem of the superselection structure in theories with long range forces, such as Abelian gauge theories with unscreened charges of electromagnetic type. In such theories there exists for each value of the charge an abundance of sectors due to the multifarious ways in which accompanying clouds of low energy massless particles can be formed. These clouds obstruct the general analysis of the superselection structure since it is difficult to disentangle their fuzzy localization properties from those of the charges one is actually interested in.

A promising step towards the solution of this problem is the observation, made in some models [33], that charges of electromagnetic type have certain clearcut localization properties in spite of their long range effects. Roughly speaking, they appear to be localized in a given Lorentz system with respect to distinguished observables, such as current densities. In the framework of AQFT this restricted localizability of sectors can be expressed by assuming that there is a subnet $\mathcal{O} \longmapsto \mathcal{B}(\mathcal{O}) \subset \mathcal{A}(\mathcal{O})$ on which the corresponding morphisms are localized,

$$\rho_{\mathcal{O}} \restriction \mathcal{B}(\mathcal{O}') = \mathrm{id}. \tag{18}$$

In contrast to the preceding class of localizable sectors, the net $\mathcal{O} \longmapsto \mathcal{B}(\mathcal{O})$ need not be maximal, however. In the case of electrically charged sectors, it will also not be stable under Lorentz transformations. So the general results of Doplicher and Roberts cannot be applied in this case. Yet there is evidence [33] that if the localizing subnet of observables is sufficiently big (it has to satisfy a condition of weak additivity) one can still establish symmetry and statistics properties for the physically interesting class of so–called simple sectors which are induced by automorphisms ρ of the algebra of observables with the above localization properties.

Another important issue which is closely related to the sector structure is the problem of symmetry breakdown. The consequences of the spontaneous breakdown of internal symmetries are well understood in AQFT in the context of localizable sectors, cf. [34] and references quoted there. They lead to a degeneracy of the vacuum state and, under more restrictive assumptions, to the appearance of massless Goldstone particles. For the class of cone–like localizable sectors the consequences of spontaneous symmetry breaking are less clear, however. One knows from model studies in gauge theories that under these circumstances there can appear a mass gap in the theory (Higgs mechanism). But the understanding of this phenomenon in the general framework of AQFT is not yet in a satisfactory state.

6 Short distance structure

Renormalization group methods have proved to be a powerful tool for the analysis of the short distance (ultraviolet) properties of field–theoretic models. They provided the basis for the discussion of pertinent physical concepts, such as the notion of parton, confinement, asymptotic freedom, etc. In view of these successes it was a natural step to transfer these methods to the abstract field theoretic setting [35]. More recently, the method has also been established in the algebraic formulation of AQFT [36]. We give here a brief account of the latter approach in which renormalization group transformations are introduced in a novel, implicit manner.

The essential idea is to consider functions $\underline{A}$ of a parameter $\lambda \in \mathbb{R}_+$, fixing the space–time scale, which have values in the algebra of observables,

$$\lambda \longmapsto \underline{A}_\lambda \in \mathcal{A}. \tag{19}$$

These functions form under the obvious pointwise defined algebraic operations a normed algebra, the scaling algebra $\underline{\mathcal{A}}$, on which the Poincaré transformations (Λ, x) act continuously by automorphisms $\underline{\alpha}_{\Lambda,x}$ according to

$$\left(\underline{\alpha}_{\Lambda,x}(\underline{A})\right)_\lambda \doteq \alpha_{\Lambda,\lambda x}(\underline{A}_\lambda). \tag{20}$$

The local structure of the original net can be lifted to $\underline{\mathcal{A}}$ by setting

$$\underline{\mathcal{A}}(\mathcal{O}) \doteq \{\underline{A} : \underline{A}_\lambda \in \mathcal{A}(\lambda\mathcal{O}),\ \lambda \in \mathbb{R}_+\}. \tag{21}$$

It is easily checked that with these definitions one obtains a local, Poincaré covariant net of subalgebras of the scaling algebra which is canonically associated with the original theory.

We refer the reader to [36] for a discussion of the physical interpretation of this formalism and only mention here that the values $\underline{A}_\lambda$ of the functions $\underline{A}$ are to be regarded as observables in the theory at space–time scale λ. So the graphs of the functions $\underline{A}$ establish a relation between observables at different space–time scales, in analogy to renormalization group transformations. Yet, in contrast to the field theoretic setting, one need not identify specific observables in the algebraic framework since the physical information is contained in the net structure. For that reason one has much more freedom in the choice of the functions $\underline{A}$ which have to satisfy only the few general constraints indicated above. Thus one arrives at a universal framework for the discussion of the short distance structure in quantum field theory.

The physical states, such as the vacuum $|0\rangle$, can be analyzed at any scale with the help of the scaling algebra and one can define a short distance (scaling) limit of the theory by the formula

$$\langle 0|\, \underline{A}_0\, \underline{B}_0 \cdots \underline{C}_0\, |0\rangle \;\doteq\; \lim_{\lambda \to 0}\, \langle 0|\, \underline{A}_\lambda\, \underline{B}_\lambda \cdots \underline{C}_\lambda\, |0\rangle. \tag{22}$$

To be precise, the convergence on the right hand side may only hold for suitable subsequences of the scaling parameter, and it should also be noticed that the limit may not be interchanged with the expectation value. The resulting correlation functions determine a pure vacuum state on the scaling algebra. So by the reconstruction theorem one obtains a net $\mathcal{O} \mapsto \mathcal{A}_0(\mathcal{O})$ of local algebras and automorphisms $\alpha^{(0)}_{\Lambda,x}$ which induce the Poincaré transformations on this net. This scaling limit net describes the properties of the underlying theory at very small space–time scales.

Based on these results the possible structure of scaling limit theories has been classified in the general framework of AQFT [36]. There appear three qualitatively different cases (classical, quantum and degenerate limits). There is evidence that they correspond to the various possibilities in the field theoretic setting of having (no, stable, unstable) ultraviolet fixed points. One can also characterize in purely algebraic terms those local nets which ought to correspond to asymptotically free theories. Yet a fully satisfactory clarification of the relation between the field theoretic renormalization group and its algebraic version requires further work.

The framework of the scaling algebra has also shed new light on the physical interpretation of the short distance properties of quantum field theories [37]. Particle like entities and symmetries appearing only at very small space–time scales, such as partons and color, can be uncovered from a local net of observables by proceeding to its scaling limit. Since the resulting net has all properties required in AQFT, the methods and results outlined in the preceding two sections can be applied to determine these structures. It is conceptionally very satisfactory that one does not need to rely on unphysical quantities in this analysis, such as gauge fields; moreover, the method has also proven to be useful as a computational tool.

There remain, however, many interesting open problems in this setting. For example, one may hope to extract in the scaling limit information about the presence of a *local* gauge group in the underlying theory. A possible strategy could be to consider the scaling limit of the theory with respect to base points $p \neq 0$ in Minkowski space and to introduce some canonical identification of the respective nets. One may then study how this identification lifts to the corresponding algebras of charged fields, which are fixed by the reconstruction theorem of Doplicher and Roberts outlined in the preceding section. It is an interesting question whether some non–trivial information about the presence of local gauge symmetries is encoded in this structure.

Another problem of physical interest is the development of methods for the determination of the particle content of a given state at short distances which, in contrast to the determination of the particle content of a given theory, has not yet been accomplished. In order to understand this problem one has to notice that there holds for any physical state $|\Phi\rangle$

$$\lim_{\lambda \to 0} \langle \Phi | \underline{A}_\lambda \underline{B}_\lambda \cdots \underline{C}_\lambda | \Phi \rangle = \langle 0 | \underline{A}_0 \underline{B}_0 \cdots \underline{C}_0 | 0 \rangle, \tag{23}$$

so all states look like the vacuum state in the scaling limit [36]. In order to extract more detailed information about the short distance structure of $|\Phi\rangle$, one has to determine also the next to leading contribution of the matrix element on the left hand side of this relation. One may expect that this term describes the particle like structures which appear in the state $|\Phi\rangle$ at small scales, but this question has not yet been explored.

7 Thermal states

The rigorous analysis of thermal states in non–relativistic quantum field theory is an old subject which is well covered in the literature. Interest in thermal states in relativistic quantum field theory arose only more recently. For this reason there is a backlog in our understanding of the structure of thermal states in AQFT and one may therefore hope that this physically relevant topic will receive increasing attention in the future.

We recall that, from the algebraic point of view, a theory is fixed by specifying a net of local algebras of observables. The thermal states correspond to distinguished positive, linear and normalized functionals $\langle\cdot\rangle$ on the corresponding global algebra $\mathcal{A}$. We will primarily discuss here the case of thermal *equilibrium* states, which can be characterized in several physically meaningful ways, the technically most convenient one being the KMS–condition, which imposes analyticity and boundary conditions on the correlation functions [5]. Having characterized the equilibrium states, it is natural to ask which properties of a local net matter if such states are to exist. The important point is that, in answering this question, one arrives at criteria which distinguish a physically reasonable class of theories and can be used for their further analysis.

It has become clear by now that phase space properties, which were already mentioned at various points in this article, are of vital importance in this context. A quantitative measure of these properties, where the relation to thermodynamical considerations is particularly transparent, has been introduced in [38]. In that approach one considers for any $\beta > 0$ and bounded space–time region $\mathcal{O}$ the linear map $\Theta_{\beta,\mathcal{O}}$ from the local algebra $\mathcal{A}(\mathcal{O})$ to the vacuum Hilbert space $\mathcal{H}$ given by

$$\Theta_{\beta,\mathcal{O}}(A) \doteq e^{-\beta H} A|0\rangle, \tag{24}$$

where H is the Hamiltonian. Roughly speaking, this map amounts to restricting the operator $e^{-\beta H}$ to states which are localized in $\mathcal{O}$, whence the sum of its eigenvalues yields the partition function of the theory for spatial volume $|\mathcal{O}|$ and inverse temperature β. This idea can be made precise by noting that for maps between Banach spaces, such as $\Theta_{\beta,\mathcal{O}}$, one can introduce a nuclear norm $||\cdot||_1$ which is the appropriate generalization of the concept of the trace of Hilbert space operators. Thus $||\Theta_{\beta,\mathcal{O}}||_1$ takes the place of the partition function and provides the desired information on the phase space properties (level density) of the theory. If the theory is to have reasonable thermodynamical properties there must hold for small β and large $|\mathcal{O}|$

$$||\Theta_{\beta,\mathcal{O}}||_1 \leq e^{c|\mathcal{O}|^m \beta^{-n}}, \tag{25}$$

where c, m and n are positive constants [38]. This condition, which can be checked in the vacuum sector $\mathcal{H}$, should be regarded as a selection criterion for theories of physical interest.

It has been shown in [39] that any local net which satisfies condition (25) admits thermal equilibrium states $\langle\cdot\rangle_\beta$ for all $\beta > 0$. As a matter of fact, these states can be reached from the vacuum sector by a quite general procedure. Namely, there holds for $A \in \mathcal{A}$

$$\langle A \rangle_\beta = \lim_{\mathcal{O}\to\mathbf{R}^4} \frac{1}{Z_\mathcal{O}} \operatorname{Tr} E_\mathcal{O}\, e^{-\beta H} E_\mathcal{O}\, A, \tag{26}$$

where $E_\mathcal{O}$ projects onto certain "local" subspaces of $\mathcal{H}$ and $Z_\mathcal{O}$ is a normalization constant [39]. This formula provides a direct description of the Gibbs ensemble in the thermodynamic limit which does not require the definition of the theory in a "finite box". It has led to a sharpened characterization of thermal equilibrium states in terms of a *relativistic* KMS–condition [40]. This condition states that for any $A, B \in \mathcal{A}$ the correlation functions

$$x \longmapsto \langle A\, \alpha_x(B) \rangle_\beta \tag{27}$$

admit an analytic continuation in all space–time variables x into the complex tube $\mathbb{R}^4 + i\, C_\beta$, where $C_\beta \subset V_+$ is a double cone of size proportional to β,

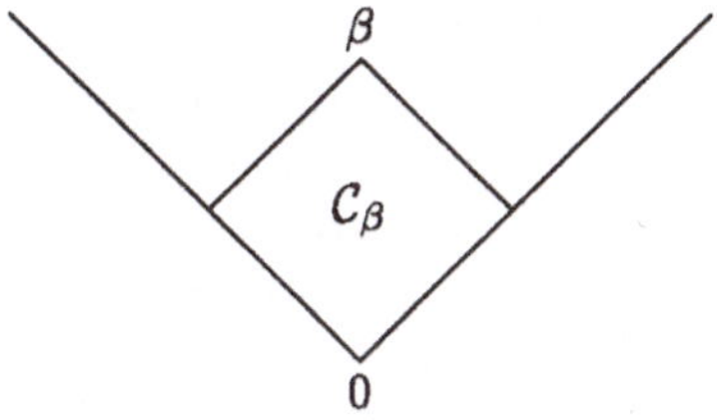

and the boundary values of these functions at the upper tip of C_β coincide with

$$x \longmapsto \langle B\, \alpha_x(A) \rangle_\beta . \tag{28}$$

The relativistic KMS–condition may be regarded as a generalization of the relativistic spectrum condition to the case of thermal equilibrium states. One recovers from it the spectrum condition for the vacuum sector in the limit $\beta \to \infty$.

The (relativistic) KMS–condition and the condition of locality lead to enlargements of the domains of analyticity of correlation functions of pointlike localized fields, in analogy to the case of the vacuum. Moreover, there exists an analogue of the Källén–Lehmann representation for thermal correlation functions which provides a basis for the discussion of the particle aspects of thermal states. These results are, however, far from being complete. We refer to [41] for a more detailed account of this topic and further references.

Amongst the many intriguing problems in thermal AQFT let us also mention the unclear status of perturbation theory [42], the still pending clarification of the relation between the Euclidean and Minkowski space formulation of the theory [43] and the characterization of non–equilibrium states. Any progress on these issues would be an important step towards the consolidation of thermal quantum field theory.

8 Curved spacetime

The unification of general relativity and quantum theory to a consistent theory of quantum gravity is an important issue which has become a major field of activity in theoretical physics. There are many stimulating proposals which are based on far–reaching theoretical ideas and novel mathematical structures. Yet the subject is still in an experimental state and has not yet reached a point where one could extract from the various approaches and results a consistent mathematical formalism with a clearcut physical interpretation. In a sense, one may compare the situation with the status of relativistic quantum field theory before the invention of AQFT.

In view of these theoretical uncertainties and lacking experimental clues it seems appropriate to treat, in an intermediate step, the effects of gravity in quantum field theory as a classical background. This idea has motivated the formulation of AQFT on curved space–time manifolds $(\mathcal{M}, g)$. As far as the algebraic aspects are concerned, this step does not require any new ideas. One still deals with nets (2) of local algebras which are assigned to the bounded space–time regions of $\mathcal{M}$. On these nets there act the isometries of $(\mathcal{M}, g)$ by automorphisms and they satisfy the principle of locality (commutativity of observables in causally disjoint regions). A novel difficulty which appears in this setting is the characterization of states of physical interest. For in general the isometry group of $(\mathcal{M}, g)$ does not contain global future directed Killing vector fields which could be interpreted as time translations and would allow for the characterization of distinguished ground states, representing the vacuum, or thermal equilibrium states.

The common strategy to overcome these difficulties is to invent local regularity conditions which distinguish subsets (folia) of physically acceptable states amongst the set of all states on the global algebra of observables. Such conditions have successfully been formulated for free field theories [6], yet their generalization turned out to be difficult and required new ideas. There are two promising proposals which can be applied to arbitrary theories, the "condition of local stability" [44], which fits well into the algebraic setting, and the "microlocal spectrum condition" [45,46], which so far has only been stated in a setting based on point fields. These conditions have proved to be useful for the discussion of prominent gravitational effects, such as the Hawking temperature [44,47], and they provided the basis for the formulation of a consistent renormalized perturbation theory on curved spacetimes [48].

In the long run, however, it appears to be inevitable to solve the problem of characterizing states describing specific physical situations. If the underlying spacetime is sufficiently symmetric in the sense that it admits Killing vector fields which are future directed on subregions of the space–time manifold (an example being de Sitter space) one can indeed identify vacuum–like states by symmetry and local stability properties [49,50]. The general situation is unclear, however. It has been suggested in [51] to distinguish the physically preferred states by a "condition of geometric modular action" which can be stated for a large class of space–time manifolds. But this proposal has so far been proven to work only for spacetimes where the preceding local stability conditions are also applicable. So there is still much work needed until our understanding of this important issue may be regarded as satisfactory.

Even more mysterious is the generalization of the particle concept to curved space–time manifolds and the description of collision processes. Since curvature gives rise to interaction with the classical background its effects have to be taken into account in the characterization of the corresponding states. These problems are of a similar nature as those occuring in the description of particle states in Minkowski space in the presence of long range forces or in thermal states. One may therefore hope that progress in the understanding of the latter problems will also provide clues to the solution of these conceptual difficulties in curved spacetime.

9 Concluding remarks

In the present survey of AQFT emphasis was put on issues which are of relevance for the discussion of relativistic quantum field theory in physical spacetime. There has been considerable progress in recent years in our understanding of the general mathematical framework and its physical interpretation. In particular, it has become clear that the modern algebraic approach is suitable for the discussion of the conceptual problems appearing in gauge theories. Yet there are still many intriguing questions which deserve further clarification.

Several topics which are presently in the limelight of major research activities had to be omitted here, such as the general analysis of low–dimensional quantum field theories. These theories provide a laboratory for the exploration of new theoretical ideas and methods and their thorough investigation brought to light novel mathematical structures which stimulated the interest of mathematicians. It was also not possible to outline here the many pertinent results and interesting perspectives which are based on the powerful techniques of modular theory. For an account of these exciting developments we refer to [52,53].

Thus, in spite of its age, AQFT is still very much alive and continues to be a valuable source of our theoretical understanding. One may therefore hope that it will eventually lead, together with the constructive efforts, to the rigorous consolidation of relativistic quantum field theory.

References

1. R. Jost, *The General Theory of Quantized Fields*, American Math. Soc. 1965
2. R.F. Streater and A.S. Wightman, *PCT, Spin and Statistics, and all that*, Benjamin 1964
3. N.N. Bogolubov, A.A. Logunov and I.T. Todorov, *Introduction to Axiomatic Quantum Field Theory*, Benjamin 1975
4. J. Glimm and A. Jaffe, *Quantum Physics: A Functional Integral Point of View*, Springer 1987
5. R. Haag, *Local Quantum Physics: Fields, Particles, Algebras*, Springer 1996
6. R.M. Wald, *Quantum Field Theory in Curved Spacetime and Black Hole Thermodynamics*, Univ. Chicago Press 1994
7. J. Dimock, "Locality in free string theory", preprint mp_arc 98-311, to appear in J. Math. Phys.
8. K.–H. Rehren, "Comment on a recent solution to Wightman's axioms", Commun. Math. Phys. **178** (1996) 453

9. H.–J. Borchers and W. Zimmermann, "On the selfadjointness of field operators", Nuovo Cimento **31** (1963) 1047
10. W. Driessler and J. Fröhlich, "The reconstruction of local observable algebras from the Euclidean Green's functions of relativistic quantum field theory", Ann. Inst. H. Poincaré **27** (1977) 221
11. H.–J. Borchers and J. Yngvason, "From quantum fields to local von Neumann algebras", Rev. Math. Phys. **Special Issue** (1992) 15
12. K. Fredenhagen and J. Hertel, "Observables and pointlike localized fields", Commun. Math. Phys. **80** (1981) 555
13. M. Wollenberg, "The existence of quantum fields for local nets of algebras of observables", J. Math. Phys. **29** (1988) 2106
14. R. Haag and I. Ojima, "On the problem of defining a specific theory within the frame of local quantum physics", Ann. Inst. H. Poincaré **64** (1996) 385
15. H. Bostelmann, "Zustandskeime in der lokalen Quantenfeldtheorie", Diploma Thesis, Univ. Göttingen 1998
16. K. Wilson and W. Zimmermann, "Operator product expansions and composite field operators in the general framework of quantum field theory", Commun. Math. Phys. **24** (1972) 87
17. D. Buchholz, C. D'Antoni and K. Fredenhagen, "The universal structure of local algebras", Commun. Math. Phys. **111** (1987) 123
18. S.J. Summers and R. Werner, "On Bell's inequalities and algebraic invariants", Lett. Math. Phys. **33** (1995) 321
19. H.W. Wiesbrock, "Half sided modular inclusions of von Neumann algebras", Commun. Math. Phys. **157** (1993) 83, Erratum Commun. Math. Phys. **184** (1997) 683
20. H.W. Wiesbrock, "Modular intersections of von Neumann algebras in quantum field theory", Commun. Math. Phys. **193** (1998) 269
21. H.–J. Borchers, "Modular groups in quantum field theory", these proceedings
22. H. Lehmann, K. Symanzik and W. Zimmermann, "Zur Formulierung quantisierter Feldtheorien", Nuovo Cimento **1** (1955) 425
23. A. Martin and F. Cheung, *Analyticity properties and bounds of the scattering amplitudes*, Documents on modern physics, Gordon and Breach 1970
24. J. Bros and H. Epstein, "Charged physical states and analyticity of scattering amplitudes in the Buchholz–Fredenhagen framework" in: *XIth International Congress of Mathematical Physics. Paris 1994*, D. Iagolnitzer ed., Internat. Press 1995
25. D. Iagolnitzer, *Scattering in quantum field theories*, Princeton Univ. Press 1992
26. D. Buchholz, "Gauss' law and the infraparticle problem", Phys. Lett. **B174** (1986) 331
27. D. Buchholz, "On the manifestations of particles" in: *Mathematical Physics Towards the 21st Century. Beer–Sheva 1993*, R.N. Sen and A. Gersten ed., Ben Gurion University Press 1994
28. D. Buchholz, M. Porrmann and U. Stein, "Dirac versus Wigner: Towards a universal particle concept in local quantum field theory", Phys. Lett. **B267** (1991) 377
29. H.–J. Borchers, *Translation Group and Particle Representations in Quantum Field Theory*, Springer 1996
30. H. Baumgärtel and M. Wollenberg, *Causal Nets of Operator Algebras*, Akademie Verlag 1992
31. D. Buchholz and K. Fredenhagen, "Locality and the structure of particle states", Commun. Math. Phys. **84** (1982) 1

32. K. Fredenhagen, K.–H. Rehren and B. Schroer, "Superselection sectors with braid group statistics and exchange algebras. 1. General theory", Commun. Math. Phys. **125** (1989) 201
33. D. Buchholz, S. Doplicher, G. Morchio, J.E. Roberts and F. Strocchi, "A model for charge of electromagnetic type" in: *Operator Algebras and Quantum Field Theory. Rome 1996*, S. Doplicher, R. Longo, J.E. Roberts and L. Zsido ed., International Press 1997
34. D. Buchholz, S. Doplicher, R. Longo and J.E. Roberts, "A new look at Goldstone's theorem", Rev. Math. Phys. **Special Issue** (1992) 47
35. W. Zimmermann, "The renormalization group of the model of the A^4–coupling in the abstract approach of quantum field theory", Commun. Math. Phys. **76** (1980) 39
36. D. Buchholz and R. Verch, "Scaling algebras and renormalizaton group in algebraic quantum field theory", Rev. Math. Phys. **7** (1995) 1195
37. D. Buchholz, "Quarks, gluons, color: Facts or fiction?", Nucl. Phys. **B469** (1996) 333
38. D. Buchholz and E.H. Wichmann, "Causal independence and the energy level density of states in local quantum field theory", Commun. Math. Phys. **106** (1986) 321
39. D. Buchholz and P. Junglas, "On the existence of equilibrium states in local quantum field theory", Commun. Math. Phys. **121** (1989) 255
40. J. Bros and D. Buchholz, "Towards a relativistic KMS–condition", Nucl. Phys. **B429** (1994) 291
41. J. Bros and D. Buchholz, "Axiomatic analyticity properties and representations of particles in thermal quantum field theory", Ann. Inst. H. Poincaré **64** (1996) 495
42. O. Steinmann, "Perturbative quantum field theory at positive temperatures: An axiomatic approach", Commun. Math. Phys. **170** (1995) 405
43. J. Fröhlich, "The reconstruction of quantum fields from Euclidean Green's functions at arbitrary temperatures", Helv. Phys. Acta **48** (1975) 355
44. R. Haag, H. Narnhofer and U. Stein, "On quantum field theory in gravitational background", Commun. Math. Phys. **94** (1984) 219
45. M. Radzikowski, "Micro–local approach to the Hadamard condition in quantum field theory in curved space–time", Commun. Math. Phys. **179** (1996) 529
46. R. Brunetti, K. Fredenhagen and M. Köhler, "The microlocal spectrum condition and Wick polynomials of free fields on curved space–times", Commun. Math. Phys. **180** (1996) 633
47. K. Fredenhagen and R. Haag, "On the derivation of the Hawking radiation associated with the formation of a black hole", Commun. Math. Phys. **127** (1990) 273
48. R. Brunetti and K. Fredenhagen, "Interacting quantum fields in curved space: Renormalizability of ϕ^4" in: *Operator Algebras and Quantum Field Theory. Rome 1996*, S. Doplicher, R. Longo, J.E. Roberts and L. Zsido ed., International Press 1997
49. J. Bros, H. Epstein and U. Moschella, "Analyticity properties and thermal effects for general quantum field theory on de Sitter space–time", Commun. Math. Phys. **196** (1998) 535
50. H.–J. Borchers and D. Buchholz, "Global properties of vacuum states in de Sitter space", preprint gr-qc/9803036, to appear in Ann. Inst. H. Poincaré
51. D. Buchholz, O. Dreyer, M. Florig and S.J. Summers, "Geometric modular action and space–time symmetry groups", preprint math-ph/9805026, to appear in Rev. Math. Phys.

52. D. Kastler, *The Algebraic Theory of Superselection Sectors and Field Theory: Introduction and Recent Results. Palermo 1989*, World Scientific 1990
53. B. Schroer, "A course on: An algebraic approach to nonperturbative quantum field theory", preprint hep-th/9805093

Operator Product Expansion, Renormalization Group and Weak Decays

Andrzej J. Buras

Technische Universität München, Physik Department,
D-85748 Garching, Germany

Abstract. A non-technical description of the Operator Product Expansion and Renormalization Group techniques as applied to weak decays of mesons is presented. We use this opportunity to summarize briefly the present status of the next-to-leading QCD corrections to weak decays and their implications for the unitarity triangle, the ratio ε'/ε, the radiative decay $B \to X_s\gamma$, and the rare decays $K^+ \to \pi^+\nu\bar{\nu}$ and $K_L \to \pi^0\nu\bar{\nu}$.

1 Preface

It is a great privilege and a great pleasure to give this talk at the symposium celebrating the 70th birthday of Wolfhart Zimmermann. The Operator Product Expansion [1] to which Wolfhart Zimmermann contributed in such an important manner [2–4] had an important impact on my research during the last 20 years. I do hope very much to give another talk on this subject in 2008 at a symposium celebrating Wolfhart Zimmermann's 80th birthday. I am convinced that OPE will play an important role in the next 10 years in the field of weak decays as it played already in almost 25 years since the pioneering applications of this very powerful method by Gaillard and Lee [5] and Altarelli and Maiani [6].

2 Operator Product Expansion

The basic starting point for any serious phenomenology of weak decays of hadrons is the effective weak Hamiltonian which has the following generic structure

$$\mathcal{H}_{eff} = \frac{G_F}{\sqrt{2}} \sum_i V^i_{\rm CKM} C_i(\mu) Q_i \ . \tag{1}$$

Here G_F is the Fermi constant and Q_i are the relevant local operators which govern the decays in question. The Cabibbo-Kobayashi-Maskawa factors V^i_{CKM} [7,8] and the Wilson Coefficients C_i [1] describe the strength with which a given operator enters the Hamiltonian.

In the simplest case of the β-decay, $\mathcal{H}_{eff}$ takes the familiar form

$$\mathcal{H}^{(\beta)}_{eff} = \frac{G_F}{\sqrt{2}} \cos\theta_c [\bar{u}\gamma_\mu(1-\gamma_5)d \otimes \bar{e}\gamma^\mu(1-\gamma_5)\nu_e] \ , \tag{2}$$

where V_{ud} has been expressed in terms of the Cabibbo angle. In this particular case the Wilson Coefficient is equal unity and the local operator, the object between the square brackets, is given by a product of two $V-A$ currents. This

P. Breitenlohner and D. Maison (Eds.): Proceedings 1998, LNP 558, pp. 65–85, 2000.

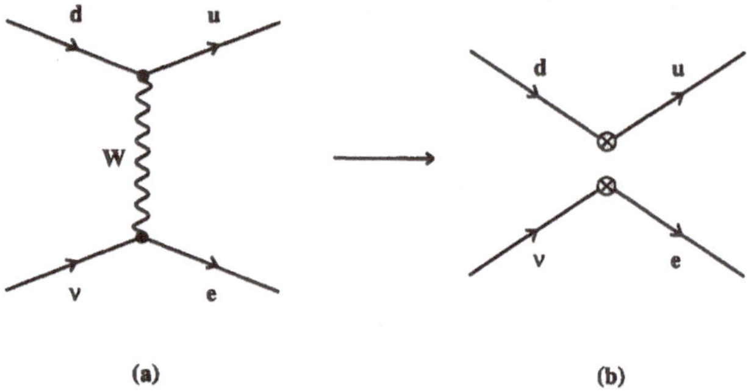

Fig. 1. β-decay at the quark level in the full (a) and effective (b) theory.

local operator is represented by the diagram (b) in Fig. 1. Equation (2) represents the Fermi theory for β-decays as formulated by Sudarshan and Marshak [9] and Feynman and Gell-Mann [10] forty years ago, except that in (2) the quark language has been used and following Cabibbo a small departure of V_{ud} from unity has been incorporated. In this context the basic formula (1) can be regarded as a generalization of the Fermi Theory to include all known quarks and leptons as well as their strong and electroweak interactions as summarized by the Standard Model. It should be stressed that the formulation of weak decays in terms of effective Hamiltonians is very suitable for the inclusion of new physics effects. We will discuss this issue briefly later on.

Now, I am aware of the fact that the formal operator language used here is hated by experimentalists and frequently disliked by more phenomenological minded theorists. Consequently the literature on weak decays, in particular on B-meson decays [11], is governed by Feynman diagram drawings with W-, Z- and top quark exchanges, rather than by the operators in (1). In the case of the β-decay we have the diagram (a) in Fig. 1. Yet such Feynman diagrams with full W-propagators, Z-propagators and top-quark propagators really represent the situation at very short distance scales $\mathcal{O}(\mathrm{M_{W,Z}}, \mathrm{m_t})$, whereas the true picture of a decaying hadron with masses $\mathcal{O}(m_{\mathrm{b}}, m_{\mathrm{c}}, m_K)$ is more properly described by effective point-like vertices which are represented by the local operators Q_i. The Wilson coefficients C_i can then be regarded as coupling constants associated with these effective vertices.

Thus $\mathcal{H}_{eff}$ in (1) is simply a series of effective vertices multiplied by effective coupling constants C_i. This series is known under the name of the operator product expansion (OPE) [1]-[4], [12]. Due to the interplay of electroweak and strong interactions the structure of the local operators (vertices) is much richer than in the case of the β-decay. They can be classified with respect to the Dirac structure, colour structure and the type of quarks and leptons relevant for a given decay. Of particular interest are the operators involving quarks only. They govern the non-leptonic decays. To be specific let us list the operators relevant for non-leptonic B–meson decays. They are:

Current–Current:

$$Q_1 = (\bar{c}_\alpha b_\beta)_{V-A}\,(\bar{s}_\beta c_\alpha)_{V-A} \qquad Q_2 = (\bar{c}b)_{V-A}\,(\bar{s}c)_{V-A} \tag{3}$$

QCD–Penguins:

$$Q_3 = (\bar{s}b)_{V-A} \sum_{q=u,d,s,c,b} (\bar{q}q)_{V-A} \qquad Q_4 = (\bar{s}_\alpha b_\beta)_{V-A} \sum_{q=u,d,s,c,b} (\bar{q}_\beta q_\alpha)_{V-A} \tag{4}$$

$$Q_5 = (\bar{s}b)_{V-A} \sum_{q=u,d,s,c,b} (\bar{q}q)_{V+A} \qquad Q_6 = (\bar{s}_\alpha b_\beta)_{V-A} \sum_{q=u,d,s,c,b} (\bar{q}_\beta q_\alpha)_{V+A} \tag{5}$$

Electroweak–Penguins:

$$\begin{aligned} Q_7 &= \frac{3}{2}\,(\bar{s}b)_{V-A} \sum_{q=u,d,s,c,b} e_q\,(\bar{q}q)_{V+A} \\ Q_8 &= \frac{3}{2}\,(\bar{s}_\alpha b_\beta)_{V-A} \sum_{q=u,d,s,c,b} e_q(\bar{q}_\beta q_\alpha)_{V+A} \end{aligned} \tag{6}$$

$$\begin{aligned} Q_9 &= \frac{3}{2}\,(\bar{s}b)_{V-A} \sum_{q=u,d,s,c,b} e_q(\bar{q}q)_{V-A} \\ Q_{10} &= \frac{3}{2}\,(\bar{s}_\alpha b_\beta)_{V-A} \sum_{q=u,d,s,c,b} e_q\,(\bar{q}_\beta q_\alpha)_{V-A}\,. \end{aligned} \tag{7}$$

Here, α and β are colour indices and e_q denotes the electrical quark charges reflecting the electroweak origin of $Q_7, \ldots, Q_{10}$. Q_2, Q_{3-6} and $Q_{7,9}$ originate in the tree level $W^\pm$-exchange, gluon penguin and (γ, Z^0)-penguin diagrams respectively. These are the diagrams a)–c) in Fig. 2. To generate Q_1, Q_8 and Q_{10} additional gluonic exchanges are needed. The operators given above have dimension six. Of interest are also operators of dimension five which are responsible for the $B \to s\gamma$ decay. They originate in the diagram d) in Fig. 2 where γ and the gluon are on-shell. They will be given in Section 7. In what follows we will neglect the higher dimensional operators as their contributions to weak decays are marginal.

Now what about the couplings $C_i(\mu)$ and the scale μ? The important point is that $C_i(\mu)$ summarize the physics contributions from scales higher than μ and due to asymptotic freedom of QCD they can be calculated in perturbation theory as long as μ is not too small. C_i include the top quark contributions and contributions from other heavy particles such as W, Z-bosons and charged Higgs particles or supersymmetric particles in the supersymmetric extensions of the Standard Model. At higher orders in the electroweak coupling the neutral Higgs may also contribute. Consequently $C_i(\mu)$ depend generally on m_t and also on the masses of new particles if extensions of the Standard Model are considered. This dependence can be found by evaluating the *box* and *penguin* diagrams with full W-, Z-, top- and new particles exchanges shown in Fig. 2 and *properly* including short distance QCD effects. The latter govern the μ-dependence of the couplings $C_i(\mu)$.

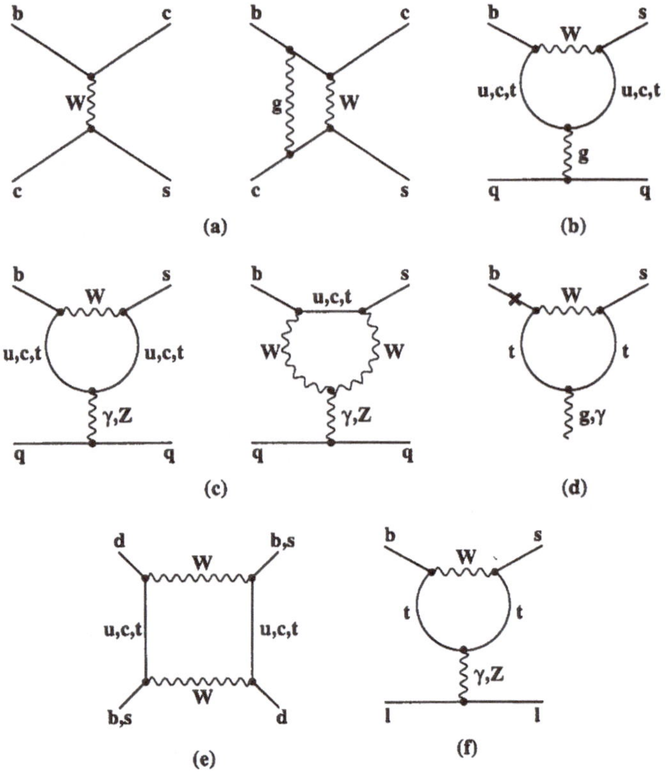

Fig. 2. Typical Penguin and Box Diagrams.

The value of μ can be chosen arbitrarily. It serves to separate the physics contributions to a given decay amplitude into short-distance contributions at scales higher than μ and long-distance contributions corresponding to scales lower than μ. It is customary to choose μ to be of the order of the mass of the decaying hadron. This is $\mathcal{O}(m_b)$ and $\mathcal{O}(m_c)$ for B-decays and D-decays respectively. In the case of K-decays the typical choice is $\mu = \mathcal{O}(1 - 2\ GeV)$ instead of $\mathcal{O}(m_K)$, which is much too low for any perturbative calculation of the couplings C_i.

Now due to the fact that $\mu \ll M_{W,Z},\ m_t$, large logarithms $\ln M_W/\mu$ compensate in the evaluation of $C_i(\mu)$ the smallness of the QCD coupling constant α_s and terms $\alpha_s^n(\ln M_W/\mu)^n$, $\alpha_s^n(\ln M_W/\mu)^{n-1}$ etc. have to be resummed to all orders in α_s before a reliable result for C_i can be obtained. This can be done very efficiently by means of the renormalization group methods [13–15]. Indeed solving the renormalization group equations for the Wilson coefficients $C_i(\mu)$ summs automatically large logarithms. The resulting *renormalization group improved* perturbative expansion for $C_i(\mu)$ in terms of the effective coupling constant $\alpha_s(\mu)$ does not involve large logarithms and is more reliable.

It should be stressed at this point that the construction of the effective Hamiltonian $\mathcal{H}_{eff}$ by means of the operator product expansion and the renormalization group methods can be done fully in the perturbative framework. The fact that the decaying hadrons are bound states of quarks is irrelevant for this construction. Consequently the coefficients $C_i(\mu)$ are independent of the particular decay considered in the same manner in which the usual gauge couplings are universal and process independent.

Having constructed the effective Hamiltonian we can proceed to evaluate the decay amplitudes. An amplitude for a decay of a given meson $M = K, B, ..$ into a final state $F = \pi\nu\bar{\nu},\ \pi\pi,\ DK$ is simply given by

$$A(M \to F) = \langle F|\mathcal{H}_{eff}|M\rangle = \frac{G_F}{\sqrt{2}} \sum_i V^i_{CKM} C_i(\mu) \langle F|Q_i(\mu)|M\rangle, \qquad (8)$$

where $\langle F|Q_i(\mu)|M\rangle$ are the hadronic matrix elements of Q_i between M and F. As indicated in (8) these matrix elements depend similarly to $C_i(\mu)$ on μ. They summarize the physics contributions to the amplitude $A(M \to F)$ from scales lower than μ.

We realize now the essential virtue of OPE: it allows to separate the problem of calculating the amplitude $A(M \to F)$ into two distinct parts: the ***short distance*** (perturbative) calculation of the couplings $C_i(\mu)$ and the ***long-distance*** (generally non-perturbative) calculation of the matrix elements $\langle Q_i(\mu)\rangle$. The scale μ, as advertised above, separates then the physics contributions into short distance contributions contained in $C_i(\mu)$ and the long distance contributions contained in $\langle Q_i(\mu)\rangle$. By evolving this scale from $\mu = \mathcal{O}(M_{\rm W})$ down to lower values one simply transforms the physics contributions at scales higher than μ from the hadronic matrix elements into $C_i(\mu)$. Since no information is lost this way the full amplitude cannot depend on μ. Therefore the μ-dependence of the couplings $C_i(\mu)$ has to cancel the μ-dependence of $\langle Q_i(\mu)\rangle$. In other words it is a matter of choice what exactly belongs to $C_i(\mu)$ and what to $\langle Q_i(\mu)\rangle$. This cancellation of μ-dependence involves generally several terms in the expansion in (8).

Clearly, in order to calculate the amplitude $A(M \to F)$, the matrix elements $\langle Q_i(\mu)\rangle$ have to be evaluated. Since they involve long distance contributions one is forced in this case to use non-perturbative methods such as lattice calculations, the 1/N expansion (N is the number of colours), QCD sum rules, hadronic sum rules, chiral perturbation theory and so on. In the case of certain B-meson decays, the ***Heavy Quark Effective Theory*** (HQET) turns out to be a useful tool. Needless to say, all these non-perturbative methods have some limitations. Consequently the dominant theoretical uncertainties in the decay amplitudes reside in the matrix elements $\langle Q_i(\mu)\rangle$.

The fact that in most cases the matrix elements $\langle Q_i(\mu)\rangle$ cannot be reliably calculated at present, is very unfortunate. One of the main goals of the experimental studies of weak decays is the determination of the CKM factors $V^i_{\rm CKM}$ and the search for the physics beyond the Standard Model. Without a reliable estimate of $\langle Q_i(\mu)\rangle$ this goal cannot be achieved unless these matrix elements can be determined experimentally or removed from the final measurable quantities by taking the ratios or suitable combinations of amplitudes or branching ratios. However, this can be achieved only in a handful of decays and generally one has to face directly the calculation of $\langle Q_i(\mu)\rangle$.

Now in the case of semi-leptonic decays, in which there is at most one hadron in the final state, the chiral perturbation theory in the case of K-decays and HQET in the case of B-decays have already provided useful estimates of the relevant matrix elements. This way it was possible to achieve satisfactory determinations of the CKM elements V_{us} and V_{cb} in $K \to \pi e\nu$ and $B \to D^* e\nu$ respectively. Similarly certain rare decays like $K \to \pi\nu\bar{\nu}$ and $B \to \mu\bar{\mu}$ can be calculated very reliably.

The case of non-leptonic decays in which the final state consists exclusively out of hadrons is a completely different story. Here even the matrix elements entering the simplest decays, the two-body decays like $K \to \pi\pi$, $D \to K\pi$ or $B \to DK$ cannot be calculated in QCD reliably at present. More promising in this respect is the evaluation of hadronic matrix elements relevant for $K^0 - \bar{K}^0$ and $B^0_{d,s} - \bar{B}^0_{d,s}$ mixings.

Returning to the Wilson coefficients $C_i(\mu)$ it should be stressed that similar to the effective coupling constants they do not depend only on the scale μ but also on the renormalization scheme used: this time on the scheme for the renormalization of local operators. That the local operators undergo renormalization is not surprising. After all they represent effective vertices and as the usual vertices in a field theory they have to be renormalized when quantum corrections like QCD or QED corrections are taken into account. As a consequence of this, the hadronic matrix elements $\langle Q_i(\mu)\rangle$ are renormalization scheme dependent and this scheme dependence must be cancelled by the one of $C_i(\mu)$ so that the physical amplitudes are renormalization scheme independent. Again, as in the case of the μ-dependence, the cancellation of the renormalization scheme dependence involves generally several terms in the expansion (8).

Now the μ and the renormalization scheme dependences of the couplings $C_i(\mu)$ can be evaluated efficiently in the renormalization group improved perturbation theory. Unfortunately the incorporation of these dependences in the non-perturbative evaluation of the matrix elements $\langle Q_i(\mu)\rangle$ remains as an important challenge and most of the non-perturbative methods on the market are insensitive to these dependences. The consequence of this unfortunate situation is obvious: the resulting decay amplitudes are μ and renormalization scheme dependent which introduces potential theoretical uncertainty in the predictions. On the other hand in certain decays these dependences can be put under control.

So far I have discussed only *exclusive* decays. It turns out that in the case of *inclusive* decays of heavy mesons, like B-mesons, things turn out to be easier. In an inclusive decay one sums over all (or over a special class) of accessible final states so that the amplitude for an inclusive decay takes the form:

$$A(B \to X) = \frac{G_F}{\sqrt{2}} \sum_{f \in X} V^i_{\rm CKM} C_i(\mu) \langle f | Q_i(\mu) | B \rangle \ . \tag{9}$$

At first sight things look as complicated as in the case of exclusive decays. It turns out, however, that the resulting branching ratio can be calculated in the expansion in inverse powers of m_b with the leading term described by the spectator model in which the B-meson decay is modelled by the decay of the b-quark:

$$\mathrm{Br}(B \to X) = \mathrm{Br}(b \to q) + \mathcal{O}(\frac{1}{m_b^2}) \; . \tag{10}$$

This formula is known under the name of the Heavy Quark Expansion (HQE) [16]-[18]. Since the leading term in this expansion represents the decay of the quark, it can be calculated in perturbation theory or more correctly in the renormalization group improved perturbation theory. It should be realized that also here the basic starting point is the effective Hamiltonian (1) and that the knowledge of the couplings $C_i(\mu)$ is essential for the evaluation of the leading term in (10). But there is an important difference relative to the exclusive case: the matrix elements of the operators Q_i can be "effectively" evaluated in perturbation theory. This means, in particular, that their μ and renormalization scheme dependences can be evaluated and the cancellation of these dependences by those present in $C_i(\mu)$ can be explicitly investigated.

Clearly in order to complete the evaluation of $Br(B \to X)$ also the remaining terms in (10) have to be considered. These terms are of a non-perturbative origin, but fortunately they are suppressed by at least two powers of m_b. They have been studied by several authors in the literature with the result that they affect various branching ratios by less than 10% and often by only a few percent. Consequently the inclusive decays give generally more precise theoretical predictions at present than the exclusive decays. On the other hand their measurements are harder. There is of course an important theoretical issue related to the validity of HQE in (10) which appear in the literature under the name of quark-hadron duality. I will not discuss it here. Recent discussions of this issue can be found in [19].

We have learned now that the matrix elements of Q_i are easier to handle in inclusive decays than in the exclusive ones. On the other hand the evaluation of the couplings $C_i(\mu)$ is equally difficult in both cases although as stated above it can be done in a perturbative framework. Still in order to achieve sufficient precision for the theoretical predictions it is desirable to have accurate values of these couplings. Indeed it has been realized at the end of the eighties that the leading term (LO) in the renormalization group improved perturbation theory, in which the terms $\alpha_s^n(\ln M_{\rm W}/\mu)^n$ are summed, is generally insufficient and the inclusion of next-to-leading corrections (NLO) which correspond to summing the terms $\alpha_s^n(\ln M_{\rm W}/\mu)^{n-1}$ is necessary. In particular, unphysical left-over μ-dependences in the decay amplitudes and branching ratios resulting from the truncation of the perturbative series are considerably reduced by including NLO corrections. These corrections are known by now for the most important and interesting decays and will be briefly reviewed below.

3 Penguin–Box Expansion and OPE

The FCNC decays, in particular rare and CP violating decays are governed by various penguin and box diagrams with internal top quark and charm quark exchanges. Some examples are shown in Fig. 2. These diagrams can be evaluated in the full theory and are summarized by a set of basic universal (process independent) $m_{\rm t}$-dependent functions $F_r(x_t)$ [20] where $x_t = m_{\rm t}^2/M_{\rm W}^2$. Explicit expressions for these functions can be found in [21–23].

It is useful to express the OPE formula (8) directly in terms of the functions $F_r(x_t)$ [25]. To this end we rewrite the $A(M \to F)$ in (8) as follows

$$A(M \to F) = \frac{G_{\rm F}}{\sqrt{2}} V_{\rm CKM} \sum_{i,k} \langle F \mid O_k(\mu) \mid M \rangle \, \hat{U}_{ki}\,(\mu, M_{\rm W})\, C_i(M_{\rm W}), \tag{11}$$

where $\hat{U}_{kj}(\mu, M_W)$ is the renormalization group transformation from $M_{\rm W}$ down to μ. Explicit formula for this transformation will be given below. In order to simplify the presentation we have removed the index "i" from $V^i_{\rm CKM}$

Now $C_i(M_{\rm W})$ are linear combinations of the basic functions $F_r(x_t)$ so that we can write

$$C_i(M_{\rm W}) = c_i + \sum_r h_{ir} F_r(x_t) \tag{12}$$

where c_i and h_{ir} are $m_{\rm t}$-independent constants. Inserting (12) into (11) and summing over i and k we find

$$A(M \to F) = P_0(M \to F) + \sum_r P_r(M \to F)\, F_r(x_t), \tag{13}$$

with

$$P_0(M \to F) = \sum_{i,k} \langle F \mid O_k(\mu) \mid M \rangle \, \hat{U}_{ki}\,(\mu, M_{\rm W}) c_i \ , \tag{14}$$

$$P_r(M \to F) = \sum_{i,k} \langle F \mid O_k(\mu) \mid M \rangle \, \hat{U}_{ki}\,(\mu, M_{\rm W}) h_{ir} \ , \tag{15}$$

where we have suppressed the overall factor $(G_F/\sqrt{2})V_{CKM}$. I would like to call (13) *Penguin-Box Expansion* (PBE) [25].

The coefficients P_0 and P_r are process dependent. This process dependence enters through $\langle F \mid O_k(\mu) \mid M \rangle$. In certain cases like $K \to \pi\nu\bar{\nu}$ these matrix elements are very simple implying simple formulae for the coefficients P_0 and P_r. In other situations, like ε'/ε, this is not the case.

Originally PBE was designed to expose the $m_{\rm t}$-dependence of FCNC processes [25]. After the top quark mass has been measured precisely this role of PBE is less important. On the other hand, PBE is very well suited for the study of the extentions of the Standard Model in which new particles are exchanged in the loops. We know already that these particles are heavier than W-bosons and consequently they can be integrated out together with the weak bosons and the top quark. If there are no new local operators the mere change is to modify the functions $F_r(x_t)$ which now acquire the dependence on the masses of new particles such as charged Higgs particles and supersymmetric particles. The process dependent coefficients P_0 and P_r remain unchanged. This is particularly useful as the most difficult part is the evaluation of $\hat{U}_{kj}(\mu, M_W)$ and of the hadronic matrix elements, both contained in these coefficients. However, if new effective operators with different Dirac and colour structures are present the values of P_0 and P_r are modified. Examples of the applications of PBE to physics beyond the Standard Model can be found in [26–28].

The universality of the functions $F_r(x_t)$ can be violated partly when QCD corrections to one loop penguin and box diagrams are included. For instance in the case of semi-leptonic FCNC transitions there is no gluon exchange in a Z^0-penguin diagram parallel to the Z^0-propagator but such an exchange takes place in non-leptonic decays in which the bottom line is a quark-line. Thus the general universality of $F_r(x_t)$ present at one loop level is reduced to two universality classes relevant for semi-leptonic and non-leptonic transitions. However, the $\mathcal{O}(\alpha_s)$ corrections to the functions $F_r(x_t)$ are generally rather small when the top quark mass $\overline{m}_t(m_t)$ is used and consequently the inclusion of QCD effects plays mainly the role in reducing various μ-dependences.

In order to see the general structure of $A(M \to F)$ more transparently let us write it as follows:

$$A(M \to F) = B_{M\to F} V_{\rm CKM} \eta_{\rm QCD} F(x_t) + \text{Charm} \tag{16}$$

where the first term represents the internal top quark contribution and "Charm" stands for remaining contributions, in particular those with internal charm quark exchanges. $F(x_t)$ represents one of the universal functions and $\eta_{\rm QCD}$ the corresponding short distance QCD corrections. The parameter $B_{M\to F}$ represents the relevant hadronic matrix element, which can only be calculated by means of non-perturbative methods. However, in certain lucky situations $B_{M\to F}$ can be extracted from well measured leading decays and when it enters also other decays, the latter are then free from hadronic uncertainties and offer very useful means for extraction of CKM parameters. One such example is the decay $K^+ \to \pi^+\nu\bar{\nu}$ for which one has

$$Br(K^+ \to \pi^+\nu\bar{\nu}) = \left[\frac{\alpha_{\rm QED}^2 Br(K^+ \to \pi^0 e^+ \nu)}{V_{us}^2 2\pi^2 \sin^4\theta_W}\right] \times \left|V_{ts}^* V_{td} \eta_{\rm QCD}^t F(x_t) + V_{cs}^* V_{cd} \eta_{\rm QCD}^c F(x_c)\right|^2 \tag{17}$$

The factor in square brackets stands for the "B-factor" in (16), which is given in terms of well measured quantities. Since V_{cs}, V_{cd} and V_{ts} are already rather well determined and $F(x_i)$ and $\eta_{\rm QCD}^i$ can be calculated in perturbation theory, the element V_{td} can be extracted from $Br(K^+ \to \pi^+\nu\bar{\nu})$ without essentially any theoretical uncertainties. We will be more specific about this in Section 7.

4 Motivations for NLO Calculations

Going beyond the LO approximation for $C_i(\mu)$ is certainly an important but a non-trivial step. For this reason one needs some motivations to perform this step. Here are the main reasons for going beyond LO:

- The NLO is first of all necessary to test the validity of the renormalization group improved perturbation theory.
- Without going to NLO the QCD scale $\Lambda_{\overline{MS}}$ [29] extracted from various high energy processes cannot be used meaningfully in weak decays.
- Due to renormalization group invariance the physical amplitudes do not depend on the scales μ present in α_s or in the running quark masses, in particular $m_t(\mu)$, $m_b(\mu)$ and $m_c(\mu)$. However, in perturbation theory this property is broken through the truncation of the perturbative series. Consequently one finds sizable scale ambiguities in the leading order, which can be reduced considerably by going to NLO.
- The Wilson Coefficients are renormalization scheme dependent quantities. This scheme dependence appears first at NLO. For a proper matching of the short distance contributions to the long distance matrix elements obtained from lattice calculations it is essential to calculate NLO. The same is true for inclusive heavy quark decays in which the hadron decay can be modeled by a decay of a heavy quark and the matrix elements of Q_i can be effectively calculated in an expansion in $1/m_b$.
- In several cases the central issue of the top quark mass dependence is strictly a NLO effect.

5 General Structure of Wilson Coefficients

We will give here a formula for the Wilson coefficient $C(\mu)$ of a single operator Q including NLO corrections. The case of several operators which mix under renormalization is much more complicated. Explicit formulae are given in [21,23].

$C(\mu)$ is given by

$$C(\mu) = U(\mu, M_W) C(M_W) \tag{18}$$

where

$$U(\mu, M_W) = \exp\left[\int_{g_s(M_W)}^{g_s(\mu)} dg' \frac{\gamma_Q(g'_s)}{\beta(g'_s)}\right] \tag{19}$$

is the evolution function, which allows to calculate $C(\mu)$ once $C(M_W)$ is known. The latter can be calculated in perturbation theory in the process of integrating out $W^\pm$, Z^0 and top quark fields. Details can be found in [21,23]. Next γ_Q is the anomalous dimension of the operator Q and $\beta(g_s)$ is the renormalization group function which governs the evolution of the QCD coupling constant $\alpha_s(\mu)$.

At NLO we have

$$C(M_W) = 1 + \frac{\alpha_s(M_W)}{4\pi} B \tag{20}$$

$$\gamma_Q(\alpha_s) = \gamma_Q^{(0)} \frac{\alpha_s}{4\pi} + \gamma_Q^{(1)} \left(\frac{\alpha_s}{4\pi}\right)^2 \tag{21}$$

$$\beta(g_s) = -\beta_0 \frac{g_s^3}{16\pi^2} - \beta_1 \frac{g_s^5}{(16\pi^2)^2} \tag{22}$$

Inserting the last two formulae into (19) and expanding in α_s we find

$$U(\mu, M_W) = \left[1 + \frac{\alpha_s(\mu)}{4\pi} J\right] \left[\frac{\alpha_s(M_W)}{\alpha_s(\mu)}\right]^d \left[1 - \frac{\alpha_s(M_W)}{4\pi} J\right] \tag{23}$$

with

$$J = \frac{d}{\beta_0}\beta_1 - \frac{\gamma_Q^{(1)}}{2\beta_0} \qquad d = \frac{\gamma_Q^{(0)}}{2\beta_0}. \tag{24}$$

Inserting (23) and (20) into (18) we find an important formula for $C(\mu)$ in the NLO approximation:

$$C(\mu) = \left[1 + \frac{\alpha_s(\mu)}{4\pi} J\right] \left[\frac{\alpha_s(M_W)}{\alpha_s(\mu)}\right]^d \left[1 + \frac{\alpha_s(M_W)}{4\pi}(B - J)\right] . \tag{25}$$

6 Status of NLO Calculations

Since the pioneering leading order calculations of Wilson coefficients for current–current [5,6] and penguin operators [30], enormous progress has been made, so that at present most of the decay amplitudes are known at the NLO level. We list all existing NLO calculations for weak decays in table 1. In addition to the calculations in the Standard Model we list the calculations in two-Higgs doublet models and supersymmetry. In table 2 we list references to calculations of two-loop electroweak contributions to rare decays. The latter calculations allow to reduce scheme and scale dependences related to the definition of electroweak parameters like $\sin^2\theta_W$, α_{QED}, etc. Next, useful techniques for three-loop calculations can be found in [77] and a very general discussion of the evanescent operators including earlier references is presented in [78]. Further details on these calculations can be found in the orignal papers, in the review [21] and in the Les Houches lectures [23]. Some of the implications of these calculations will be analyzed briefly in subsequent sections.

7 Applications: News

7.1 Preliminaries

There is a vast literature on the applications of NLO calculations listed in table 1. As they are already reviewed in detail in [21–23] there is no point to review them here again. I will rather discuss briefly some of the most important applications

Table 1. References to NLO Calculations

Decay	Reference
$\Delta F = 1$ Decays	
current-current operators	[31,32]
QCD penguin operators	[33,35–38]
electroweak penguin operators	[34–37]
magnetic penguin operators	[39,40]
$Br(B)_{SL}$	[31,41–43]
inclusive $\Delta S = 1$ decays	[44]
Particle-Antiparticle Mixing	
η_1	[45]
$\eta_2,\ \eta_B$	[46,47]
η_3	[48]
Rare K- and B-Meson Decays	
$K_L^0 \to \pi^0\nu\bar{\nu}$, $B \to l^+l^-$, $B \to X_s\nu\bar{\nu}$	[49–52]
$K^+ \to \pi^+\nu\bar{\nu}$, $K_L \to \mu^+\mu^-$	[53,52]
$K^+ \to \pi^+\mu\bar{\mu}$	[54]
$K_L \to \pi^0e^+e^-$	[55]
$B \to X_s\mu^+\mu^-$	[56,57]
$B \to X_s\gamma$	[58]-[65]
$\Delta\Gamma_{B_s}$	[67]
inclusive B $\to$ Charmonium	[68]
Two-Higgs Doublet Models	
$B \to X_s\gamma$	[64,66,65]
Supersymmetry	
ΔM_K and ε_K	[69,70]
$B \to X_s\gamma$	[71]

Table 2. Electroweak Two-Loop Calculations

Decay	Reference
$K_L^0 \to \pi^0\nu\bar{\nu}$, $B \to l^+l^-$, $B \to X_s\nu\bar{\nu}$	[72]
$B \to X_s\gamma$	[73–75]
$B^0 - \bar{B}^0$ mixing	[76]

in general terms. This will also give me the opportunity to update some of the numerical results presented in [23]. This update is related mainly to the improved experimental lower bound on $B_s^0 - \bar{B}_s^0$ mixing ($(\Delta M)_s > 12.4/ps$) and a slight increase in $|V_{ub}|/|V_{cb}|$: 0.091 ± 0.016, both presented at the last Rochester Conference in Vancouver [79].

7.2 Unitarity Triangle

The standard analysis of the unitarity triangle (see Fig. 3) uses the values of $|V_{us}|$, $|V_{cb}|$, $|V_{ub}/V_{cb}|$ extracted from tree level K- and B- decays, the indirect

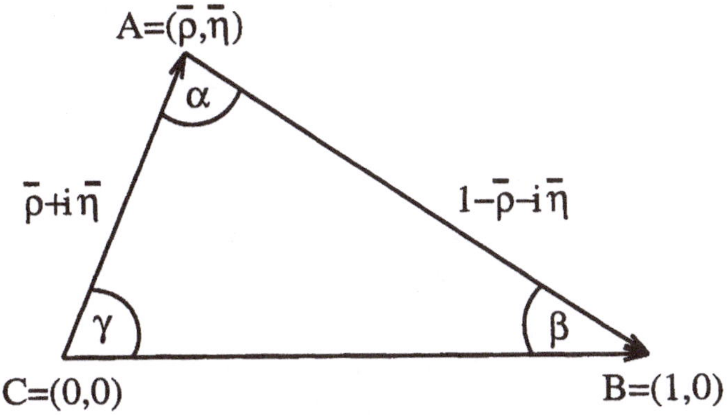

Fig. 3. Unitarity Triangle.

CP-violation in $K_L \to \pi\pi$ represented by the parameter ε and the $B^0_{d,s} - \bar{B}^0_{d,s}$ mixings described by the mass differences $(\Delta M)_{d,s}$. From this analysis follows the allowed range for $(\bar{\varrho}, \bar{\eta})$ describing the apex of the unitarity triangle. Here [81]

$$\bar{\varrho} = \varrho(1 - \frac{\lambda^2}{2}), \qquad \bar{\eta} = \eta(1 - \frac{\lambda^2}{2}). \tag{26}$$

where λ, ϱ and η are Wolfenstein parameters [82] with $|V_{us}| = \lambda = 0.22$. We have in particular

$$V_{ub} = \lambda|V_{cb}|(\varrho - i\eta), \qquad V_{td} = \lambda|V_{cb}|(1 - \bar{\varrho} - i\bar{\eta}). \tag{27}$$

$\eta \neq 0$ is responsible for CP violation in the Standard Model.

The allowed region for $(\bar{\varrho}, \bar{\eta})$ is presented in Fig. 4. It is the shaded area on the right hand side of the solid circle which represents the upper bound for $(\Delta M)_d/(\Delta M)_s$. The hyperbolas give the constraint from ε and the two circles centered at $(0,0)$ the constraint from $|V_{ub}/V_{cb}|$. The white areas between the lower ε-hyperbola and the shaded region are excluded by $B^0_d - \bar{B}^0_d$ mixing. We observe that the region $\bar{\varrho} < 0$ is practically excluded. The main remaining theoretical uncertainties in this analysis are the values of non-perturbative parameters: B_K in ε, $F_{B_d}\sqrt{B_d}$ in $(\Delta M)_d$ and $\xi = F_{B_s}\sqrt{B_s}/F_{B_d}\sqrt{B_d}$ in $(\Delta M)_d/(\Delta M)_s$. I have used $B_K = 0.80 \pm 0.15$, $F_{B_d}\sqrt{B_d} = 200 \pm 40\,\mathrm{MeV}$ and $\xi_{\max} = 1.2$. On the experimental side $|V_{ub}/V_{cb}|$ and $(\Delta M)_s$ should be improved.

From this analysis we extract

$$|V_{td}| = (8.6 \pm 1.6) \cdot 10^{-3}, \qquad \mathrm{Im}(V^*_{ts}V_{td}) = (1.38 \pm 0.33) \cdot 10^{-4} \tag{28}$$

and

$$\sin 2\beta = 0.71 \pm 0.13, \qquad \sin\gamma = 0.83 \pm 0.17 \tag{29}$$

7.3 ε'/ε

ε'/ε is the ratio of the direct and indirect CP violation in $K_L \to \pi\pi$. A measurement of a non-vanishing value of ε'/ε would give the first signal for the direct

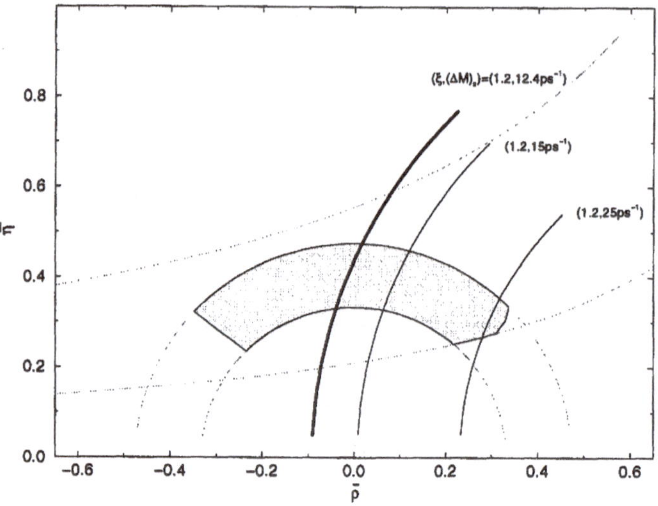

Fig. 4. Unitarity Triangle 1998.

CP violation ruling out the superweak models [83]. In the Standard Model ε'/ε is governed by QCD penguins and electroweak (EW) penguins. The corresponding operators are given in (4)-(7). With increasing $m_{\rm t}$ the EW penguins become increasingly important [84,85], and entering ε'/ε with the opposite sign to QCD penguins suppress this ratio for large $m_{\rm t}$. For $m_{\rm t} \approx 200$ GeV the ratio can even be zero [85]. Because of this strong cancellation between two dominant contributions and due to uncertainties related to hadronic matrix elements of the relevant local operators, a precise prediction of ε'/ε is not possible at present.

A very simplified formula (not to be used for any serious numerical analysis) which exhibits main uncertainties is given as follows

$$\frac{\varepsilon'}{\varepsilon} = 15\cdot 10^{-4}\left[\frac{\eta\lambda|V_{cb}|^2}{1.3\cdot 10^{-4}}\right]\left[\frac{120\ MeV}{m_s(2\ \text{GeV})}\right]^2\left[\frac{\Lambda^{(4)}_{\overline{\rm MS}}}{300\ \text{MeV}}\right]^{0.8}[B_6 - Z(x_t)B_8] \quad (30)$$

where $Z(x_t) \approx 0.18(m_{\rm t}/M_{\rm W})^{1.86}$ represents the leading $m_{\rm t}$-dependence of EW penguins. B_6 and B_8 represent the hadronic matrix elements of the dominant QCD-penguin operator Q_6 and the dominant electroweak penguin operator Q_8 (see (5) and (6)) respectively. Together with $m_{\rm s}(2\,\text{GeV})$ the values of these parameters constitute the main theoretical uncertainty in evaluating ε'/ε. Present status of $m_{\rm s}$, B_6 and B_8 is reviewed in [86,23]. Roughly one has $B_6 = 1.0 \pm 0.2$ and $B_8 = 0.7 \pm 0.2$. Taking these values, η of Fig. 4 and $|V_{cb}| = 0.040 \pm 0.003$, I find:

$$\varepsilon'/\varepsilon = \begin{cases} (5.7 \pm 3.6)\cdot 10^{-4}\,, & m_{\rm s}(2\,\text{GeV}) = 130 \pm 20\,\text{MeV} \\ (9.1 \pm 5.7)\cdot 10^{-4}\,, & m_{\rm s}(2\,\text{GeV}) = 110 \pm 20\,\text{MeV}. \end{cases} \quad (31)$$

where the chosen values for $m_{\rm s}$ are in the ball park of various QCD sum rules and lattice estimates [86]. This should be compared with the result of NA31 collaboration at CERN which finds $(\varepsilon'/\varepsilon) = (23 \pm 7) \cdot 10^{-4}$ [87] and the value of E731 at Fermilab, $(\varepsilon'/\varepsilon) = (7.4 \pm 5.9) \cdot 10^{-4}$ [88].

The Standard Model expectations are closer to the Fermilab result, but due to large theoretical and experimental errors no firm conclusion can be reached at present. The new improved data from CERN and Fermilab in 1999 and later from DAΦNE should shed more light on ε'/ε. In this context improved estimates of B_6, B_8 and m_s are clearly desirable.

7.4 $B \to X_s\gamma$

A lot of efforts have been put into predicting the branching ratio for the inclusive decay $B \to X_s\gamma$ including NLO QCD corrections and higher order electroweak corrections. The relevant references are given in table 1 and in [23], where details can be found. The final result of these efforts can be summarized by

$$Br(B \to X_s\gamma)_{\rm th} = (3.30 \pm 0.15({\rm scale}) \pm 0.26({\rm par})) \cdot 10^{-4} \tag{32}$$

where the first error represents residual scale dependences and the second error is due to uncertainties in input parameters. The main achievement is the reduction of the scale dependence through NLO calculations, in particular those given in [61] and [40]. In the leading order the corresponding error would be roughly ± 0.6 [89,90].

The theoretical result in (32) should be compared with experimental data:

$$Br(B \to X_s\gamma)_{\rm exp} = \begin{cases} (3.15 \pm 0.35 \pm 0.41) \cdot 10^{-4}\,, & \text{CLEO} \\ (3.11 \pm 0.80 \pm 0.72) \cdot 10^{-4}\,, & \text{ALEPH,} \end{cases} \tag{33}$$

which implies the combined branching ratio:

$$Br(B \to X_s\gamma)_{\rm exp} = (3.14 \pm 0.48) \cdot 10^{-4} \tag{34}$$

Clearly, the Standard Model result agrees well with the data. In order to see whether any new physics can be seen in this decay, the theoretical and in particular experimental errors should be reduced. This is certainly a very difficult task. Most recent analyses of $B \to X_s\gamma$ in supersymmetric models and two–Higgs doublet models are listed in table 1.

7.5 $K_L \to \pi^0\nu\bar\nu$ and $K^+ \to \pi^+\nu\bar\nu$

$K_L \to \pi^0\nu\bar\nu$ and $K^+ \to \pi^+\nu\bar\nu$ are the theoretically cleanest decays in the field of rare K-decays. $K_L \to \pi^0\nu\bar\nu$ is dominated by short distance loop diagrams (Z-penguins and box diagrams) involving the top quark. $K^+ \to \pi^+\nu\bar\nu$ receives additional sizable contributions from internal charm exchanges. The great virtue of $K_L \to \pi^0\nu\bar\nu$ is that it proceeds almost exclusively through direct CP violation [91] and as such is the cleanest decay to measure this important phenomenon. It also offers a clean determination of the Wolfenstein parameter η and in particular offers the cleanest measurement of ${\rm Im}V_{ts}^*V_{td}$ [92]. $K^+ \to \pi^+\nu\bar\nu$ is CP conserving and offers a clean determination of $|V_{td}|$. Due to the presence of the charm contribution and the related m_c dependence it has a small scale uncertainty absent in $K_L \to \pi^0 \nu\bar{\nu}$.

The next-to-leading QCD corrections [49,50,53,51,52] to both decays considerably reduced the theoretical uncertainty due to the choice of the renormalization scales present in the leading order expressions, in particular in the charm contribution to $K^+ \to \pi^+ \nu\bar{\nu}$. Since the relevant hadronic matrix elements of the weak currents entering $K \to \pi\nu\bar{\nu}$ can be related using isospin symmetry to the leading decay $K^+ \to \pi^0 e^+ \nu$, the resulting theoretical expressions for Br($K_L \to \pi^0 \nu\bar{\nu}$) and Br($K^+ \to \pi^+ \nu\bar{\nu}$) are only functions of the CKM parameters, the QCD scale $\Lambda_{\overline{\rm MS}}$ and the quark masses m_t and m_c. The isospin braking corrections have been calculated in [93]. The long distance contributions to $K^+ \to \pi^+ \nu\bar{\nu}$ have been considered in [94] and found to be very small: a few percent of the charm contribution to the amplitude at most, which is safely neglegible. The long distance contributions to $K_L \to \pi^0 \nu\bar{\nu}$ are negligible as well [95].

The explicit expressions for $Br(K^+ \to \pi^+ \nu\bar{\nu})$ and $Br(K_L \to \pi^0 \nu\bar{\nu})$ can be found in [21–23]. Here we give approximate expressions in order to exhibit various dependences:

$$Br(K^+ \to \pi^+ \nu\bar{\nu}) = 0.7 \cdot 10^{-10} \left[\left[\frac{|V_{td}|}{0.010}\right]^2 \left[\frac{|V_{cb}|}{0.040}\right]^2 \left[\frac{\overline{m}_t(m_t)}{170 \text{ GeV}}\right]^{2.3} + \text{cc} + \text{tc} \right] \tag{35}$$

$$Br(K_L \to \pi^0 \nu\bar{\nu}) = 3.0 \cdot 10^{-11} \left[\frac{\eta}{0.39}\right]^2 \left[\frac{\overline{m}_t(m_t)}{170\, GeV}\right]^{2.3} \left[\frac{|V_{cb}|}{0.040}\right]^4 \tag{36}$$

where in (35) we have shown explicitly only the pure top contribution.

The impact of NLO calculations is the reduction of scale uncertainties in $Br(K^+ \to \pi^+ \nu\bar{\nu})$ from ±23% to ±7%. This corresponds to the reduction in the uncertainty in the determination of $|V_{td}|$ from ±14% to ±4%. The remaining scale uncertainties in $Br(K_L \to \pi^0 \nu\bar{\nu})$ and in the determination of η are fully negligible.

Updating the analysis of [23] one finds [52]:

$$\begin{aligned} Br(K^+ \to \pi^+ \nu\bar{\nu}) &= (8.2 \pm 3.2) \cdot 10^{-11} , \\ Br(K_L \to \pi^0 \nu\bar{\nu}) &= (3.1 \pm 1.3) \cdot 10^{-11} , \end{aligned} \tag{37}$$

where the errors come dominantly from the uncertainties in the CKM parameters.

As stressed in [92] simultaneous measurements of $Br(K^+ \to \pi^+ \nu\bar{\nu})$ and $Br(K_L \to \pi^0 \nu\bar{\nu})$ should allow a clean determination of the unitarity triangle as shown in Fig. 5. In particular the measurements of these branching ratios with an error of ±10% will determine $|V_{td}|$, $\text{Im} V_{ts}^* V_{td}$ and $\sin 2\beta$ with an accuraccy of ±10%, ±5% and ±0.05 respectively. The comparision of this determination of

$\sin 2\beta$ with the one by means of the CP-asymmetry in $B \to \psi K_S$ should offer a very good test of the Standard Model.

Experimentally we have [96]

$$BR(K^+ \to \pi^+ \nu\bar{\nu}) = (4.2{+}9.7{-}3.5) \cdot 10^{-10} \tag{38}$$

and the bound [97]

$$BR(K_{\rm L} \to \pi^0 \nu\bar{\nu}) < 1.6 \cdot 10^{-6}. \tag{39}$$

Moreover from (38) and isospin symmetry one has [98] $BR(K_{\rm L} \to \pi^0 \nu\bar{\nu}) < 6.1 \cdot 10^{-9}$.

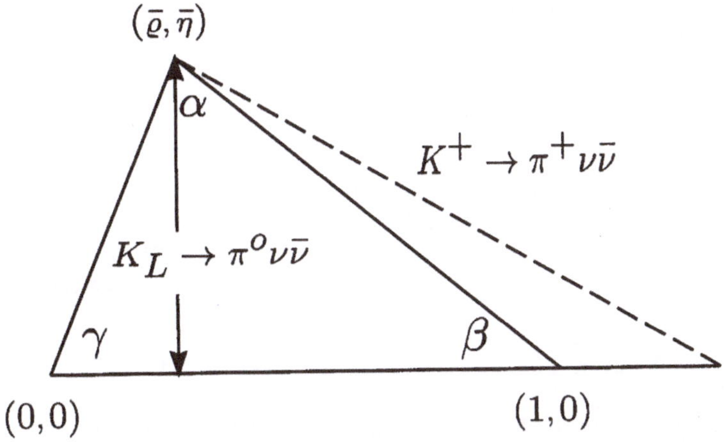

Fig. 5. Unitarity triangle from $K \to \pi\nu\bar{\nu}$.

The central value in (38) is by a factor of 4 above the Standard Model expectation but in view of large errors the result is compatible with the Standard Model. The analysis of additional data on $K^+ \to \pi^+ \nu\bar{\nu}$ present on tape at BNL787 should narrow this range in the near future considerably. In view of the clean character of this decay a measurement of its branching ratio at the level of $2 \cdot 10^{-10}$ would signal the presence of physics beyond the Standard Model [52]. The Standard Model sensitivity is expected to be reached at AGS around the year 2000 [99]. Also Fermilab with the Main Injector could measure this decay [100].

The present upper bound on $Br(K_{\rm L} \to \pi^0 \nu\bar{\nu})$ is about five orders of magnitude above the Standard Model expectation (37). FNAL-E799 expects to reach the accuracy $\mathcal{O}(10^{-8})$ and a very interesting new experiment at Brookhaven (BNL E926) [99] expects to reach the single event sensitivity $2 \cdot 10^{-12}$ allowing a 10% measurement of the expected branching ratio. There are furthermore plans to measure this gold-plated decay with comparable sensitivity at Fermilab [101] and KEK [102].

8 Summary

We have given a general description of OPE and Renormalization Group techniques as applied to weak decays of mesons. Further details can be found in [21–23,103]. One of the outstanding and important challanges for theorists in this field is a quantitative description of non–leptonic meson decays. In the field of K–decays this is in particular the case of the $\Delta I = 1/2$ rule for which some progress has been made in [104]. In the field of B–decays progress in a quantitative description of two–body decays is very desirable in view of forthcoming B-physics experiments at Cornell, SLAC, KEK, DESY, FNAL and later at LHC. Recent reviews on non-leptonic two-body decays are given in [105–107] where further references can be found.

I would like to thank Peter Breitenlohner, Dieter Maison and Julius Wess for inviting me to such a pleasant symposium.

References

1. K.G. Wilson, Phys. Rev. **179** (1969) 1499;
2. W. Zimmermann, in *Proc. 1970 Brandeis Summer Institute in Theor. Phys*, (eds. S. Deser, M. Grisaru and H. Pendleton), MIT Press, 1971, p.396;
3. K.G. Wilson and W. Zimmermann, Commun. Math. Phys. **24** (1972) 87.
4. W. Zimmermann, Ann. Phys. **77** (1973) 570.
5. M.K. Gaillard and B.W. Lee, Phys. Rev. Lett. **33** (1974) 108.
6. G. Altarelli and L. Maiani, Phys. Lett. **B 52** (1974) 351.
7. N. Cabibbo, Phys. Rev. Lett. **10** (1963) 531.
8. M. Kobayashi and K. Maskawa, Prog. Theor. Phys. **49** (1973) 652.
9. E.C.G. Sudarshan and R.E. Marshak, Proc. Padua-Venice Conf. on Mesons and Recently Discovered Particles (1957).
10. R.P. Feynman and M. Gell-Mann, Phys. Rev. **109** (1958) 193.
11. D. Zeppenfeld, Z. Phys. **C 8** (1981) 77; L.L. Chau, Phys. Rev. **D 43** (1991) 2176; M. Gronau, J.L. Rosner and D. London, Phys. Rev. Lett. **73** (1994) 21; O.F. Hernandez, M. Gronau, J.L. Rosner and D. London, Phys. Lett. **B 333** (1994) 500, Phys. Rev. **D 50** (1994) 4529.
12. E. Witten, Nucl. Phys. **B 120** (1977) 189.
13. E.C.G. Stueckelberg and A. Petermann, Helv. Phys. Acta **26** (1953) 499; M. Gell–Mann and F.E. Low, Phys. Rev. **95** (1954) 1300; N.N. Bogoliubov and D.V. Shirkov, Dokl. Acad. Nauk SSSR **103** (1955) 203, 391; L.V. Ovsyannikov, Dokl. Acad. Nauk SSSR **109** (1956) 1112; K. Symanzik, Commun. Math. Phys. **18** (1970) 227; C.G. Callan Jr, Phys. Rev. **D 2** (1970) 1541.
14. G. 't Hooft, Nucl. Phys. **B 61** (1973) 455.
15. S. Weinberg, Phys. Rev. **D 8** (1973) 3497.
16. J. Chay, H. Georgi and B. Grinstein, Phys. Lett. **B 247** (1990) 399.
17. I.I. Bigi, N.G. Uraltsev and A.I. Vainshtein, Phys. Lett. **B 293** (1992) 430 [E: **B 297** (1993) 477]. I.I. Bigi, M.A. Shifman, N.G. Uraltsev and A.I. Vainshtein, Phys. Rev. Lett. **71** (1993) 496; B. Blok, L. Koyrakh, M.A. Shifman and A.I. Vainshtein, Phys. Rev. **D 49** (1994) 3356 [E: **D 50** (1994) 3572].
18. A.V. Manohar and M.B. Wise, Phys. Rev. **D 49** (1994) 1310.

19. Z. Ligeti and A.V. Manohar, Phys. Lett. **B 433** (1998) 396; I. Bigi, M. Shifman, N. Uraltsev and A. Vainshtein, hep-ph/9805241; B.Grinstein and R.F. Lebed, hep-ph/9805404; N. Isgur, hep-ph/9811377.
20. T. Inami and C.S. Lim, Progr. Theor. Phys. **65** (1981) 297.
21. G. Buchalla, A.J. Buras and M. Lautenbacher, Rev. Mod. Phys. **68** (1996) 1125.
22. A.J. Buras and R. Fleischer, hep-ph/9704376, page 65 in [24].
23. A.J. Buras, hep-ph/9806471, to appear in *Probing the Standard Model of Particle Interactions*, eds. F. David and R. Gupta (Elsevier Science B.V., Amsterdam, 1998).
24. A.J. Buras and M. Lindner, Heavy Flavours II, World Scientific, 1998.
25. G. Buchalla, A.J. Buras and M.K. Harlander, Nucl. Phys. **B 349** (1991) 1.
26. G. Buchalla, A.J. Buras, M.K. Harlander, M.E. Lautenbacher and C. Salazar, Nucl. Phys. **B 355** (1991) 305.
27. P. Cho, M. Misiak and D. Wyler, Phys. Rev. **D 54** (1996) 3329.
28. A. Ali, Th. Mannel and Ch. Greub, Zeit. Phys. **C 67** (1995) 417.
29. W.A. Bardeen, A.J. Buras, D.W. Duke and T. Muta, Phys. Rev. **D 18** (1978) 3998.
30. A.I. Vainshtein, V.I. Zakharov, and M.A. Shifman, JTEP **45** (1977) 670; F. Gilman and M.B. Wise, Phys. Rev. **D 20** (1979) 2392; B. Guberina and R. Peccei, Nucl. Phys. **B 163** (1980) 289.
31. G. Altarelli, G. Curci, G. Martinelli and S. Petrarca, Nucl. Phys. **B 187** (1981) 461.
32. A.J. Buras and P.H. Weisz, Nucl. Phys. **B 333** (1990) 66.
33. A.J. Buras, M. Jamin, M.E. Lautenbacher and P.H. Weisz, Nucl. Phys. **B 370** (1992) 69; **B 400** (1993) 37.
34. A.J. Buras, M. Jamin and M.E. Lautenbacher, Nucl. Phys. **B 400** (1993) 75.
35. A.J. Buras, M. Jamin and M.E. Lautenbacher, Nucl. Phys. **B 408** (1993) 209.
36. M. Ciuchini, E. Franco, G. Martinelli and L. Reina, Phys. Lett. **B 301** (1993) 263.
37. M. Ciuchini, E. Franco, G. Martinelli and L. Reina, Nucl. Phys. **B 415** (1994) 403.
38. K. Chetyrkin, M. Misiak and Münz, Nucl. Phys. **B520** (1998) 279.
39. M.Misiak and M. Münz, Phys. Lett. **B344** (1995) 308.
40. K.G. Chetyrkin, M. Misiak and M. Münz, Phys. Lett. **B400** (1997) 206; Erratum ibid. **B425** (1998) 414.
41. G. Buchalla, Nucl. Phys. **B 391** (1993) 501.
42. E. Bagan, P. Ball, V.M. Braun and P.Gosdzinsky, Nucl. Phys. **B 432** (1994) 3; E. Bagan et al., Phys. Lett. **B 342** (1995) 362; **B 351** (1995) 546.
43. A. Lenz, U. Nierste and G. Ostermaier, Phys. Rev. **D 56** (1997) 7228.
44. M. Jamin and A. Pich, Nucl. Phys. **B425** (1994) 15.
45. S. Herrlich and U. Nierste, Nucl. Phys. **B419** (1994) 292.
46. A.J. Buras, M. Jamin, and P.H. Weisz, Nucl. Phys. **B347** (1990) 491.
47. J. Urban, F. Krauss, U.Jentschura and G. Soff, Nucl. Phys. **B523** (1998) 40.
48. S. Herrlich and U. Nierste, Phys. Rev. **D52** (1995) 6505; Nucl. Phys. **B476** (1996) 27.
49. G. Buchalla and A.J. Buras, Nucl. Phys. **B 398** (1993) 285.
50. G. Buchalla and A.J. Buras, Nucl. Phys. **B 400** (1993) 225.
51. M. Misiak and J. Urban, hep-ph/9901278.
52. G. Buchalla and A.J. Buras, hep-ph/9901288.
53. G. Buchalla and A.J. Buras, Nucl. Phys. **B 412** (1994) 106.

54. G. Buchalla and A.J. Buras, Phys. Lett. **B 336** (1994) 263.
55. A. J. Buras, M. E. Lautenbacher, M. Misiak and M. Münz, Nucl. Phys. **B423** (1994) 349.
56. M. Misiak, Nucl. Phys. **B393** (1993) 23; Erratum, Nucl. Phys. **B439** (1995) 461.
57. A.J. Buras and M. Münz, Phys. Rev. **D 52** (1995) 186.
58. A. Ali, and C. Greub, Z.Phys. **C49** (1991) 431; Phys. Lett. **B259** (1991) 182; **B361** (1995) 146.
59. K. Adel and Y.P. Yao, Modern Physics Letters **A8** (1993) 1679; Phys. Rev. **D 49** (1994) 4945.
60. N. Pott, Phys. Rev. **D 54** (1996) 938.
61. C. Greub, T. Hurth and D. Wyler, Phys. Lett. **B380** (1996) 385; Phys. Rev. **D 54** (1996) 3350;
62. C. Greub and T. Hurth, Phys. Rev. **D 56** (1997) 2934; hep-ph/9809468.
63. A.J. Buras, A. Kwiatkowski and N. Pott, Phys. Lett. **B 414** (1997) 157; Nucl. Phys. **B 517** (1998) 353.
64. M. Ciuchini, G. Degrassi, P. Gambino and G.F. Giudice, Nucl. Phys. **B 527** (1998) 21.
65. F.M. Borzumati and Ch. Greub, Phys. Rev. **D 58** (1998) 07004; hep-ph/9809438.
66. P. Ciafaloni, A. Romanino, and A. Strumia, Nucl. Phys. **B 524** (1998) 361.
67. M. Beneke, G. Buchalla, C. Greub, A. Lenz and U. Nierste, hep-ph/9808385.
68. M. Beneke, F. Maltoni and I.Z. Rothstein, hep-ph/9808360.
69. M. Ciuchini, E. Franco, V. Lubicz, G. Martineli, I. Scimeni and L. Silvestrini, Nucl. Phys. **B 523** (1998) 501.
70. M. Ciuchini, et al. JHEP **9810** (1998) 008.
71. M. Ciuchini, G. Degrassi, P. Gambino and G.F. Giudice, Nucl. Phys. **B 534** (1998) 3.
72. G. Buchalla and A.J. Buras, Phys. Rev. **D57** (1998) 216.
73. A. Czarnecki and W.J. Marciano, Phys. Rev. Lett. **81** (1998) 277.
74. A. Strumia, Nucl. Phys. **B532** (1998) 28.
75. A.L. Kagan and M. Neubert, hep-ph/9805303.
76. P. Gambino, A. Kwiatkowski and N. Pott, hep-ph/9810400.
77. K. Chetyrkin, M. Misiak and Münz, Nucl. Phys. **B518** (1998) 473.
78. S. Herrlich and U. Nierste, Nucl. Phys. **B 445** (1995) 39.
79. J. Alexander; D. Karlen; P. Rosnet; talks given at the 29th Intl. Conf. on High Energy Physics (ICHEP 98), Vancouver, Canada, 23-29 July 1998, to appear in the proceedings.
80. Particle Data Group, Phys. Rev. **D 54** (1996) 1.
81. A.J. Buras, M.E. Lautenbacher and G. Ostermaier, Phys. Rev. **D 50** (1994) 3433.
82. L. Wolfenstein, Phys. Rev. Lett. **51** (1983) 1945.
83. L. Wolfenstein, Phys. Rev. Lett. **13** (1964) 562.
84. J. M. Flynn and L. Randall, Phys. Lett. **B224** (1989) 221; erratum Phys. Lett. **B235** (1990) 412.
85. G. Buchalla, A. J. Buras, and M. K. Harlander, Nucl. Phys. **B337** (1990) 313.
86. R. Gupta, hep-ph/9801412; R.S. Sharpe, hep-lat/9811006.
87. G. D. Barr et al., Phys. Lett. **B317** (1993) 233.
88. L. K. Gibbons et al., Phys. Rev. Lett. **70** (1993) 1203.
89. A. Ali, and C. Greub, Z.Phys. **C60** (1993) 433.
90. A. J. Buras, M. Misiak, M. Münz and S. Pokorski, Nucl. Phys. **B424** (1994) 374.
91. L. Littenberg, Phys. Rev. **D39** (1989) 3322.
92. G. Buchalla and A.J. Buras, Phys. Lett. **B333** (1994) 221; Phys. Rev. **D54** (1996) 6782.

93. W. Marciano and Z. Parsa, Phys. Rev. **D53**, R1 (1996).
94. D. Rein and L.M. Sehgal, Phys. Rev. **D39** (1989) 3325; J.S. Hagelin and L.S. Littenberg, Prog. Part. Nucl. Phys. **23** (1989) 1; M. Lu and M.B. Wise, Phys. Lett. **B324** (1994) 461; S. Fajfer, Nuovo Cim. **110A** (1997) 397; C.Q. Geng, I.J. Hsu and Y.C. Lin, Phys. Rev. **D54** (1996) 877.
95. G. Buchalla and G. Isidori, hep-ph/9806501.
96. S. Adler et al., Phys. Rev. Lett. **79**, (1997) 2204.
97. J. Adams et al., hep-ex/9806007.
98. Y. Grossman, Y. Nir and R. Rattazzi, [hep-ph/9701231] in [24]; Y. Grossman and Y. Nir, Phys. Lett. **B398** (1997) 163;
99. L. Littenberg and J. Sandweiss, eds., AGS2000, Experiments for the 21st Century, BNL 52512.
100. P. Cooper, M. Crisler, B. Tschirhart and J. Ritchie (CKM collaboration), EOI for measuring $Br(K^+ \to \pi^+ \nu\bar{\nu})$ at the Main Injector, Fermilab EOI 14, 1996.
101. K. Arisaka et al., KAMI conceptual design report, FNAL, June 1991.
102. T. Inagaki, T. Sato and T. Shinkawa, Experiment to search for the decay $K_L \to \pi^0 \nu\bar{\nu}$ at KEK 12 GeV proton synchrotron, 30 Nov. 1991.
103. R. Fleischer, Int. J. of Mod. Phys. **A12** (1997) 2459.
104. W. A. Bardeen, A. J. Buras and J.-M. Gérard, Phys. Lett. **B192** (1987) 138; A. Pich and E. de Rafael, Nucl. Phys. **B358** (1991) 311; M. Neubert and B. Stech, Phys. Rev. **D 44** (1991) 775; M. Jamin and A. Pich, Nucl. Phys. **B425** (1994) 15; J. Kambor, J. Missimer and D. Wyler, Nucl. Phys. **B346** (1990) 17; Phys. Lett. **B261** (1991) 496; V. Antonelli, S. Bertolini, M. Fabrichesi, and E.I. Lashin, Nucl. Phys. **B469** (1996) 181; J. Bijnens and J. Prades, hep-ph/9811472.
105. M. Neubert and B. Stech, [hep-ph/9705292], page 294 in [24]; B. Stech [hep-ph/9706384]; M.Neubert, Nucl. Phys. **B** (Proc. Suppl.) 64 (1998) 474, [hep-ph/9801269].
106. A.J. Buras and L. Silvestrini, hep-ph/9812392.
107. A. Ali, hep-ph/9812434.

The Quantum Noether Condition in Terms of Interacting Fields

Tobias Hurth[1] and Kostas Skenderis[3]

[1] Max-Planck-Institute for Physics, Werner-Heisenberg-Institute, Föhringer Ring 6, D-80805 Munich, Germany
[2] Spinoza Institute, University of Utrecht, Leuvenlaan 4, 3584 CE Utrecht, The Netherlands

Abstract. We review our recent work, hep-th/9803030, on the constraints imposed by global or local symmetries on perturbative quantum field theories. The analysis is performed in the Bogoliubov-Shirkov-Epstein-Glaser formulation of perturbative quantum field theory. In this formulation the S-matrix is constructed directly in the asymptotic Fock space with only input causality and Poincaré invariance. We reformulate the symmetry condition proposed in our earlier work in terms of interacting Noether currents.

1 Introduction

The relation between symmetries and quantum theory is an important and fundamental issue. For instance, symmetry relations among correlation functions (Ward identities) are often used in order to prove that a quantum field theory is unitary and renormalizable. Conversely, the violation of a classical symmetry at the quantum level (anomalies) often indicates that the theory is inconsistent. Furthermore, in recent years symmetries (such as supersymmetry) have been instrumental in uncovering non-perturbative aspects of quantum theories (see, for example, [1]). It is, thus, desirable to understand the interplay between symmetries and quantization in a manner which is free of the technicalities inherent in the conventional Lagrangian approach (regularization/renormalization) and in a way which is model independent as much as possible.

In a recent paper [2] we have presented a general method, the Quantum Noether Method, for constructing perturbative quantum field theories with global symmetries. Gauge theories are within this class of theories, the global symmetry being the BRST symmetry [3]. The method is established in the causal approach to quantum field theory introduced by Bogoliubov and Shirkov [4] and developed by Epstein and Glaser [5,6]. This explicit construction method rests directly on the axioms of relativistic quantum field theory. The infinities encountered in the conventional approach are avoided by a proper handling of the correlation functions as operator-valued distributions. In particular, the well-known problem of ultraviolet (UV) divergences is reduced to the mathematically well-defined problem of splitting an operator-valued distribution with causal support into a distribution with retarded and a distribution with advanced support or, alternatively [6,7], to the continuation of time-ordered products to coincident points. Implicitly, every consistent renormalization scheme solves this problem.

P. Breitenlohner and D. Maison (Eds.): Proceedings 1998, LNP 558, pp. 86–105, 2000.

Thus, the explicit Epstein-Glaser (EG) construction should not be regarded as a special renormalization scheme but as a general framework in which the conditions posed by the fundamental axioms of quantum field theory (QFT) on any renormalization scheme are built in by construction. In this sense our method is independent from the causal framework. Any renormalization scheme can be used to work out the consequences of the general symmetry conditions proposed in [2].

In the EG approach the S-matrix is directly constructed in the Fock space of free asymptotic fields in a form of formal power series. The coupling constant is replaced by a tempered test function $g(x)$ (i.e. a smooth function rapidly decreasing at infinity) which switches on the interaction. Instead of evaluating the S-matrix by first computing off-shell Greens functions by means of Feynman rules and then applying the LSZ formalism, the S-matrix is directly obtained by imposing causality and Poincaré invariance. The method can be regarded as an "inverse" of the cutting rules. One builds n-point functions out of m-point functions ($m < n$) by suitably "gluing" them together. The precise manner in which this is done is dictated by causality and Poincaré invariance (see appendix A for details). One shows, that this process uniquely fixes the S-matrix up to local terms (which we shall call "local normalization terms"). At tree level these local terms are nothing but the Lagrangian of the conventional approach [2].

The problem we set out to solve in [2] was to determine how to obtain a quantum theory which, on top of being causal and Poincaré invariant, is also invariant under a global symmetry. For linear symmetries such as global internal symmetries or discrete C, P, T symmetries the solution is well-known: one implements the symmetry in the asymptotic Fock space space by means of an (anti-) unitary transformation. The focus of our investigation in [2] was symmetries that are non-linear in the Lagrangian formulation. The prime examples are BRST symmetry and supersymmetry (in the absence of auxiliary fields). The main puzzle is how a theory formulated in terms of asymptotic fields only knows about the inherent non-linear structure.

The solution to the problem is rather natural. One imposes that the Noether current that generates the asymptotic symmetry is conserved at the quantum level, i.e. inside correlation functions. This condition, the Quantum Noether Condition (QNC), constrains the local normalization terms left unspecified by causality and Poincaré invariance. At tree-level one finds that the asymptotic Noether current renormalizes such that it generates the full non-linear transformation rules. At the quantum level the same condition yields the corresponding Ward identities. The way the methods works is analogous to the classical Noether method [8,9], hence its name. In addition, we have shown that the QNC is equivalent to the condition that the S-matrix is invariant under the symmetry under question (i.e. the S-matrix commutes with the generator of the asymptotic symmetry).

Quantum field theory, however, is usually formulated in terms of interacting fields. In the Lagrangian formulation, the symmetries of the theory are the symmetries of the action (or more generally of the field equations) that survive at the quantum level. These symmetries are generated by interacting Noether currents. It will, thus, be desirable to express the QNC in terms of the latter. As we shall see, this is indeed possible. The QNC in term of the interacting current is given in (22). If the symmetry is linear then the condition is that the interacting current is conserved (as expected). If the symmetry, however, is non-linear the interacting current is only conserved in the adiabatic limit ($g \to$ const.).

One important example is Yang-Mills theory. In this case, the corresponding Noether current is the BRST current. Because there are unphysical degrees of freedom present in gauge theories, one needs a subsidiary condition in order to project out the unphysical states. The subsidiary condition should remain invariant under time evolution. This means that it should be expressed in terms of a conserved charge. The appropriate charge for gauge theories is the BRST charge [10]. The subsidiary condition is that physical states should be annihilated by the BRST charge Q_{int} (and not be Q_{int}-exact).

The considerations in [10], however, (implicitly) assumed the naive adiabatic limit. For pure gauge theories this limit seem not to exist. Then from the Quantum Noether Condition (22) follows that the interacting BRST current is not conserved before the adiabatic limit. We stress, however, that the Quantum Noether Condition allows one to work out all consequences of non-linear symmetries for time-ordered operator products before the adiabatic limit is taken. As we shall see, one can even identify the non-linear transformation rules.

We organize this paper as follows: In the next section we shortly review the Quantum Noether Method. In section 3 we express the Quantum Noether Condition in terms of the interacting Noether current. Section 4 contains a discussion of future directions. In the appendix we present the main formulae of the causal framework and our conventions.

2 The Quantum Noether Method

In the EG approach one starts with a set of free fields in the asymptotic Fock space. These fields satisfy their (free) field equations and certain commutation relations. To define the theory one still needs to specify T_1, the first term in the S-matrix. (Actually, as we shall see, even T_1 is not free in our construction method but is also constrained by the Quantum Noether Condition). Given T_1 one can, in a well defined manner, construct iteratively the perturbative S-matrix. The requirements of causality and Poincaré invariance completely fix the S-matrix up to local terms. The additional requirement that the theory is invariant under a global and/or local symmetry imposes constraints on these local terms.

To construct a theory with global and/or local symmetry we introduce the coupling $g_\mu j_0^\mu$ in the theory, where j_0^μ is the Noether current that generates the asymptotic (linear) symmetry transformations, and we impose the condition that "the Noether current is conserved at the quantum level"

$$\partial_\mu \mathscr{T}_n^\mu(x_1, \cdots, x_n; \hbar) = 0, \tag{1}$$

where we introduce the notation (we use the abbreviation $\partial/\partial x_l^\mu = \partial_\mu^l$)

$$\partial_\mu \mathcal{T}_n^\mu(x_1, \cdots, x_n; \hbar) = \sum_{l=1}^{n} \partial_\mu^l \mathcal{T}_{n/l}^\mu, \tag{2}$$

and

$$\mathcal{T}_{n/l}^\mu = T[T_1(x_1) \cdots j_0^\mu(x_l) \cdots T_1(x_n)]. \tag{3}$$

(for $n = 1$, $\mathcal{T}_1^\mu(x_1) = j_0^\mu(x_1)$). In other words we consider an n-point function with one insertion of the current j_0^μ at the point x_l. Notice that since the left hand side of (1) is a formal Laurent series in $\hbar$, this condition is actually a set of conditions.

One may apply the inductive EG construction to work out the consequences of (1). This may be done by first working out $T[j_0 T_1 ... T_1]$ and then constructing (2). However, there is an alternative route [2]. One relaxes the field equations of the fields ϕ^A. Then the inductive hypothesis takes the form: for $m < n$,

$$\sum_{l=1}^{m} \partial_\mu^l \mathcal{T}_{m/l}^\mu = \sum_A R^{A;m}(\hbar) \mathcal{K}_{AB} \phi^B \delta(x_1, \ldots, x_m), \tag{4}$$

where

$$\mathcal{K}_{AB} \phi^B = \partial^\mu \frac{\partial \mathcal{L}_0}{\partial(\partial^\mu \phi^A)} - \frac{\partial \mathcal{L}_0}{\partial \phi^A} \tag{5}$$

are the free field equations ($\mathcal{L}_0$ is the free Lagrangian that yields (5); the present formulation assumes that such a Lagrangian exists). The coefficients $R^{A;m}(\hbar)$ are defined by (4) and are formal series in $\hbar$.

Clearly, if we impose the field equation we go back to (1). The converse is also true. Once one relaxes the field equations in the inductive step, (1) implies (4) as was shown in [2]. The advantage of the off-shell formulation is that it makes manifest the non-linear structure: the coefficients $R^{A;m}(\hbar)$ are just the order m part of the non-linear transformation rules. In addition, the calculation of local on-shell terms arising from tree-level graphs simplifies:

We now discuss the condition (1) at tree-level. For the analysis at loop level we refer to [2]. At tree-level we only need the $\hbar^0$ part of (4). Let us define

$$s_{(m-1)} \phi^A = \frac{1}{m!} R^{A;m}(\hbar^0). \tag{6}$$

Depending on the theory under consideration the quantities $R^{A;m}(\hbar^0)$ may be zero after some value of m. Without loss of generality we assume that they are zero for $m > k + 1$, for some integer k (which may be infinity; the same applies for k' below.). One shows that

$$s\phi^A = \sum_{m=0}^{k} g^m s_m \phi^A \tag{7}$$

are symmetry transformation rules that leave the Lagrangian,

$$\mathscr{L} = \sum_{m=0}^{k'} g^m \mathscr{L}_m, \tag{8}$$

invariant (up to total derivatives), where k' is also an integer (generically not equal to k). The Lagrangian $\mathscr{L}$ will be determined from the tree-level normalization conditions as follows,

$$\mathscr{L}_m = \frac{\hbar}{i} \frac{N_m}{m!}, \quad \text{for} \quad m > 1, \tag{9}$$

where N_m denotes the local normalization ambiguity of $T_m[T_1(x_1)...T_1(x_m)]$ in tree graphs defined with respect to the naturally split solution (i.e. the Feynman propagator is used in tree-graphs). For $m = 1$, $\mathscr{L}_1 = (\hbar/i)T_1$. The factor $m!$ reflects the fact that $T_m[...]$ appears in (A.55) with a combinatorial factors $m!$ while the factor $\hbar/i$ is there to cancel the overall factor $i/\hbar$ that multiplies the action in the tree-level S-matrix. Notice that we regard (9) as definition of $\mathscr{L}_m$. Let us further define j_n^μ as the local normalization ambiguity of $T_n[j_0 T_1 ... T_1]$,[1]

$$T_n[j_0^\mu(x_1)T_1(x_2)\cdots T_1(x_n)] = T_{c,n}[j_0^\mu(x_1)T_1(x_2)\cdots T_1(x_n)] + j_{n-1}^\mu \delta^{(n)} \tag{10}$$

where $T_{n,c}$ denotes the naturally splitted solution. We shall see that the normalization terms j_n complete the asymptotic current j_0 to the Noether current that generates the non-linear symmetry transformations (7).

We wish to calculate the tree-level terms at nth order. The causal distribution $\sum_{l=1}^n \partial_\mu^l \mathscr{D}_{n/l}^\mu$ at the nth order consists of a sum of terms each of these being a tensor product of $T_m[T_1...T_1\partial{\cdot}j_0 T_1...T_1)$ $(m < n)$ with T-products that involve only T_1 vertices according to the general formulae (A.61,A.62,A.68). By the off-shell induction hypothesis, we have for all $m < n$

$$\sum_{l=1}^{m} \partial_\mu^l \mathscr{T}_{m/l}^\mu = \sum_A (m! s_{m-1}\phi^A)\mathscr{K}_{AB}\phi^B \delta^{(m)}. \tag{11}$$

As explained in detail in [2], at order n one obtains all local on-shell terms by performing the so-called "relevant contractions", namely the contractions between the ϕ^B in the right hand side of (11) and ϕ in local terms. In this manner we get the following general formula for the local term $A_{c,n}$ arising through tree-level contractions at level n,

$$A_{c,n}(tree) = \sum_{\pi \in \Pi^n} \sum_{m=1}^{n-1} \partial_\mu \mathscr{T}_m^\mu(x_{\pi(1)}, \ldots, x_{\pi(m)}) N_{n-m} \delta(x_{\pi(k+1)}, \ldots, x_{\pi(n)}) \tag{12}$$

[1] We use the following abbreviations for the delta function distributions: $\delta^{(n)} = \delta(x_1, \ldots, x_n) = \delta(x_1 - x_2) \cdots \delta(x_{n-1} - x_n)$.

where it is understood that in the right hand side only "relevant contractions" are made. The factors N_{n-m} are tree-level normalization terms of the T-products that contain $n-m$ T_1 vertices.

In [2] we have provided a detailed analysis of (12) for any n (under the assumption that the Quantum Noether Method is not obstructed). In the next section, we will need these results in order to show that condition (22) is equivalent to condition (1). We therefore list them here without proofs.

The $n=1$ case is trivial. One just gets that $R^{A;1}(\hbar^0) = s_0\phi^A$. For $2 \leq n \leq k+1$, the condition (4) at tree-level yields the following constraint on the local normalization terms of the T_m, $m<n$,

$$s_0\mathscr{L}_{n-1} + s_1\mathscr{L}_{n-2} + \cdots + s_{n-2}\mathscr{L}_1 = \partial_\mu \mathscr{L}^\mu_{n-1} + s_{n-1}\phi^A \mathscr{K}_{AB}\phi^B \tag{13}$$

and, furthermore, determines j^μ_{n-1},

$$j^\mu_{n-1} = -n!\mathscr{L}^\mu_{n-1} + (n-1)!\sum_{l=0}^{n-2}(l+1)\frac{\partial \mathscr{L}_{n-1-l}}{\partial(\partial_\mu \phi^A)} s_l \phi^A. \tag{14}$$

For $n > k+1$ we obtain,

$$s_0\mathscr{L}_{n-1} + s_1\mathscr{L}_{n-2} + \cdots + s_k\mathscr{L}_{n-1-k} = \partial_\mu \mathscr{L}^\mu_{n-1}, \tag{15}$$

and

$$j^\mu_{n-1} = -n!\mathscr{L}^\mu_{n-1} + (n-1)!\sum_{l=1}^{k} l \frac{\partial \mathscr{L}_{n-l}}{\partial(\partial_\mu \phi^A)} s_{l-1}\phi^A. \tag{16}$$

Depending on the theory under consideration the $\mathscr{L}_n$'s will be zero for $n > k'$, for some integer k'. Given the integers k and k', there is also an integer k'' (determined from the other two) such that $\mathscr{L}^\mu_n = 0$, for $n > k''$.

Summing up the necessary and sufficient conditions (13), (15) for the Quantum Noether method to hold at tree level we obtain,

$$s\sum_{l=1}^{k'} g^l \mathscr{L}_l = \sum_{l=1}^{k''} \partial_\mu \mathscr{L}^\mu_l + (\sum_{l=1}^{k} g^l s_l \phi^A)\mathscr{K}_{AB}\phi^B \tag{17}$$

Using $s_0\mathscr{L}_0 = \partial_\mu k^\mu_0$ and for $l \leq k$

$$s_l\phi^A \mathscr{K}_{AB}\phi^B = \partial_\mu\left(\frac{\partial \mathscr{L}_0}{\partial(\partial_\mu \phi^A)} s_l \phi^A\right) - s_l\mathscr{L}_0 \tag{18}$$

we obtain,

$$s\mathscr{L} = \partial_\mu\left(\sum_{l=0}^{k''} g^l k^\mu_l\right) \tag{19}$$

where, for $1 < l \leq k$,

$$k^\mu_l = \mathscr{L}^\mu_l + \frac{\partial \mathscr{L}_0}{\partial(\partial_\mu \phi^A)} s_l \phi^A \tag{20}$$

and for $l > k$, $k_l^\mu = \mathscr{L}_l^\mu$. We therefore find that $\mathscr{L}$ is invariant under the symmetry transformation,

$$s\phi^A = \sum_{l=0}^{k} g^l s_l \phi^A. \tag{21}$$

According to Noether's theorem there is an associated Noether current. One may check that the current normalization terms j_m^μ (14), (16) are in one-to-one correspondence with the terms in the Noether current. Therefore the current j_0 indeed renormalizes to the full non-linear current.

3 Conservation of the Interacting Noether Current

The Quantum Noether Condition (1) can be reformulated in terms of interacting fields. Let $j_{0,\text{int}}^\mu$ and $\tilde{j}_{1,\text{int}}^\mu$ be the interacting currents corresponding to free field operators j_0^μ and $\tilde{j}_1^\mu$, respectively, perturbatively constructed according to (A.82). $\tilde{j}_1^\mu$ is equal to $-\mathscr{L}_1^\mu$ (defined in (13)) as will see below. Then the general Ward identity

$$\partial_\mu j_{0,\text{int}}^\mu = \partial_\mu g \tilde{j}_{1,\text{int}}^\mu \tag{22}$$

is equivalent to condition (1). According to condition (22) the interacting Noether current $j_{0,\text{int}}^\mu$ is conserved only if it generates a linear symmetry, i.e. $\tilde{j}_1^\mu$ vanishes, or otherwise in the adiabatic limit $g(x) \to 1$, provided this limit exists. In the following we shall show that the condition (22) yields the same conditions on the the time-ordered products $T_n[T_1...T_1]$ as the Quantum Noether condition (1). In this sense the two general symmetry conditions are considered equivalent.

Because Poincaré invariance and causality already fix the time-ordered products $T_n[T_1...T_1]$ up to the local normalization ambiguity N_n, we only have to show that these local normalization terms N_n are constrained in the same way by both conditions, (22) and (1).

First, we translate the condition (22) to a condition on time-ordered products using the formulae given in the appendix:

The perturbation series for the interacting field operator j_{int}^μ of a free field operator j^μ is given by the advanced distributions of the corresponding expansion of the S-matrix (see (A.82)):

$$j_{\text{int}}^\mu(g,x) = j^\mu(x) + \sum_{n=1}^{\infty} \frac{1}{n!} \int d^4x_1 \dots d^4x_n$$
$$Ad_{n+1}\left[T_1(x_1)\dots T_1(x_n); j^\mu(x)\right] g(x_1)\dots g(x_n), \tag{23}$$

where Ad_{n+1} denotes the advanced operator-valued distribution with n vertices T_1 and one vertex $j^\mu(x)$ at the $(n+1)$th position. This distribution is only symmetric in the first n variables $x_1, \dots, x_n$. The support properties are defined with respect to the unsymmetrized variable x.

With the help of (23), we rewrite the left hand side of equation (22)

$$\partial_\mu^x j^\mu_{0,\mathrm{int}}(x) = \partial_\mu^x j_0^\mu(x) + \sum_{n=1}^{\infty} \frac{1}{n!} \int d^4x_1 \dots d^4x_n$$
$$\partial_\mu^x Ad_{n+1}\left[T_1(x_1)\dots T_1(x_n); j_0^\mu(x)\right] g(x_1)\dots g(x_n) \qquad (24)$$

and the right hand side of (22)

$$\tilde{j}^\mu_{1,\mathrm{int}}(x)\partial_\mu g(x) = \sum_{n=0}^{\infty} \frac{1}{n!} \int d^4x_1 \dots d^4x_n d^4x_{n+1}$$
$$Ad_{n+1}\left[T_1(x_1)\dots T_1(x_n); \tilde{j}_1^\mu(x)\right] \delta(x - x_{n+1})$$
$$g(x_1)\dots g(x_n)\partial_\mu^{x_{n+1}} g(x_{n+1}) \qquad (25)$$

After partial integration, symmetrization of the integrand in the variables (x_1, $\dots, x_{n+1}$) and shifting the summation index, the right hand side of (22) can be further rewritten as

$$\tilde{j}^\mu_{1,\mathrm{int}}(x)\partial_\mu g(x) = -\sum_{n=1}^{\infty} \frac{1}{n!} \int d^4x_1 \dots d^4x_n$$
$$\sum_{j=1}^{n} \left\{ Ad_n \left[T_1(x_1)\dots \widehat{T(x_j)} \dots T(x_n); \tilde{j}_1^\mu(x)\right] \partial_\mu^{x_j} \delta(x_j - x) \right\}$$
$$g(x_1)\dots g(x_n) \qquad (26)$$

where the hat indicates that this coupling has to be omitted. Equation (22) reads then

$$\partial_\mu j_0^\mu = 0, \qquad (n = 0)$$
$$\partial_\mu^x Ad_{n+1}\left[T_1(x_1)\dots T(x_n); j_0^\mu(x)\right]$$
$$+ \sum_{j=1}^{n} Ad_n \left[T_1(x_1)\dots \widehat{T(x_j)} \dots T(x_n); \tilde{j}_1^\mu(x)\right] \partial_\mu^{x_j} \delta(x_j - x) = 0, \; (n > 0) \qquad (27)$$

where the local normalization terms of the *Ad*-distributions with respect to a specified splitting solution will be given below.

In the following we discuss the equivalent condition of the time-ordered distributions instead of the advanced ones in order to compare the unsymmetrized condition (22) with the symmetrized Quantum Noether Condition (1). We get instead of (27)

$$\partial_\mu^x T_{n+1}\left[T_1(x_1)\dots T_1(x_n); j_0^\mu(x)\right]$$
$$= -\sum_{j=1}^{n} T_n \left[T_1(x_1)\dots \widehat{T_1(x_j)} \dots T_1(x_n); \tilde{j}_1^\mu(x)\right] \partial_\mu^{x_j} \delta(x_j - x) \qquad (28)$$

These distributions get smeared out by $g(x_1)\dots g(x_n)\tilde{g}(x)$, where the test function $\tilde{g}$ differs from g. One easily verifies the left hand side of (28) is just the Quantum Noether Condition (1) but without the symmetrization; the missing symmetrization produces the extra terms on the right hand side of (28) as we shall see.

We shall use the same off-shell procedure in order to fix the local on-shell obstruction terms (which is explained in detail in [2], section 4.2). The starting point ($n = 0$) of both conditions is the same

$$\partial_\mu j_0^\mu(x) = s_0\phi^A \mathcal{K}_{AB}\phi^B \tag{29}$$

We have now for $n = 1$,

$$\partial_\mu^x \left(T_{2,c}[T_1(x_1)j_0^\mu(x)] + j_1^\mu \delta(x_1 - x)\right) = -\tilde{j}_1^\mu(x)\partial_\mu^{x_1}\delta(x_1 - x) \tag{30}$$

Working out the left hand side (and using $T_1 = \frac{i}{\hbar}\mathcal{L}_1$) we obtain,

$$\begin{aligned}\partial_\mu^x \left(j_1^\mu \delta(x_1 - x)\right) + s_0\mathcal{L}_1\delta(x_1 - x) - \partial_\mu^x \left(\frac{\partial \mathcal{L}_1}{\partial(\partial_\mu \phi^A)} s_0\phi^A \delta(x_1 - x)\right)\\ = \tilde{j}_1^\mu(x)\partial_\mu^x\delta(x_1 - x)\end{aligned} \tag{31}$$

This condition fixes the local renormalization of j_0^μ at order g, denoted by j_1^μ (defined with respect to the natural splitting solution $T_{2,c}$) and also $\tilde{j}_1^\mu$ in condition (22). The latter term, proportional to the derivative of the δ-distribution, is left over in our new unsymmetrized condition. Note that in the symmetrized case, we reduced these kind of terms to ones proportional to the δ-distribution with the help of distributional identities.

The condition (31) can be fulfilled for some local operators j_1^μ and $\tilde{j}_1^\mu$ if and only if $s_0\mathcal{L}_1$ is a divergence up to field equation terms,

$$s_0\mathcal{L}_1 = \partial_\mu \mathcal{L}_1^\mu + s_1\phi^A \mathcal{K}_{AB}\phi^B. \tag{32}$$

In the absence of real obstructions this equation has solutions and we get

$$j_1^\mu = -\mathcal{L}_1^\mu + \frac{\partial \mathcal{L}_1}{\partial(\partial_\mu \phi^A)} s_0\phi^A \tag{33}$$

as local renormalization of $j_{0,\mathrm{int}}^\mu$ at order g^1 and

$$\tilde{j}_1^\mu = -\mathcal{L}_1^\mu. \tag{34}$$

Equation (33) should be compared with the analogous formulae (14) for $n = 2$[2]. We finally have

$$\partial_\mu^x T_2\left[T_1(x_1)j_0^\mu(x)\right] + \tilde{j}_1^\mu(x)\partial_\mu^{x_1}\delta(x_1 - x) = s_1\phi^A \mathcal{K}_{AB}\phi^B\delta(x_1 - x). \tag{35}$$

[2] Notice that n in the present section should be compared with $n + 1$ in section 2.

The off-shell term on the right hand side of (35) is responsible for local obstruction terms at the next order, $n = 2$. We get (taking special care of derivative terms and advantage of our off-shell procedure):

$$\begin{aligned}\partial^x_\mu T_{3,c}\left[T_1(x_1)T_1(x_2)j^\mu_0(x)\right] + \left(T_{2,c}\left[T_1(x_1)\tilde{j}^\mu_1(x)\right]\partial^{x_2}_\mu\delta(x_2-x) + [x_1 \leftrightarrow x_2]\right)\\ = \frac{\hbar}{i}\Bigg[2s_1T_1\delta^{(3)} - \left(2\partial^x_\mu + \partial^{x_1}_\mu + \partial^{x_2}_\mu\right)\left(\frac{\partial T_1}{\partial(\partial_\mu\phi^A)}s_1\phi^A\delta^{(3)}\right)\\ +s_0N_2\delta^{(3)} - \partial^x_\mu\left(\frac{\partial N_2}{\partial(\partial_\mu\phi^A)}s_0\phi^A\delta^{(3)}\right)\Bigg] \qquad (36)\end{aligned}$$

where N_2 denotes the tree-normalization term of $T_2[T_1T_1]$ which is uniquely defined with respect to the natural splitting solution $T_{2,c}[T_1T_1]$. Now we include also the normalization ambiguity of the other distributions involved:

$$\begin{aligned}T_3\left[T_1(x_1)T_1(x_2)j^\mu_0(x)\right] &= T_{3,c}\left[T_1(x_1)T_1(x_2)j^\mu_0(x)\right] + j^\mu_2(x)\delta(x_1,x_2,x) \qquad (37)\\ T_2\left[T_1(x_i)\tilde{j}^\mu_1(x)\right] &= T_{2,c}\left[T_1(x_i)\tilde{j}^\mu_1(x)\right] + \tilde{j}^\mu_2\delta(x_i - x)\end{aligned}$$

According to (13) the Quantum Noether Condition (1) at order $n = 3$ is fulfilled if and only if

$$s_1\mathscr{L}_1 + s_0\mathscr{L}_2 = \partial_\mu\mathscr{L}^\mu_2 + s_2\phi^A\mathscr{K}_{AB}\phi^B \qquad (38)$$

where the definition $\mathscr{L}_n = (\hbar/i)(N_n/n!)$ is used. Now the same is true for condition (36). Only if (38) holds one can absorb the local terms on the right hand side of (36) in the normalization terms $j^\mu_2(x)$ and $\tilde{j}^\mu_2(x)$ given in (37). The reasoning is again slightly different from the one in the symmetrized case. The distributions are only symmetric in the variables x_i, but x is a distinguished variable. This means that the two local operator-valued distributions [3]

$$\hat{A}_0\delta(x_1,x_2,x); \quad \sum_{i=1}^{2}\partial_{x_i}\left(\hat{A}_1\delta(x_1,x_2,x)\right), \qquad (39)$$

where $\hat{A}_0(x)$ and $\hat{A}_1(x)$ are local operators, are independent (on the test functions $\tilde{g}(x_1,x_2,x) := g(x_1)g(x_2)\tilde{g}(x)$ with $g \neq \tilde{g}$)[4].

So if and only if (38) is true the condition (22) can be fulfilled at order $n = 2$ and the local normalization terms of the interacting currents, $j^\mu_{0,\text{int}}$ and $\tilde{j}^\mu_{1,\text{int}}$, get fixed to

$$\begin{aligned}j^\mu_2 &= 2!\left(-\mathscr{L}^\mu_2 + \frac{\partial\mathscr{L}_2}{\partial(\partial_\mu\phi^A)}s_0\phi^A + \frac{\partial\mathscr{L}_1}{\partial(\partial_\mu\phi^A)}s_1\phi^A\right)\\ \tilde{j}^\mu_2 &= -2!\mathscr{L}^\mu_2 + \frac{\partial\mathscr{L}_1}{\partial(\partial_\mu\phi^A)}s_1\phi^A \qquad (40)\end{aligned}$$

[3] One could also choose as a basis $\hat{A}'_0\delta(x_1,x_2,x); \partial^x\left(\hat{A}'_1\delta(x_1,x_2,x)\right)$.

[4] In the symmetrized case, where one smears out with totally symmetric test functions $g(x_1,x_2,x_3) := g(x_1)g(x_2)g(x_3)$, one has $\sum_{i=1}^2\partial_{x_i}\left(\hat{A}_1\delta(x_1,x_2,x)\right) = (2/3)\partial\hat{A}_1\delta(x_1,x_2,x)$.

Note the different symmetry factors in j_2^μ compared with the symmetrized case (14). With these normalizations we get

$$\partial_\mu^x T_3\left[T_1(x_1)T_1(x_2)j_0^\mu(x)\right] + \left(T_2\left[T_1(x_1)\tilde{j}_1(x)\right]\partial_\mu^{x_2}\delta(x_2-x) + [x_1 \leftrightarrow x_1]\right) \quad (41)$$
$$= 2! s_2 \phi^A \mathscr{K}_{AB}\phi^B \delta(x_1, x_2, x)$$

This corresponds to (22) at order $n = 2$:

$$\partial_\mu^x j_{0,\mathrm{int}}^\mu(x)\Big|_{g^2} = \tilde{j}_{1,\mathrm{int}}^\mu(x)\Big|_{g^1}\partial_\mu g(x) + 2! s_2 \phi^A \mathscr{K}_{AB}\phi^B(x). \quad (42)$$

From these first two steps of the inductive construction, one already realizes that in general the additional terms proportional to $\partial_\mu g$ in (22) correspond to terms proportional to $\partial_\mu \delta^n$ which are now independent. In the former condition (1) we got rid of these terms by symmetrization and moding out the general formula $\sum_{l=1}^n \partial^l \delta^n = 0$. This formula is a direct consequence of translation invariance. Regardless this slight technical difference both conditions, (1) and (22), pose the same consistency conditions on the physical normalization ambiguity.

For $0 < n \leq k$, (where k is the minimal integer such that $\forall m > k, s_m = 0$), condition (28) yields

$$\partial_\mu^x(j_n \delta^{(n+1)}) + n!\left(\sum_{l=0}^{n-1} s_l \mathscr{L}_{n-l}\right)\delta^{(n+1)} -$$
$$-\sum_{l=0}^{n-1}\left(n!\,\partial_\mu^x + l\,(n-1)!\sum_{i=1}^{n}(\partial_\mu^{x_i})\right)\left(\frac{\partial \mathscr{L}_{n-l}}{\partial(\partial_\mu \phi^A)} s_l \phi^A \delta^{(n+1)}\right)$$
$$= \tilde{j}_n^\mu(x)\partial_\mu^x \delta^{(n+1)} \quad (43)$$

where j_n^μ and $\tilde{j}_n^\mu$ are defined by analogous to (37) formulae. The sufficient and necessary condition for this equation to have a solution is

$$s_0 \mathscr{L}_n + \cdots + s_{n-1}\mathscr{L}_1 = \partial_\mu \mathscr{L}_n^\mu + s_n \phi^A \mathscr{K}_{AB}\phi^B. \quad (44)$$

This agrees with (13) (we remind the reader that n in present section corresponds to $n+1$ in section 2). Then the current normalization terms are given by

$$j_n^\mu = n!\left(-\mathscr{L}_n^\mu + \sum_{l=0}^{n-1}\frac{\partial \mathscr{L}_{n-l}}{\partial(\partial_\mu \phi^A)} s_l \phi^A\right) \quad (45)$$

$$\tilde{j}_n^\mu = -n!\mathscr{L}_n^\mu + (n-1)!\sum_{l=0}^{n-1} l\,\frac{\partial \mathscr{L}_{n-l}}{\partial(\partial_\mu \phi^A)} s_l \phi^A \quad (46)$$

and we have

$$\partial_\mu^x j_{0,\mathrm{int}}^\mu(x)\Big|_{g^n} = \tilde{j}_{1,\mathrm{int}}^\mu\Big|_{g^{n-1}}\partial_\mu g(x) + n! s_n \phi^A \mathscr{K}_{AB}\phi^B(x) \quad (47)$$

For $n > k$, equation (28) yields

$$\partial_\mu^x (j_n \delta^{(n+1)}) + n! \left(\sum_{l=0}^{k} s_l \mathscr{L}_{n-l} \right) \delta^{(n+1)} -$$
$$- \sum_{l=0}^{k} \left(n! \partial_\mu^x - l\,(n-1)! \sum_{i=1}^{n} (\partial_\mu^{x_i}) \right) \left(\frac{\partial \mathscr{L}_{n-l}}{\partial(\partial_\mu \phi^A)} s_l \phi^A \delta^{(n+1)} \right)$$
$$= \tilde{j}_n^\mu(x) \partial_\mu^x \delta^{(n+1)} \tag{48}$$

This equation now implies

$$s_0 \mathscr{L}_n + \cdots + s_k \mathscr{L}_{n-k} = \partial_\mu \mathscr{L}_n^\mu. \tag{49}$$

We further obtain for the current normalization terms,

$$j_n^\mu = n! \left(-\mathscr{L}_n^\mu + \sum_{l=0}^{k} \frac{\partial \mathscr{L}_{n-l}}{\partial(\partial_\mu \phi^A)} s_l \phi^A \right) \tag{50}$$

$$\tilde{j}_n^\mu = -n! \mathscr{L}_n^\mu + (n-1)! \sum_{l=0}^{k} l \frac{\partial \mathscr{L}_{n-l}}{\partial(\partial_\mu \phi^A)} s_l \phi^A \tag{51}$$

Therefore,

$$\partial_\mu^x j_{0,\text{int}}^\mu(x) \Big|_{g^n} = \tilde{j}_{1,\text{int}}^\mu \Big|_{g^{n-1}} \partial_\mu g(x) \tag{52}$$

without using the free field equations.

In exactly the same way as in section 2, we deduce that the sum of all tree-level local normalization terms consitute a Lagrangian which is invariant (up to a total derivative) under the symmetry transformation $s\phi^A = \sum s_i \phi^A$. Inserting now the local normalization terms (45) and (50) into (23) we obtain,

$$j_{0,\text{int}}^\mu = \frac{\partial \mathscr{L}}{\partial(\partial_\mu \phi^A)} s\phi^A - k^\mu \tag{53}$$

where we have used the definitions (8), (7), and (20). The combinatorial factor $n!$ in (45) and (50) exactly cancels the same factor in (23). We, therefore, see that the interacting free current exactly becomes the full non-linear current.

We have, thus, found that going from condition (1) to condition (22) just corresponds to a different technical treatment of the $\partial_\mu \delta^{(n)}$ terms which has no influence on the fact that both conditions pose the same conditions on the normalization ambiguity of the physical T_n distributions, namely the consistency conditions of the classical Noether method. Our analysis of the condition (1) at the loop level is also independent of this slight technical rearrangement of the derivative terms. Thus, the issue of stability can be analyzed in exactly the same way as before (see section 4.3 of [2]). One shows (under the assumption that the Wess-Zumino consistency condition has only trivial solutions) that condition

(22) at loop level also implies that the normalization ambiguity at the loop level, $N_n(\hbar)$, is constrained in the same way as the tree-level normalizations, $N_n(\hbar^0)$. Once the stability has been established the equivalence of (1) and (22) at loop level follows.

Summing up, we have shown that conditions (1)-(22) yield all consequences of non-linear symmetries for time-ordered products before the adiabatic limit. So at that level currents seem to be sufficient. As mentioned in the introduction, however, if one wants to identify the physical Hilbert space, one may need to use the Noether charge $Q_{\text{int}} = \int d^3x j^0_{\text{int}}(x)$. As our Quantum Noether Condition (22) shows, only in the adiabatic limit (provided the latter exists) the interacting Noether current is conserved. Moreover, there is an additional technical obstacle. In the construction of the BRST charge a volume divergence occurs. In [11] a resolution was proposed for the case of QED. It was also described there how the analysis of Kugo-Ojima may hold locally. One may expect more technical problems in the construction of the BRST charge in the case non-abelian gauge theories where the free non-interacting Noether current includes two quantum fields. However, at least for the implementation of the symmetry transformations in correlation functions, such an explicit construction of the BRST charge is not necessary, as we have shown. Symmetries are implemented with the help of Noether currents only.

4 Discussion

We have presented a general method for constructing perturbative quantum field theories with global and/or local symmetries. The analysis was performed in the Bogoliubov-Shirkov-Epstein-Glaser approach. In this framework the perturbative S-matrix is directly constructed in the asymptotic Fock space with only input causality and Poincaré invariance. The construction directly yields a finite perturbative expansion without the need of intermediate regularization. The invariance of the theory under a given symmetry is imposed by requiring that the asymptotic Noether current is conserved at the quantum level.

The novel feature of the present discussion with respect to the usual approach is that our results are manifestly scheme independent. In addition, in the conventional approach one implicitly assumes the naive adiabatic limit. Our construction is done before the adiabatic limit is taken. The difference between the two approaches is mostly seen when the symmetry condition is expressed in terms of the interacting Noether current. If the interacting current generates non-linear symmetries, it is not conserved before the adiabatic limit is taken. An important example is pure gauge theory. In this case, the global symmetry is BRST symmetry. The interacting BRST current is not conserved before the adiabatic limit. Nevertheless, one may still construct correlation functions that satisfy the expected Ward identities.

In the present contribution and in [2] we analyzed the symmetry conditions assuming that there are no true tree-level or loop-level obstructions. The algebra of the symmetry transformation imposes integrability conditions on the possible form of these obstructions [12]. Therefore, to analyze the question of anomalies in the present context one would have to understand how to implement the algebra of symmetry transformations in this framework. This is expected to be encoded in multi-current correlation functions. We will report on this issue in a future publication [13].

The Quantum Noether Condition (1) or (22) leads to specific constraints (equations (13), (15)) that the local normalization terms should satisfy. We have seen that these conditions are equivalent to the condition that one has an invariant action. So, one may infer the most general solution of equations (13), (15) from the most general solution of the problem of finding an action invariant under certain symmetry transformation rules.

For the particular case of gauge theories the global symmetry used in the construction is BRST symmetry. In EG one always works with a gauged fixed theory since one needs to have propagators for all fields. Therefore, the symmetry transformation rules are the gauged fixed ones. Physics, however, should not depend on the particular gauge fixing chosen. The precise connection between the results of the gauge invariant cohomology (which may be derived with the help of the antifield formalism [15,16]) and the present gauged-fixed formulation will be presented elsewhere [14].

The symmetry condition we proposed involves the (Lorentz invariant) condition of conservation of the Noether current. There are cases, however, where one has a charge that generates the symmetry but not a Noether current (for this to happen the theory should not possess a Lagrangian). A more fundamental formulation that will also cover these cases may be to demand that the charge that generates the symmetry is conserved at the quantum level (i.e. inside correlation functions). A precise formulation of this condition may require a Hamiltonian reformulation of the EG approach. Such a reformulation may be interesting on its own right.

Acknowledgements

We thank Klaus Fredenhagen and Raymond Stora for discussions. KS is supported by the Netherlands Organization for Scientific Research (NWO).

5 Appendix

In this appendix we give the basic conventions and formulae of the causal framework, in particular the definition of the interacting field. A self-contained introduction to the EG construction may be found in section 3 of [2]. For further technical details we refer the reader to the literature [5,6,17,18].

We describe the construction for the case of a massive scalar field. The very starting point is the Fock space $\mathcal{F}$ of the massive scalar field (based on a representation space H^m_s of the Poincaré group) with the defining equations

$$(\Box + m^2)\varphi = 0 \quad \text{(a)}, \quad [\varphi(x), \varphi(y)] = i\hbar D_m(x-y) \quad \text{(b)}, \tag{A.54}$$

where $D_m(x-y) = \frac{-i}{(2\pi)^3}\int dk^4 \delta(k^2-m^2)\mathrm{sgn}\,(k^0)\exp(-ikx)$ is the Pauli-Jordan distribution. In contrast to the Lagrangian approach, the S-matrix is directly constructed in this Fock space in the form of a formal power series

$$S(g) = 1 + \sum_{n=1}^{\infty} \frac{1}{n!} \int dx_1^4 \cdots dx_n^4 \quad T_n(x_1, \cdots, x_n; \hbar) \quad g(x_1) \cdots g(x_n). \qquad (A.55)$$

The coupling constant g is replaced by a tempered test function $g(x) \in \mathscr{S}$ (i.e. a smooth function rapidly decreasing at infinity) which switches on the interaction.

The central objects are the n-point operator-valued distributions $T_n \in \mathscr{S}'$, where $\mathscr{S}'$ denotes the space of functionals on $\mathscr{S}$. They should be viewed as mathematically well-defined (renormalized) time-ordered products,

$$T_n(x_1, \cdots, x_n; \hbar) = T\left[T_1(x_1) \cdots T_1(x_n)\right], \qquad (A.56)$$

of a given specific coupling, say $T_1 = \frac{i}{\hbar} : \Phi^4 :$ (c), which is the third defining equation in order to specify the theory in this formalism. Notice that the expansion in (A.55) is *not* a loop expansion. Each T_n in (A.55) can receive tree-graph and loop-contributions. One can distinguish the various contributions from the power of $\hbar$ that multiplies them.

Epstein and Glaser present an explicit inductive construction of the most general perturbation series in the sense of (A.55) which is compatible with the fundamental axioms of relativistic quantum field theory, causality and Poincaré invariance.

The main guiding principle is the property of causal factorization which can be stated as follows:

• Let g_1 and g_2 be two tempered test functions. Then causal factorization means that

$$S(g_1 + g_2) = S(g_2)S(g_1) \quad \text{if} \quad \mathrm{supp} g_1 \precsim \mathrm{supp} g_2 \qquad (A.57)$$

the latter notion means that the support of g_1 and the support of g_2, two closed subsets of $\mathbf{R}^4$, can be separated by a space like surface.

It is well-known that the heuristic solution for (A.57), namely

$$T_n(x_1, \ldots, x_n; \hbar) \qquad (A.58)$$
$$= \sum_{\pi} T_1(x_{\pi(1)}) \ldots T_1(x_{\pi(n)}) \Theta(x^0_{\pi(1)} - x^0_{\pi(2)}) \ldots \Theta(x^0_{\pi(n-1)} - x^0_{\pi(n)}),$$

is, in general, affected by ultra-violet divergences (π runs over all permutations of $1, \ldots, n$). The reason for this is that the product of the discontinuous Θ-step function with Wick monomials like T_1 which are operator-valued distributions is ill-defined. One can handle this problem by using the usual regularization and renormalization procedures and finally end up with the renormalized time-ordered products of the couplings T_1.

Epstein and Glaser suggest another path which leads directly to well-defined T-products without any intermediate modification of the theory using the fundamental property of causality (A.57) as a guide. They translate the condition

(A.57) into an induction hypothesis, $H_m, m < n$, for the T_m-distribution which reads

$$H_m : \begin{cases} T_m(X \cup Y) = T_{m_1}(X)\, T_{m-m_1}(Y) \\ \qquad \text{if} \quad X \succeq Y, \quad X, Y \neq \emptyset, \quad 0 < m_1 < m \\ [T_{m_1}(X), T_{m_2}(Y)] = 0 \\ \qquad \text{if} \quad X \sim Y \ (\Leftrightarrow X \succeq Y \wedge X \preceq Y), \quad \forall m_1, m_2 \leq m \end{cases} \tag{A.59}$$

Here we use the short-hand notation $T_m(x_1, \ldots, x_m; \hbar) = T(X);\ | X |= m$. Besides other properties they also include the Wick formula for the T_m distributions into the induction hypothesis. This is most easily done by including the so-called Wick submonomials of the specific coupling $T_1 = (i/\hbar) : \Phi^4 :$ as additional couplings in the construction $T_1^j := (i/\hbar)(4!/(4-j)!) : \Phi^{4-j} :, 0 < j < 4$. Then the Wick formula for the T_n products can be written as

$$\begin{aligned} &T_m[T_1^{j_1}(x_1) \cdots T_1^{j_m}(x_m)] \\ &\quad = \sum_{s_1, \ldots, s_m} \langle 0 \mid T[T_1^{j_1+s_1}(x_1) \cdots T_1^{j_m+s_m}(x_m)] \mid 0 \rangle : \prod_{i=1}^{m} [\frac{\Phi^{s_i}(x_1)}{s_i!}] : \end{aligned} \tag{A.60}$$

That such a quantity is a well-defined operator-valued distribution in Fock space is assured by distribution theory (see Theorem O in [5], 2. p. 229). Note also that the coefficients in the Wick expansion are now represented as vacuum expectation values of operators.

Now let us assume that T_m distributions with all required properties are successfully constructed for all $m < n$. Epstein and Glaser introduce then the retarded and the advanced n-point distributions (from now on we suppress the $\hbar$ factor in our notation):

$$R_n(x_1, \ldots, x_n) = T_n(x_1, \ldots, x_n) + R'_n, \quad R'_n = \sum_{P_2} T_{n-n_1}(Y, x_n) \tilde{T}_{n_1}(X) \tag{A.61}$$

$$A_n(x_1, \ldots, x_n) = T_n(x_1, \ldots, x_n) + A'_n, \quad A'_n = \sum_{P_2} \tilde{T}_{n_1}(X) T_{n-n_1}(Y, x_n). \tag{A.62}$$

The sum runs over all partitions $P_2 : \{x_1, \ldots x_{n-1}\} = X \cup Y, \quad X \neq \emptyset$ into disjoint subsets with $| X |= n_1 \geq 1, | Y | \leq n-2$. The $\tilde{T}$ are the operator-valued distributions of the inverse S-matrix:

$$S(g)^{-1} = 1 + \sum_{n=1}^{\infty} \frac{1}{n!} \int d^4x_1 \ldots d^4x_n \tilde{T}_n(x_1, \ldots x_n) g(x_1) \ldots g(x_n) \tag{A.63}$$

The distributions $\tilde{T}$ can be computed by formal inversion of $S(g)$:

$$S(g)^{-1} = (\mathbf{1} + \mathrm{T})^{-1} = \mathbf{1} + \sum_{\mathrm{n}=1}^{\infty} (-\mathrm{T})^{\mathrm{r}} \tag{A.64}$$

$$\tilde{T}_n(X) = \sum_{r=1}^{n} (-)^r \sum_{P_r} T_{n_1}(X_1) \ldots T_{n_r}(X_r), \tag{A.65}$$

where the second sum runs over all partitions P_r of X into r disjoint subsets $X = X_1 \cup \ldots \cup X_r, \quad X_j \neq \emptyset, \quad | X_j | = n_j$.

We stress the fact that all products of distributions are well-defined because the arguments are disjoint sets of points so that the products are tensor products of distributions. We also remark that both sums, R'_n and A'_n, in contrast to T_n, contain T_j's with $j \leq n-1$ only and are therefore known quantities in the inductive step from $n-1$ to n. Note that the last argument x_n is marked as the reference point for the support of R_n and A_n. The following crucial support property is a consequence of the causality conditions (A.59):

$$\mathrm{supp} R_m(x_1, \ldots, x_m) \subseteq \Gamma^+_{m-1}(x_m), \quad m < n \tag{A.66}$$

where Γ^+_{m-1} is the $(m-1)$-dimensional closed forward cone,

$$\Gamma^+_{m-1}(x_m) = \{(x_1, \ldots, x_{m-1}) \mid (x_j - x_m)^2 \geq 0, x_j^0 \geq x_m^0, \forall j\}. \tag{A.67}$$

In the difference

$$D_n(x_1, \ldots, x_n) \stackrel{\mathrm{def}}{=} R'_n - A'_n \tag{A.68}$$

the unknown n-point distribution T_n cancels. Hence this quantity is also known in the inductive step. With the help of the causality conditions (A.59) again, one shows that D_n has causal support

$$\mathrm{supp} D_n \subseteq \Gamma^+_{n-1}(x_n) \cup \Gamma^-_{n-1}(x_n) \tag{A.69}$$

Thus, this crucial support property is preserved in the inductive step from $n-1$ to n.

Given this fact, the following inductive construction of the n-point distribution T_n becomes possible: Starting off with the known $T_m(x_1, \ldots, x_n)$, $m \leq n-1$, one computes A'_n, R'_n and $D_n = R'_n - A'_n$. With regard to the supports, one can decompose D_n in the following way:

$$D_n(x_1, \ldots, x_n) = R_n(x_1, \ldots, x_n) - A_n(x_1, \ldots, x_n) \tag{A.70}$$

$$\mathrm{supp} R_n \subseteq \Gamma^+_{n-1}(x_n), \quad \mathrm{supp} A_n \subseteq \Gamma^-_{n-1}(x_n) \tag{A.71}$$

Having obtained these quantities we define T'_n as

$$T'_n = R_n - R'_n = A_n - A'_n \tag{A.72}$$

Symmetrizing over the marked variable x_n, we finally obtain the desired T_n,

$$T_n(x_1, \ldots x_n) = \sum_\pi \frac{1}{n!} T'_n(x_{\pi(1)}, \ldots x_{\pi(n)}) \tag{A.73}$$

One can verify that the T_n satisfy the conditions (A.59) and all other further properties of the induction hypothesis [5].

Summing up, with the help of the corresponding causal factorization property of the T_m-distribution one is able to reduce the problem of constructing well-defined time-ordered products to the following splitting problem of distributions:

Given an operator-valued tempered distribution $D_n \in \mathscr{S}'(\mathbf{R^{4n}})$ with causal support,

$$\mathrm{supp} D_n \subseteq \Gamma^+_{n-1}(x_n) \cup \Gamma^-_{n-1}(x_n). \tag{A.74}$$

one has to find a pair (R, A) of tempered distributions on $\mathbf{R^{4n}}$ with the following characteristics:

$$\bullet \quad R, A \in \mathscr{S}'(\mathbf{R^{4n}}) \qquad \mathbf{(A)} \tag{A.75}$$

$$\bullet \quad \mathrm{supp} R \subset \Gamma^+(x_n), \quad \mathrm{supp} A \subset \Gamma^-(x_n) \qquad \mathbf{(B)} \tag{A.76}$$

$$\bullet \quad R - A = D \qquad \mathbf{(C)} \tag{A.77}$$

A general solution of this problem was given by the mathematician Malgrange some time ago [19]. As mentioned already, every renormalization scheme solves this problem implicitly. The advantage of the Epstein-Glaser formulation is that it separates the purely technical details (which are essential for explicit calculations) from the simple physical structure of the theory.

The singular behavior of the distributions d_n for $x \to 0$ is crucial for the splitting problem because $\Gamma^+_{n-1}(0) \cap \Gamma^-_{n-1}(0) = \{0\}$. One therefore has to classify the singularities of distributions in this region. This can be characterized in terms of the singular order ω of the distribution under consideration which turns out to be identical with the usual power-counting degree. For details on the theory of distribution splitting we refer to the literature [5,17].

One has to ask whether the splitting solution of a given numerical distribution d with singular order $\omega(d)$ is unique. Let $r_1 \in \mathscr{S}'$ and $r_2 \in \mathscr{S}'$ be two splitting solutions of the given distribution $d \in \mathscr{S}'$. By construction r_1 and r_2 have their support in Γ^+ and agree with d on $\Gamma^+ \setminus \{0\}$, from which follows that $(r_1 - r_2)$ is a tempered distribution with point support and with singular order $\omega \leq \omega(d)$:

$$\mathrm{supp}(r_1 - r_2) \subset \{0\}, \quad \omega(r_1 - r_2) = \omega(d), \quad (r_1 - r_2) \in \mathscr{S}' \tag{A.78}$$

According to a well-known theorem in the theory of distributions, we have

$$r_1 - r_2 = \sum_{|a|=0}^{\omega_0} C_a \partial^a \delta(x). \tag{A.79}$$

In the case $\omega(d) < 0$ which means that d_n is regular at the zero point, the splitting solution is thus unique. In the case $\omega(d) \geq 0$ the splitting solution is only determined up to a local distribution with a fixed maximal singular degree $\omega_0 = \omega(d)$. The demands of causality (A.57) and translational invariance leave the constants C_a in (A.79) undetermined. They have to be fixed by additional normalization conditions.

One shows that, besides this normalization ambiguity, the T_n distributions are already fixed at all orders by the fundamental axioms of QFT and the defining equations of the specific theory under consideration which includes the definition of the specific coupling T_1.

Having constructed the most general S-matrix one can construct interacting field operators (compatible with causality and Poincaré invariance) (second reference in [5], section 8) as follows:

One starts with an extended first order S-matrix

$$S(g,g_1,g_2,\ldots) = \int d^4x\{T_1(x)g(x) + \frac{i}{\hbar}(\Phi_1(x)g_1(x) + \Phi_2(x)g_2(x) + \ldots)\} \quad (A.80)$$

where Φ_i represent certain Wick monomials like φ or $:\varphi^3:$. Following Bogoliubov and Shirkov [4], Epstein and Glaser defined the corresponding interacting fields Φ_i^{int} as functional derivatives of the extended S-matrix:

$$\Phi_i^{int}(g,x) = \frac{\hbar}{i} S^{-1}(g,g_1,\ldots) \frac{\delta S(g,g_1,\ldots)}{\delta g_i(x)}\bigg|_{g_i=0} \quad (A.81)$$

One shows that the perturbation series for the interacting fields is given by the advanced distributions of the corresponding expansion of the S-matrix, namely

$$\Phi_i^{int}(g,x) = \Phi_i(x) + \sum_{n=1}^{\infty} \frac{1}{n!} \int d^4x_1 \ldots d^4x_n A_{n+1/n+1}(x_1,\ldots,x_n;x), \quad (A.82)$$

where $A_{n+1/n+1}$ denotes the advanced distributions with n original vertices T_1 and one vertex Φ_i at the $(n+1)$th position; symbolically we may write:

$$A_{n+1/n+1}(x_1,\ldots,x_n;x) = Ad_{n+1}\left[T_1(x_1)\ldots T_1(x_n);\Phi_i(x)\right] \quad (A.83)$$

One shows that the perturbative defined object Φ_i^{int} fulfills the properties like locality and field equations in the sense of formal power series. The definition can be regarded as a direct construction of renormalized composite operators. Epstein and Glaser showed that the adiabatic limit $g \to 1$ exists only in the weak sense of expectation values in massive theories. The limit possesses all the expected properties of a Green's function such as causality, Lorentz covariance and the spectral condition.

References

1. N. Seiberg and E. Witten, *Monopole Condensation, And Confinement In $N = 2$ Supersymmetric Yang-Mills Theory*, Nucl.Phys. **B426** (1994) 19, hep-th/9407087; N. Seiberg, *Electric-Magnetic Duality in Supersymmetric Non-Abelian Gauge Theories*, Nucl.Phys. **B435** (1995) 129, hep-th/9411149.
2. T. Hurth and K. Skenderis, *Quantum Noether Method*, Nucl. Phys. **B541** (1999) 566-614, hep-th/9803030.
3. C. Becchi, A. Rouet, and R. Stora, *Renormalization of the Abelian Higgs-Kibble Model*, Comm. Math. Phys. **42** (1975) 127; *Renormalizable Models with Broken Symmetries*, in *Renormalization Theory*, ed. by G. Velo and A.S. Wightman (Reidel, Dordrecht, 1976); Ann. Phys. **98** (1976) 287; I.V. Tyutin, Lebenev Institute preprint N39 (1975).
4. N.N. Bogoliubov and D.V. Shirkov, *Introduction to the Theory of Quantized Fields*, New York, Interscience, 1959.

5. H. Epstein and V. Glaser, in: *Statistical Mechanics and Quantum Field Theory*, Proceedings of the 1970 Summer School of Les Houches, eds. C. de Witt and R. Stora, New York, Gordon and Breach, 1971; H. Epstein and V. Glaser, *The Role of Locality in Perturbation Theory*, Ann. Inst. Poincaré **29** (1973) 211; H. Epstein and V. Glaser, *Adiabatic Limit in Perturbation Theory*, in G. Velo, A.S. Wightman (eds.), *Renormalization Theory*, D. Reidel Publishing Company, Dordrecht 1976, 193; O. Piguet and A. Rouet, *Symmetries in Perturbative Quantum Field theory*, Phys. Rep. **76** (1981) 1.
6. H. Epstein, V. Glaser and R. Stora, *General Properties of the n-Point Functions in Local Quantum Field Theory*, in J. Bros, D. Jagolnitzer (eds.), Les Houches Proceedings 1975; G. Popineau and R. Stora, *A Pedagogical Remark on the Main Theorem of Perturbative Renormalization Theory*, unpublished; R. Stora, *Differential Algebras*, ETH-lectures (1993) unpublished.
7. R. Brunetti and K. Fredenhagen, *Interacting Quantum Fields in Curved Space: Renormalizability of* ϕ^4, hep-th/9701048.
8. S. Deser, *Self-interaction and Gauge Invariance*, Gen. Rel. and Grav. **I,1** (1970) 9.
9. D.Z. Freedman, P. van Nieuwenhuizen and S. Ferrara, *Progress towards a Theory of Supergravity*, Phys. Rev. **D14** (1976) 912; S. Deser and B. Zumino, *Consistent Supergravity*, Phys. Lett. **B62** (1976) 335; S. Ferrara, F. Gliozzi, J. Scherk and P. van Nieuwenhuizen, *Matter Couplings in Supergravity Theory*, Nucl. Phys. **B117** (1976) 333; P. van Nieuwenhuizen, *Supergravity*, Phys. Rep. **68** (1981) 191.
10. T. Kugo and I. Ojima, *Local covariant operator formalism of non-abelian gauge theories and quark confinement problem*, Suppl. Progr. Theor. Phys. **66** (1979) 1.
11. M. Dütsch and K. Fredenhagen, *A local (perturbative) construction of observables in gauge theories: the example of QED*, hep-th/9807078; *Deformation stability of BRST-quantization*, hep-th/9807215.
12. J. Wess and B. Zumino, *Consequences of Anomalous Ward Identities*, Phys. Lett. **B37** (1971) 95.
13. T. Hurth and K. Skenderis, *Analysis of Anomalies in the Quantum Noether Method*, in preparation.
14. G. Barnich, M. Henneaux, T. Hurth and K. Skenderis, *Cohomological analysis of gauged-fixed gauge theories*, hep-th/9910201.
15. I. A. Batalin and G. A. Vilkovisky, *Gauge algebra and quantization*, Phys. Lett. **B102**; *Quantization of Gauge Theories with Linearly Dependent Generators*, (1981) 27, Phys. Rev. **D28** (1983) 2567.
16. M. Henneaux and C. Teitelboim, *Quantization of Gauge Systems*, Princeton University Press, Princeton, 1992; J. Gomis, J. Paris and S. Samuel, *Antibracket, Antifields and Gauge-Theory Quantization*, Phys. Rep. **259** (1995) 1.
17. G. Scharf, *Finite Quantum Electrodynamics, The Causal Approach*, Text and Monographs in Physics, Springer, Berlin 1995.
18. T. Hurth, *Nonabelian Gauge Theories: The Causal Approach*, Annals of Phys. **244** (1995) 340, hep-th/9411080.
19. B. Malgrange, Seminaire Schwartz **21** (1960).

Applications of the Reduction of Couplings

Jisuke Kubo

Institute for Theoretical Physics,
Kanazawa University, Kanazawa 920-1192, Japan

Abstract. Applications of the principle of reduction of couplings to the standard model and supersymmetric grand unified theories are reviewed. Phenomenological applications of renormalization group invariant sum rules for soft supersymmetry-breaking parameters are also reviewed.

1 Application to the Standard Model

High energy physicists have been using renormalizability as the predictive tool, and also to decide whether or not a quantity is calculable. As we have learned in the previous talk by Professor Oehme, it is possible, using the method of reduction of couplings [1–3], to renormalize a theory with fewer number of counter terms then usually counted, implying that the traditional notion of renormalizability should be generalized in a certain sense [1]. Consequently, the notion of the predictability and the calculability [5] may also be generalized with the help of reduction of couplings. Of course, whether the generalizations of these notions have anything to do with nature is another question. The question can be answered if one applies the idea of reduction of couplings to realistic models, make predictions that are specific for reduction of couplings, and then wait till experimentalists find positive results [2].

In 1984 Professor Zimmermann, Klaus Sibold and myself [8] began to apply the idea of reduction of couplings to the $SU(3)_C \times SU(2)_L \times U(1)_Y$ gauge model for the strong and electroweak interactions. As it is known, this theory has a lot of free parameters, and at first sight it seemed there exists no guiding principle how to reduce the couplings in this theory. There were two main problems associated with this program. The one was that it is not possible to assume a common asymptotic behavior for all couplings, and the other one is how to increase the predictive power of the model without running into the contradiction with the experimental knowledge (of that time) such as the masses of the known fermions.

[1] Earlier references related to the idea of reduction of couplings are given in [4]. Professor Shirkov who is present here was one of the authors who considered such theoretical possibilities. For reviews, see [6].

[2] Here I would like to restrict myself to phenomenological applications of reduction of couplings. See [7,9] for the other applications. Professor Oehme reminded me that in his seminar talk given at the Max-Planck-Institute early 1984, professor Peccei suggested phenomenological applications of this idea .

P. Breitenlohner and D. Maison (Eds.): Proceedings 1998, LNP 558, pp. 106–126, 2000.

Professor Zimmermann suggested to use asymptotic freedom as a guiding principle, and assumed that QCD is most fundamental among the interactions of the standard model (SM). Since pure QCD is asymptotically free, we tried to switch on as many SM interactions as possible while keeping asymptotic freedom and added them to QCD. The result was almost unique: There exist two possibilities (or two asymptotically free (AF) surfaces in the space of couplings). It turned out that on the first surface, the $SU(2)_L$ gauge coupling α_2 is bigger than the QCD coupling α_3, and on the second surface, α_2 has to identically vanish. We decided to choose the second possibility, because we found out that it is possible to include the $SU(2)_L$ gauge coupling α_2 as a certain kind of "perturbation" into the AF system. Since the perturbating couplings should be regarded as free parameters, the reduction of couplings in this case can be achieved only partially ("partial reduction"). For the first case, it was not possible.

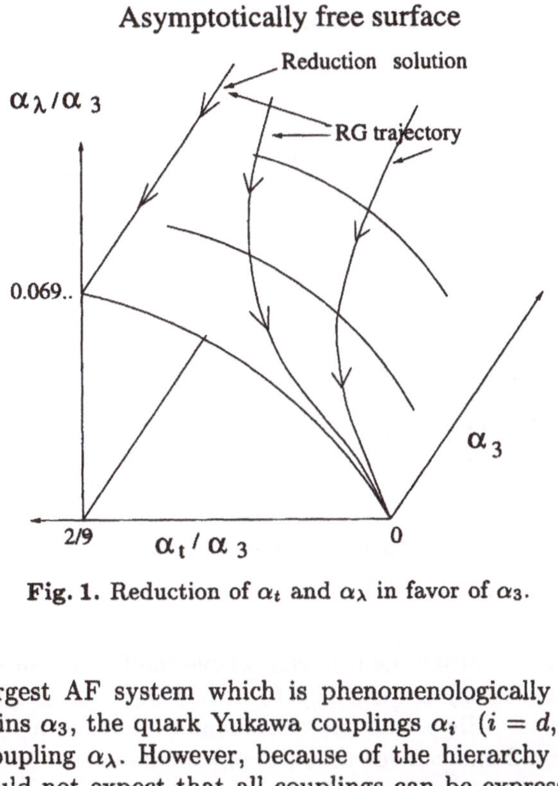

Fig. 1. Reduction of α_t and α_λ in favor of α_3.

Thus, the largest AF system which is phenomenologically acceptable (at that time) contains α_3, the quark Yukawa couplings α_i $(i = d, s, b, u, c, t)$ and the Higgs self coupling α_λ. However, because of the hierarchy of the Yukawa couplings, we could not expect that all couplings can be expressed in terms of a power series of α_3 without running into the contradiction with that hierarchy. So we decided to apply to the reduction of couplings only to the system with α_3 , α_t and α_λ, and to regard the other couplings as perturbations like α_2. Fig. 1 shows the AF surface in the space of α_3, α_{t/α_3} and α_λ/α_3. The reduction of the top Yukawa and Higgs couplings in favor of the QCD coupling corresponds to the border line on the surface, i.e., the line defined by

$$\alpha_t/\alpha_3 = \frac{2}{9}\ , \ \alpha_\lambda/\alpha_3 = \frac{\sqrt{689}-25}{18} \simeq 0.0694 \qquad (1)$$

in the one-loop approximation. This border line was already known as the Pendleton-Ross infrared (IR) fixed point (line) [10]. Note that the existence of the AF surface (shown in Fig. 1) at least for α_3 closed to the origin is mathematically ensured (see also [11]), while the line for large α_3, Pendleton-Ross infrared IR line, can be an one-loop artifact which was pointed out by Professor Zimmermann. He showed explicitly in the two-loop approximation that this is indeed the case [12].

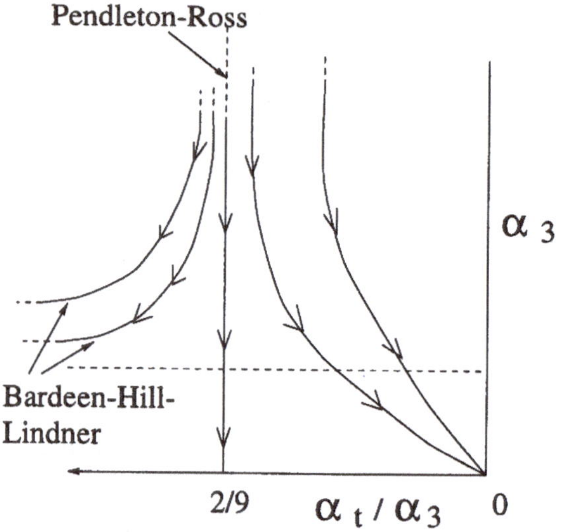

Fig. 2. Asymptotically free surface in the $\alpha_3 - \alpha_t/\alpha_3$ space.

worthwhile to mention that the branches above the Pendleton-Ross IR line (the lines left to it in Fig. 2) are used by Professor Bardeen and his collaborators [13] to interpret the Higgs particle as a bound state of the top and anti-top quarks. From Fig. 2 one can see that the higher the energy scale where the top Yukawa coupling diverges (the horizontal dotted line in Fig. 2 will be lowered), the similar is the prediction of the top mass in two methods. However, I would like to emphasize that how to include the corrections to this lowest order system (especially those due to the non-vanishing $SU(2)_L$ and $U(1)_Y$ gauge couplings) depends on the ideas behind, so that the actual predictions are different. We included these corrections within the one-loop approximation and calculated α_t/α_3 and α_h/α_3 in terms of α_3 and the perturbating free couplings. Then we

An asymptotically free renormalization group (RG) trajectory lies exactly on the surface. Fig. 2 shows trajectories projected on the $\alpha_3-\alpha_t/\alpha_3$ plane. It may be used the formulae

$$M_t^2/M_Z^2 \ = \ 2\cos^2\theta_{\mathrm{W}}\alpha_t/\alpha_2 \ , \ M_h^2/M_Z^2 \ = \ 2\cos^2\theta_{\mathrm{W}}\alpha_h/\alpha_2 \ , \tag{2}$$

to calculate the top quark and Higgs masses, M_t and M_h, from the known values of the parameters such as the Z boson mass M_Z and the Weinberg mixing angle θ_{W}. We obtained [8]

$$M_t \simeq 81 \ \ \mathrm{GeV} \ , M_h \simeq 61 \ \ \mathrm{GeV} \ . \tag{3}$$

Later I included higher order corrections such as two-loop corrections and found that the earlier predictions (3) become $M_t = 98.6 \pm 9.2$ GeV and $M_h = 64.5 \pm 1.5$ GeV, which should be compared with the present knowledge [14]

$$M_t = 173.8 \pm 5.2 \ \ \mathrm{GeV} \ , M_h \gtrsim 77.5 \ \ \mathrm{GeV} \ . \tag{4}$$

The failure of our prediction was disappointing in fact. However, this failure relieved Professor Zimmermann from a self-contradicting feeling. As we know he likes low-energy supersymmetry and also good wines. If our prediction would have been confirmed by an experiment, it would be very unlikely that low-energy supersymmetry is realized in nature, which would imply that he would loose *again* a lot of bottles of wines. So, the decision of nature was welcome at the same time. Fig. 3 summarizes.

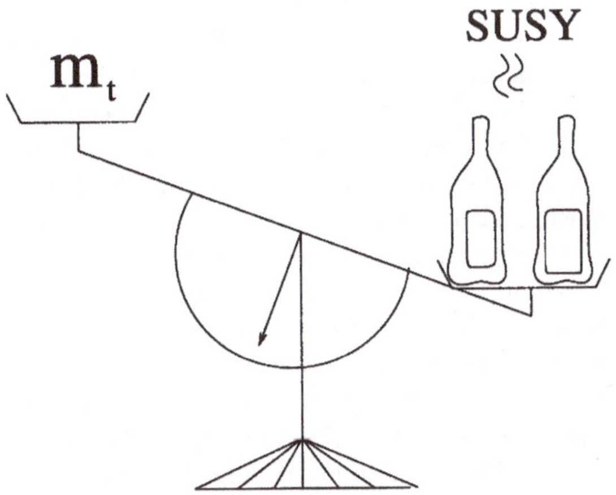

Fig. 3. Enttäuschung und Hoffnung.

2 Why Is Supersymmetry as Ideal Place for Application?

2.1 Naturalness and supersymmetry

Let me now come to the application of reduction of parameters to supersymmetric theories. I do not know why Professor Zimmermann likes low energy supersymmetry. But let me assume that he likes the usual argument for low energy supersymmetry, which is based on the naturalness notion of 't Hooft [15]. I would like to spend few minutes for that. (Let me allow to do so, although for the superexperts in the audience it might be superboring.) 't Hooft [15] said that there exist a natural scale in a given theory, and that the natural energy scale of spontaneously broken gauge theories which contain the SM is usually less than few TeV. The argument is the following. Suppose the scale at which the SM goes over to a more fundamental theory is Λ. That is, there are in the fundamental theory particles with masses of this order. Now consider the propagator $\Delta(p^2)$ of a boson field with the physical mass $m_{\rm B}$ much smaller than Λ, and suppose that it is normalized at Λ so that the propagator assumes a simple form at $p^2 = -\Lambda^2$:

$$\lim_{p^2 \to -\Lambda^2} \Delta(p^2) \to \frac{iZ(\Lambda^2)}{p^2 - m_{\rm B}^2(\Lambda^2)} , \tag{5}$$

where Z is the normalization constant for the wave function. The physical mass squared $m_{\rm B}^2$ can be expressed as

$$m_{\rm B}^2 = m_{\rm B}^2(\Lambda^2) + \delta m_{\rm B}^2 . \tag{6}$$

Then we ask ourselves how accurate we have to tune the value of $m_{\rm B}^2(\Lambda^2)$ to obtain a desired accuracy in the physical mass squared $m_{\rm B}^2$. This depends on $\delta m_{\rm B}^2$, of course. 't Hooft said that for a theory to be natural the ratio $m_{\rm B}^2(\Lambda^2)/m_{\rm B}^2$ should be of $O(1)$, which implies that $|\delta m_{\rm B}^2| < m_{\rm B}^2$. If quadratic divergences are involved in the theory, the correction $\delta m_{\rm B}^2$ will be proportional not only to the masses of the light fields, but also to the masses of the heavy fields, and so $\delta m_{\rm B}^2$ can be of the order $(\alpha/4\pi)\Lambda^2$, where α is some generic coupling. Since the Higgs mass should not exceed few hundred GeV in the SM, the natural scale of the fundamental theory, which contains the standard model Higgs and also involves quadratic divergences, is at best few TeV. So according 't Hooft, ordinary Grand Unified Theories (GUTs), for instance, are unnatural [15].

Supersymmetry, thanks to its very renormalization property known as non-renormalization theorem [16,17], can save the situation. The cancellation of the quadratic divergences, which was first observed by Professors Wess and Zumino [16], is exact if the masses of the bosonic and fermionic superpartners are the same. However, supersymmetry is unfortunately broken in nature, so that the cancellation is not exact. The mass squared difference, $m_{\rm B}^2 - m_{\rm F}^2$, characterizes the energy scale of supersymmetry breaking. To make compatible supersymmetry breaking with the naturalness notion of 't Hooft, we must impose the constraint on the supersymmetry-breaking scale $M_{\rm SUSY}$. A simple calculation yields that $M_{\rm SUSY}$ should be less than few TeV.

2.2 Soft supersymmetry-breaking parameters

Since the pioneering works by Professor Iliopolos (who could not participate in this meeting) with P. Fayet [18] and the others in late 70's, a lot of attempts to understand supersymmetry-breaking mechanism have been done. However, unfortunately, we still do not know how supersymmetry is really broken in nature. It, therefore, may be reasonable at this moment to pick up the common feature of supersymmetry breaking which effect the SM. The so-called minimal supersymmetric standard model (MSSM) is "defined" along this line of thought. The MSSM contains the ordinary gauge bosons and fermions together with their superpartners, and two supermultiplets for the Higgs sector. (With one supermultiplet in the Higgs sector, it is not possible to give masses to all the fermions of the MSSM.)

It is expected that the common effect of supersymmetry breaking is to add the so-called soft supersymmetry-breaking terms (SSB) to the symmetry theory. The SSB terms are defined as those which do not change the infinity structure of the parameters of the symmetric theory. So they are additional terms in the Lagrangian that do not change the RG functions such as the β- and γ-functions of the symmetric theory. (More precisely, there exists a renormalization scheme in which the RG functions are not altered by the SSB terms.) There exist four types of such terms [19].

$$\begin{aligned}&1.\ \text{Soft scalar mass terms}: \quad (m^2)^j_i \phi_j \phi^{*i}\ ,\\ &2.\ B-\text{terms}: \quad B^{ij}\phi_j\phi_i + \text{ H.C}\ ,\\ &3.\ \text{Gaugino mass terms}: \quad M\lambda\lambda + \text{ H.C}\ ,\\ &4.\ \text{Trilinear scalar couplings}: \quad h^{ijk}\phi_i\phi_j\phi_k + \text{ H.C}\ ,\end{aligned} \tag{7}$$

where ϕ_j and λ denote the scalar component in a chiral supermultiplet and the gaugino (the fermionic component) in a gauge supermultiplet, respectively.

If one insists only renormalizability for the MSSM, the number of the SSB parameters amounts to about 100, which is about five times of that of the SM. The commonly made assumption to reduce this number is the assumption of universality of the SSB terms, which is often justified by saying that supersymmetry breaking occurs in a flavor blind sector [20]. That is, it is assumed that the soft scalar masses and the trilinear scalar couplings are universal or flavor blind at the scale where supersymmetry breaking takes place. The so-called constrained MSSM contains thus only four independent massive parameters. But we could easily imagine that nature might not be so universal as one wants. In fact it possible to construct a lot of models with non-universal SSB terms [21] (even in models in which supersymmetry-breaking occurs in the so-called hidden sector which does not interact directly with the observable sector), and once we deviate from the universality, there will be chaotic varieties.

The application of reduction of couplings in the SSB sector is based on the assumption that the SSB terms organize themselves into a most economic structure that is consistent with renormalizability. I will come to discuss this later. I have spent a lot of time for low energy supersymmetry, because I wanted to argue that supersymmetric theories offer an ideal place where the reduction method, especially for massive parameters, can be applied and tested experimentally. It is worthwhile to mention that the current research program of Professor Zimmermann is the reduction of massive parameters [22].

3 Supersymmetric Gauge-Yukawa Unification

Before I come to discuss the SSB sector, I would like to stay in the sector of the dimensionless couplings in realistic supersymmetric GUTs and tell about certain phenomenological successes of reduction of parameters in these theories. I would like to emphasize that in contrast to the SM, supersymmetric GUTs can be asymptotically free or even finite.

3.1 Unification of the gauge and Yukawa couplings based on the principle of reduction of couplings

Few year ago, Professor Zimmermann and I were trying to apply the reduction method in the dimensionless sector of the MSSM, but we had no success. The main reason was that the power series solution to the reduction equation seemed to diverge. So we stopped to continue. About the same time, George Zoupanos (who unfortunately could not come here today) visited the Max-Planck-Institute, and told me that he obtains a top quark mass of about 180 GeV in a finite $SU(5)$ GUT [23]. Although the top quark was not found at that time (it was end of 1993, so just before we heard the rumor from Fermilab), 180 GeV for the top quark mass was a reasonable value. Finite theories have attracted many theorists. By a finite theory we mean a theory with the vanishing β-functions and anomalous dimensions. As we know, the $N = 4$ supersymmetric Yang-Mills theory is a well-known example [24]. And there were many attempts to construct $N = 1$ supersymmetric finite theories [23,25,26]. Klaus Sibold and his collaborators [27] gave an elegant existence proof of finite $N = 1$ supersymmetric theories, where I would like to recall that their proof is strongly based on the Adler-Bardeen non-renormalization theorem of chiral anomaly [28] (about which Professor Bardeen talked yesterday)[3]. The reduction of Yukawa couplings in favor of the gauge coupling is one of the necessary condition for a theory to be finite in perturbation theory. So in a finite theory, Gauge-Yukawa unification is achieved. Since Gauge-Yukawa unification results from the reduction of Yukawa couplings in favor of the gauge coupling, it can be achieved not only in finite theories but also in non-finite theories, as Myriam Mondragón, George Zoupanos and myself explicitly showed [30]. Relations among the gauge and Yukawa couplings, which are missing in ordinary GUTs, could be a consequence of a further unification provided by a more fundamental theory. And so Gauge-Yukawa unification is a natural extension to the ordinary GUT idea. This idea of unification relies on a symmetry principle as well as on the principle of reduction of couplings. The latter principle requires the existence of RG invariant relations among couplings, which do not necessarily result from a symmetry, but nevertheless preserve perturbative renormalizability or even finiteness as I mentioned.

[3] It is currently studied how to extend their theorem; for instance a non-perturbative extension has also been proposed in [29].

3.2 The double-role of $\tan\beta$

Before I come to discuss Gauge-Yukawa unification more in details, I would like to talk about an important parameter, $\tan\beta$, in the MSSM. It is a very popular parameter among SUSY physicists, but let me allow to spend few minutes for this parameter, because it plays also an important role for Gauge-Yukawa unification. As I mentioned the MSSM contains two Higgs supermultiplets. The most general form of the Higgs potential which is consistent with renormalizability and with the softness of the SSB parameters can be written as

$$V = (m_{H_d}^2 + |\mu_H|^2)\,\hat{H}_d^\dagger \hat{H}_d + (m_{H_u}^2 + |\mu_H|^2)\,\hat{H}_u^\dagger \hat{H}_u + (\,B\hat{H}_d\hat{H}_u + \text{ H.C.}\,)$$
$$+\frac{\pi}{2}(3\alpha_1^2/5 + \alpha_2^2)(\,\hat{H}_d^\dagger \hat{H}_d - \hat{H}_u^\dagger \hat{H}_u\,)^2 \;, \quad (8)$$

where μ_H is the only massive parameter in the supersymmetric limit, while $m_{H_u}^2$, $m_{H_d}^2$ and B are the SSB parameters in this sector. ($m_{H_u}^2$, $m_{H_d}^2$ are real while μ_H and B may be complex parameters.) Here $\hat{H}_{u,d}$ denote the scalar components of the two Higgs supermultiplets. There are four independent massive parameters in this sector as we can see in (8). These parameters should give the only one independent mass parameter of the SM, for instance the mass of Z. Now instead of regarding these parameters as independent one can regard also the ratio of the vacuum expectation values [32]

$$\tan\beta \equiv \frac{<\hat{H}_u>}{<\hat{H}_d>} \quad (9)$$

as independent. ($\tan\beta$ can be assumed to be real.) Usually one regards $|\mu_H|$ and B as dependent [4]. So the Higgs sector in the tree approximation is characterized by the parameters

$$\tan\beta \;,\; m_{H_1}^2 \;,\; m_{H_2}^2 \;. \quad (10)$$

The crucial point for Gauge-Yukawa unification is that $\tan\beta$ plays a double-role. On one hand, it is a parameter in the Higgs potential as we have seen above, and on the other hand it it is a mixing parameter to define the standard model Higgs field out of the two Higgs fields of the MSSM. That is, $\tan\beta$ appears also in the dimensionless sector, and in fact it can be fixed through Gauge-Yukawa unification with the knowledge of the tau mass M_τ, as I would like to explain it more in detail below.

3.3 How to predict M_t from Gauge-Yukawa Unification

The consequence of a Gauge-Yukawa unification in a GUT is that the gauge and Yukawa couplings are related above the GUT scale M_{GUT}. In the following discussions we consider only the Gauge-Yukawa unification in the third generation

[4] If $\tan\beta$ is real as we assume here, B can become complex starting in one-loop order [31].

sector [5]:

$$g_i = \kappa_i\, g \sum_{n=1} (\, 1 + \kappa_i^{(n)} g^{2n} \,) \quad (i = 1, 2, 3, t, b, \tau) \,, \tag{11}$$

where g denotes the unified gauge coupling, g_i denote the gauge and Yukawa couplings of the MSSM. Note that the constants κ_i's can be explicitly calculated from the principle of reduction of couplings. Once $\tan\beta$ and the Yukawa couplings are known, the fermion masses can be calculated as one can easily see from the tree level mass formulae

$$M_t = \sqrt{2}\frac{M_Z}{g_2} \sin\beta \cos\theta_W\, g_t \,, \quad M_{b,\tau} = \sqrt{2}\frac{M_Z}{g_2} \cos\beta \cos\theta_W\, g_{b,\tau} \,, \tag{12}$$

where M_t, M_b and M_τ are the masses of the top and bottom quarks and tau, respectively. Assume that we use the tau mass M_τ as input and also that below M_{SUSY} ($> M_t$) the effective theory of the GUT is the SM. At M_{SUSY} the couplings of the SM and MSSM have to satisfy the matching conditions [6]

$$\begin{aligned} \alpha_t^{SM} &= \alpha_t \sin^2\beta \,, \ \alpha_b^{SM} = \alpha_b \cos^2\beta \,, \ \alpha_\tau^{SM} = \alpha_\tau \cos^2\beta \,, \\ \alpha_\lambda &= \frac{1}{4}(\frac{3}{5}\alpha_1 + \alpha_2) \cos^2 2\beta \quad (\alpha_i = \frac{g_i^2}{4\pi}) \,, \end{aligned} \tag{13}$$

where α_i^{SM} $(i = t, b, \tau)$ are the SM Yukawa couplings and α_λ is the Higgs coupling. It is now easy to see that there is no longer freedom for $\tan\beta$ because with a given set of the input parameters, especially $M_\tau = 1.777$ GeV and $M_Z = 91.187$ GeV, the matching conditions (13) at M_{SUSY} and the Gauge-Yukawa unification boundary condition (11) at M_{GUT} can be simultaneously satisfied only if we have a specific value of $\tan\beta$. In this way Gauge-Yukawa unification enables us to predict the top and bottom masses in supersymmetric GUTs.

Table 1 shows the predictions in the case of a finite $SU(5)$ GUT [26], in which the one-loop reduction solution is given by

$$g_t^2 = \frac{4}{5} g^2 \,, \ g_b^2 = g_\tau^2 = \frac{3}{5} g^2 \,. \tag{14}$$

The experimental value of M_t, M_b and $\alpha_3(M_Z)$ are [14]

$$\alpha_3(M_Z) = 0.119 \pm 0.002 \,, \ M_t = 173.8 \pm 5.2 \ \text{GeV} \,, \ M_b = 5.2 \pm 0.2 \ \text{GeV} \tag{15}$$

We see that the predictions of the model reasonably agree with the experimental values[7]. This means among other things that the top-bottom hierarchy could

[5] A naive extension to include other generations into this scheme fails phenomenologically.

[6] There are MSSM threshold corrections to the matching conditions [33,34], which are ignored here.

[7] The correction to M_b coming from the MSSM superpartners can be as large as 50% for very large values of $\tan\beta$ [33,34]. In Table 1 we have not not included these corrections because they depend on the SSB parameters. The GUT threshold correction are ignored too.

Table 1. The predictions for different M_{SUSY} for the finite $SU(5)$ model.

M [GeV]	$\alpha_3(M_Z)$	$\tan\beta$	M_{GUT} [GeV]	M_b [GeV]	M_t [GeV]
800	0.118	48.2	1.3×10^{16}	5.4	173
10^3	0.117	48.1	1.2×10^{16}	5.4	173
1.2×10^3	0.117	48.1	1.1×10^{16}	5.4	173

be explained to a certain extent in this Gauge-Yukawa unified model, which should be compared with how the hierarchy of the gauge couplings of the SM can be explained if one assumes the existence of a unifying gauge symmetry at M_{GUT} [35]. More details on the different gauge-Yukawa unified models and their predictions can be found in [7,36,37].

4 Reduction of Massive Parameters: Application to the Soft Supersymmetry-Breaking Sector

To formulate reduction of massive parameters, one first has to formulate reduction of dimensionless parameters in a massive theory, which was initiated by Klaus Sibold and his collaborator Piguet [38], about ten years ago. To keep the generality of the formulation in the massive case is much more involved than in the massless case, because the RG functions now can depend on the ratios of mass parameters in a complicated way. In the massless case they are just power series in coupling constants (at least in perturbation theory). For phenomenological and also practical applications of the reduction method, it is therefore most convenient to work in a mass independent renormalization scheme, such as the dimensional renormalization scheme. There exists a transformation of one scheme to another one, which was in fact proven first by Dieter Maison in the ϕ^4 theory as far as I am informed, but not published. As I mentioned, the current research program of Professor Zimmermann is to include into the reduction program the massive parameters. He has already succeeded to carry out the program in the most general case and is able to show the renormalization scheme independence of the reduction method [22]. Consequently, there exist a transformation of a set of the reduction solutions in a mass-dependent renormalization scheme into a set of the reduction solutions in a mass-independent renormalization scheme, which generalizes the unpublished result of Dieter Maison. Thus, the naive treatment on the massive parameters (which was performed in phenomenological analyses [39,40]) can now be justified by his theorem [8].

4.1 Application to the minimal model

Now I would like to come to the SSB sector of a supersymmetric GUT. Recall that the Higgs potential (8) (in the tree approximation) is completely charac-

[8] It is assumed in the theorem that the β-functions in a mass-dependent renormalization scheme have a sufficiently smooth behavior in the massless limit [22].

terized by the soft scalar masses $m^2_{H_u}$, $m^2_{H_d}$ and $\tan\beta$, where $\tan\beta$ is fixed through Gauge-Yukawa unification as we have seen before. We [40] applied the the reduction methode of massive parameters to the SSB sector of the minimal supersymmetric $SU(5)$ GUT with Gauge-Yukawa unification in the third generation ($g_t^2 = (2533/2605)g^2$, $g_b^2 = g_\tau^2 = (1491/2605)g^2$) [30], and obtained the reduction solution

$$h_t = -g_t M \ , \ h_b = -g_b M \ , \tag{16}$$

$$\begin{aligned} m^2_{H_u} &= -\frac{569}{521}M^2 \ , \ m^2_{H_d} = -\frac{460}{521}M^2 \ , \\ m^2_{\tilde{b}_R} &= m^2_{\tilde{\tau}_L} = m^2_{\tilde{\nu}_\tau} = \frac{436}{521}M^2 \ , \\ m^2_{\tilde{d}_R} &= m^2_{\tilde{e}_L} = m^2_{\tilde{\nu}_e} = m^2_{\tilde{s}_R} = m^2_{\tilde{\mu}_L} = m^2_{\tilde{\nu}_\mu} = \frac{8}{5}M^2 \ , \\ m^2_{\tilde{t}_L} &= m^2_{\tilde{b}_L} = m^2_{\tilde{t}_R} = m^2_{\tilde{\tau}_R} = \frac{545}{521}M^2 \ , \\ m^2_{\tilde{u}_L} &= m^2_{\tilde{d}_L} = m^2_{\tilde{u}_R} = m^2_{\tilde{e}_R} = m^2_{\tilde{c}_L} = m^2_{\tilde{s}_L} = m^2_{\tilde{c}_R} = m^2_{\tilde{\mu}_R} = \frac{12}{5}M^2 \end{aligned} \tag{17}$$

in the one-loop approximation, where h_i's are the trilinear scalar couplings, m_i's are the soft scalar masses, and M is the unified gaugino mass. We found moreover that we can consistently regard μ_H and B as free parameters. As we can see from (16) and (17) the unified gaugino mass parameter M plays a similar role as the gravitino mass $m_{2/3}$ in supergravity coupled to a GUT and characterizes the scale of the supersymmetry-breaking [9]. Note that the reduction solution for the soft scalar masses (17) is not of the universal form while those for the trilinear couplings (16) are universal in the one-loop approximation.

Regarding the reduction solutions (16) and (17) as boundary conditions at $M_{\rm GUT}$ in the minimal supersymmetric GUT with Gauge-Yukawa unification in the third generation, we can compute the spectrum of the superpartners of the MSSM, which is shown in Table 2, where we have used the unified gaugino mass $M = 0.5$ TeV. The mass values [10] in Table 2 are the running masses at $M_{\rm SUSY}$ which is ~ 0.95 TeV [11] for $M = 0.5$ TeV. The prediction above depends basically only on the unified gaugino mass M, and so the model has an extremely strong predictive power. Note also that $m^2_{H_u}$, $m^2_{H_d}$ and $\tan\beta$ (see the Higgs potential (8) and the definition (9)) are now fixed outside of the Higgs sector, so that there is no guaranty that the Higgs potential (8) yields the desired symmetry breaking of $SU(2)_{\rm L} \times U(1)_{\rm Y}$ gauge symmetry. Surprisingly, in the case at hand it does! (If the sign of $m^2_{H_u}$ in (17) were different, for instance, it would not do.) In Table 3 I give the predictions from the dimensionless sector of the model. At last but not least we would like to emphasize that the reduction solutions (16) and (17) do *not* lead to the flavor changing neural current (FCNC) problem. This is not

[9] See for instance [20].

[10] For the mass of the lightest Higgs, the RG improved corrections [41] are included.

[11] $M_{\rm SUSY}$ is no longer an independent parameter and we use $M^2_{\rm SUSY} = (m^2_{\tilde{t}_1} + m^2_{\tilde{t}_2})/2$, where $m_{\tilde{t}_{1,2}}$ are the masses of the superpartners of the top quark.

Table 2. The prediction of the superpartner spectrum for $M = 0.5$ TeV in the minimal gauge-Yukawa unified model. The mass unit is TeV.

m_{χ_1}	0.22	$m_{\tilde{s}_1} = m_{\tilde{d}_1}$	1.18
m_{χ_2}	0.42	$m_{\tilde{s}_2} = m_{\tilde{d}_2}$	1.30
m_{χ_3}	0.90	$m_{\tilde{\tau}_1}$	0.42
m_{χ_4}	0.91	$m_{\tilde{\tau}_2}$	0.59
$m_{\chi_1^\pm}$	0.42	$m_{\tilde{\nu}_\tau}$	0.54
$m_{\chi_2^\pm}$	0.91	$m_{\tilde{\mu}_1} = m_{\tilde{e}_1}$	0.72
$m_{\tilde{t}_1}$	0.87	$m_{\tilde{\mu}_2} = m_{\tilde{e}_2}$	0.80
$m_{\tilde{t}_2}$	1.03	$m_{\tilde{\nu}_\mu} = m_{\tilde{\nu}_e}$	0.72
$m_{\tilde{b}_1}$	0.87	m_A	0.33
$m_{\tilde{b}_2}$	1.01	$m_{H^\pm}$	0.34
$m_{\tilde{c}_1} = m_{\tilde{u}_1}$	1.26	m_H	0.33
$m_{\tilde{c}_2} = m_{\tilde{u}_2}$	1.30	m_h	0.124
M_3	1.16		

Table 3. The predictions from the dimensionless sector of the minimal model. ($M = 0.5$ TeV)

$\alpha_3(M_Z)$	$\tan\beta$	M_{GUT} [GeV]	M_b [GeV]	M_t [GeV]
0.119	48.8	1.47×10^{16}	5.4	177

something put ad hoc by hand; it is a consequence of the principle of reduction of couplings.

5 Sum Rules for the Soft Supersymmetry-Breaking Parameters

5.1 Renormalization group invariant sum rules

Now I would like to come the next topic. To proceed I recall the result of the reduction of the SSB parameters in favor of the unified gaugino mass M in the minimal SUSY $SU(5)$ model which I have discussed just above. As we have seen, the reduction solutions for the trilinear couplings are universal while those for the soft scalar masses are not (see (16) and 17)). However, if one adds the soft scalar mass squared in an appropriate way, one finds something interesting [42]. For instance,

$$M^2 = m_{t_L}^2 + m_{t_R}^2 + m_{H_u}^2 = m_{b_L}^2 + m_{b_R}^2 + m_{H_d}^2 \,. \tag{18}$$

This is not an accidental coincidence. One can in fact show that the sum rules in this form are RG invariant at one-loop [42].

In last years there have been continues developments [43]–[47] in computing the RG functions in softly broken supersymmetric Yang-Mills theories, and the well-known result on the QCD β-function obtained by Professor Zakharov and his collaborators [48] [12] has been generalized so as to include to the SSB sector [43]–[47], which is based on a clever spurion superfield technique along with power counting [13]. Using this result, it is possible to find a closed form of the sum rules that are RG invariant to all orders in perturbation theory [44]-[47].

To be specific, we consider a softly broken supersymmetric theory described by the superpotential

$$W = \frac{1}{6} Y^{ijk}\, \Phi_i \Phi_j \Phi_k + \frac{1}{2}\, \mu^{ij}\, \Phi_i \Phi_j \,, \tag{19}$$

along with the Lagrangian for the SSB terms,

$$-\mathcal{L}_{\rm SB} = \frac{1}{6} h^{ijk}\, \phi_i \phi_j \phi_k + \frac{1}{2} b^{ij}\, \phi_i \phi_j + \frac{1}{2}\, (m^2)^j_i\, \phi^{*\,i} \phi_j + \frac{1}{2} M\, \lambda\lambda + \text{H.c.} \,, \tag{20}$$

where Φ_i stands for a chiral superfield with its scalar component ϕ_i, and λ is the gaugino field. It has been found [45] that the expressions [14]

$$\begin{aligned} b^{ij} &= -M \mu^{ij} \frac{d \ln \mu^{ij}(g)}{d \ln g} \,, \\ h^{ijk} &= -M \frac{d Y^{ijk}(g)}{d \ln g} \,, \qquad (21) \\ m_i^2 &= \frac{1}{2} |M|^2 (g/\beta_g) \frac{d\gamma_i(g)}{d \ln g} \qquad (22) \end{aligned}$$

are RG invariant to all orders in perturbation theory in a certain class of renormalization schemes, which are the higher order results for the one-loop reduction solutions (16) and (17). Similarly, the sum rule (18) in higher orders becomes [46]

$$\begin{aligned} m_i^2 + m_j^2 + m_k^2 = |M|^2 \{ & \frac{1}{1 - g^2 C(G)/(8\pi^2)} \frac{d \ln Y^{ijk}}{d \ln g} + \frac{1}{2} \frac{d^2 \ln Y^{ijk}}{d (\ln g)^2} \} \\ & + \sum_\ell \frac{m_\ell^2 T(R_\ell)}{C(G) - 8\pi^2/g^2} \frac{d \ln Y^{ijk}}{d \ln g} \,, \end{aligned} \tag{23}$$

in the renormalization scheme which corresponds to that of [48]. Here $C(G)$ is the quardratic Casimir in the adjoint representation, $T(R)$ stands for the Dynkin index of the representation R, β_g is the β-function of the gauge coupling g, γ_i is the anomalous dimension of Φ_i. These expressions look slightly complicated.

[12] Klaus Sibold pointed out that there is some correction to this β-function. See [27] for the argument.

[13] It is not clear at the moment in which class of renormalization schemes exactly the result is valid; a renormalization scheme independent investigation of this result is certainly desirable.

[14] The Yukawa couplings Y^{ijk} and μ^{ij} are assumed to be functions of the gauge coupling g.

But if one uses the freedom of reparametrization [3] (as discussed in the previous talk of Professor Oehme), they can be transformed into a more simple form $((d\ln Y^{ijk}/d\ln g) = 1)$:

$$h^{ijk} = -Y^{ijk}(g)M \ , \tag{24}$$

$$m_i^2 + m_j^2 + m_k^2 = |M|^2 \frac{1}{1 - g^2 C(G)/(8\pi^2)} + \sum_\ell \frac{m_\ell^2 T(R_\ell)}{C(G) - 8\pi^2/g^2} \ , \tag{25}$$

It is exactly this form which coincides with the results obtained in certain orbifold models of superstrings [26] [15]. I believe that this coincidence is not accidental, and I also believe that target-space duality invariance [51], which is supposed to be an exact symmetry of compactified superstring theories [16], is most responsible for the coincidence. In fact there exist already some indications for that. I hope I can report on the true reason of this interesting coincidence in near future.

5.2 Finiteness and sum rules

At this stage it may be worthwhile to mention that the reduction solution (21) and the sum rules (23) ensure the finiteness of the SSB sector in a finite theory [17]. For the $N = 4$ supersymmetric Yang Mills theory written in terms of $N = 1$ superfields, for instance, we have $\sum_\ell m_\ell^2 T(R_\ell) = (m_i^2 + m_j^2 + m_k^2)C(G)$ so that the all order sum rule (25) assumes the tree level form $m_i^2 + m_j^2 + m_k^2 = |M|^2$. Applied to the finite $SU(5)$ model [26] which I discussed in the previous section (Table 1 presents the prediction from the dimensionless sector), it means that the sum rules [26]

$$m_{H_u}^2 + 2m_{10}^2 = M^2 \ , \ m_{H_d}^2 - 2m_{10}^2 = -\frac{M^2}{3} \ , \ m_{\overline{5}}^2 + 3m_{10}^2 = \frac{4M^2}{3} \tag{26}$$

should be satisfied at and above $M_{\rm GUT}$ for the two-loop finiteness of the SSB sector requires that, where

$$m_{10} = m_{t_L} = m_{b_L} = m_{t_R} = m_{\tau_R} \ , \ m_{\overline{5}} = m_{b_R} = m_{\tau_L} = m_{\nu_\tau} \ . \tag{27}$$

In his casewe have an additional free parameter, m_{10}, in the SSB sector. It turned out that the mass of a superpartner of the tau (s-tau) tends to become very light in this model. Consequently, in order to obtain a neutral lightest superparticle (LSP) (because we assume that R- parity is intact), we have to have a large unified gravitino mass $M \gtrsim 0.8$ TeV. For $M = 1$ TeV, only the window 0.62 TeV $< m_{10} <$ 0.66 TeV is allowed. In Table 4 we give the prediction of the superpartner spectrum of the model for $m_{10} = 0.62/0.66$ TeV and $M = 1$ TeV. We have assumed the universal soft masses for the first two generations. But this assumption does not change practically our prediction of the spectrum expect for those that are directly of the first two generations.

[15] Tree-level sum rules (like (18) in string theories are found in [42], [49]-[50]

[16] See [52], for instance, for target-space duality.

[17] There exists a fine difference in the opinions about this point. See, for instance, [44,45].

Table 4. The predictions of the superpartner spectrum for the finite $SU(5)$ model. $M = 1$ TeV and $m_{10} = 0.62/0.66$ TeV.

m_{χ_1}	0.45/0.45	$m_{\tilde{s}_1} = m_{\tilde{d}_1}$	1.95/1.95
m_{χ_2}	0.84/0.84	$m_{\tilde{s}_2} = m_{\tilde{d}_2}$	2.06/2.05
m_{χ_3}	1.29/1.32	$m_{\tilde{\tau}_1}$	0.46/0.46
m_{χ_4}	1.29/1.32	$m_{\tilde{\tau}_2}$	0.73/0.66
$m_{\chi_1^\pm}$	0.84/0.84	$m_{\tilde{\nu}_\tau}$	0.70/0.57
$m_{\chi_2^\pm}$	1.29/1.32	$m_{\tilde{\mu}_1} = m_{\tilde{e}_1}$	0.70/0.71
$m_{\tilde{t}_1}$	1.50/1.51	$m_{\tilde{\mu}_2} = m_{\tilde{e}_2}$	0.89/0.89
$m_{\tilde{t}_2}$	1.72/1.74	$m_{\tilde{\nu}_\mu} = m_{\tilde{\nu}_e}$	0.89/0.88
$m_{\tilde{b}_1}$	1.51/1.46	m_A	0.63/0.77
$m_{\tilde{b}_2}$	1.70/1.71	$m_{H^\pm}$	0.63/0.77
$m_{\tilde{c}_1} = m_{\tilde{u}_1}$	1.96/1.96	m_H	0.63/0.77
$m_{\tilde{c}_2} = m_{\tilde{u}_2}$	2.05/2.05	m_h	0.127/0.127
M_3	2.21/2.21		

5.3 Sum rules in the superpartner spectrum

The sum rules (18) or (25) can be translated into the sum rules of the superpartner spectrum of the MSSM [53] as I will show now. To be specific we assume an $SU(5)$ type Gauge-Yukawa unification in the third generation of the form (11). For a given model, the constants κ's are fixed, but here we consider them as free parameters. As before we use the tau mass M_τ as an input parameter, and we go from the parameter space $(\kappa_t\ ,\ \kappa_b)$ to another one $(\kappa_t\ ,\ \tan\beta)$, because in this analysis we use the physical top quark M_t, too, as an input parameter. Then the unification conditions of the gauge and Yukawa couplings of the MSSM (i.e., $g = g_1 = g_2 = g_3\ ,\ g_b = g_\tau$) fixes the allowed region (line) in the $\kappa_t - \tan\beta$ space for a given value of the unified gaugino mass M. The parameter space in the SSB sector at $M_{\rm GUT}$ is constrained due to unification:

$$\begin{aligned} M &= M_1 = M_2 = M_3\ , \\ m_{t_R}^2 &= m_{t_L}^2 = m_{b_L}^2 = m_{\tau_R}^2\ ,\ m_{b_R}^2 = m_{\tau_L}^2 = m_{\nu_\tau}^2\ , \end{aligned} \tag{28}$$

where M_i $(i = 1,2,3)$ are the gaugino masses for $U(1)_{\rm Y}$ (bino), $SU(2)_{\rm L}$ (wino) and $SU(3)_{\rm C}$ (gluino). And the one-loop sum rules at $M_{\rm GUT}$ yield

$$h_t = -M\ ,\ h_b = h_\tau = -M\, g_b\ ,\ M^2 = m_{\Sigma(t)}^2 = m_{\Sigma(b)}^2 = m_{\Sigma(\tau)}^2\ , \tag{29}$$

where

$$m_{\Sigma(t)}^2 \equiv m_{t_R}^2 + m_{t_L}^2 + m_{H_u}^2\ ,\ m_{\Sigma(b,\tau)}^2 \equiv m_{b_R,\tau_R}^2 + m_{b_L,\tau_L}^2 + m_{H_d}^2\ . \tag{30}$$

(The above equations are the same as (16) and (18), respectively.) I would like to emphasize that in the one-loop RG evolution of m_Σ^2's in the MSSM only the same combinations of the sum of m_i^2's enter. Therefore, as far as we are interested in the evolution of m_Σ^2's, we have only one additional parameter $M_{\rm SUSY}$. To derive the announced sum rules for the superpartner spectrum, we define

$$s_i \equiv m_{\Sigma(i)}^2/M_3^2 \quad (i=t,b,\tau) \;\; \text{at} \;\; Q = M_{\rm SUSY}. \tag{31}$$

The parameters s_i's do not depend on the value of the unified gaugino mass M, but they do on $\tan\beta$. This dependence is shown in Fig. 4. We then express the

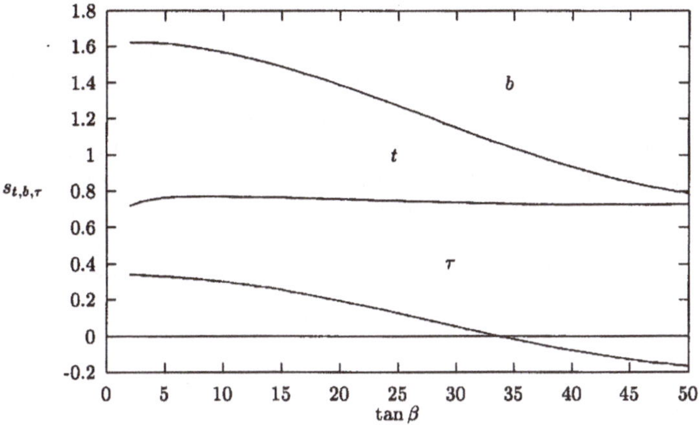

Fig. 4. s_t, s_b, s_τ against $\tan\beta$.

masses of the superpartners in terms of the soft scalar masses and the masses of the ordinary particles to obtain the sum rules [53],

$$\begin{aligned} -\cos 2\beta \; m_A^2 &= (s_b - s_t)M_3^2 + 2(\hat{m}_t^2 - m_t^2) - 2(\hat{m}_b^2 - m_b^2) \\ &= (s_\tau - s_t)M_3^2 + 2(\hat{m}_t^2 - m_t^2) - 2(\hat{m}_\tau^2 - m_\tau^2), \end{aligned} \tag{32}$$

where m_A^2 is the neutral pseudoscalar Higgs mass squared, and $\hat{m}_i^2$ stands for the arithmetic mean of the two corresponding scalar superparticle mass squared. Since we have assumed an $SU(5)$-type supersymmetric GUT with a gauge-Yukawa unification in the third generation, the result (32) is not a direct consequence of a superstring model, although the form of the sum rules in both kinds of unification schemes might coincide with each other as I mentioned (see footnote 11). However, under the following circumstances (only rough), the sum rules (32) could be a consequence of a superstring model: (i) The Yukawa coupling of the third generation is field-independent in the corresponding effective $N=1$ supergravity. (ii) Below the string scale an $SU(5)$-type gauge-Yukawa unification is realized so that the sum rules are RG invariant below the string scale and are satisfied down to $M_{\rm GUT}$. (iii) Below $M_{\rm GUT}$ the effective theory is the MSSM.

The sum rules (32) could be experimentally tested if the superpartners are found in future experiments, e.g., at LHC. In any event, an experimental verification of the sum rules of the SSB parameters would give an interesting information on physics beyond the GUT scale.

6 Conclusion

Now I will come to conclusion. Professor Zimmermann, obviously an interesting feature is coming. So please keep staying in physics and experience new developments in physics with us.

Acknowledgment

I would like to thank the Theory Group of the Max-Planck-Institute for Physics, Munich for their warm hospitality which my family and I have been enjoying since 1984. I also would like to thank my collaborators, Yoshiharu Kawamura, Tatsuo Kobayashi, Myriam Mondragón, Marek Olechowski, Klaus Sibold, Nicholas Tracas, George Zoupanos and Professor Wolfhart Zimmermann. Most of the results I presented here were obtained in collaborations. I thank Professor Reinhard Oehme for instructive discussions. At last but not least I would like to thank Professor Wolfhart Zimmermann for his encouragement of my research programs over the years.

References

1. W. Zimmermann, Commun. Math. Phys. **97** (1985) 211;
   R. Oehme and W. Zimmermann, Commun. Math. Phys. **97** (1985) 569.
2. R. Oehme, K. Sibold and W. Zimmermann, Phys. Lett. **B147** (1984) 117; **B153** (1985) 142.
3. R. Oehme, Prog. Theor. Phys. Suppl. **86** (1986) 215.
4. N.-P. Chang, Phys. Rev. **D10** (1974) 2706;
   N.-P. Chang, A. Das and J. Perez-Mercader, Phys. Rev. **D22** (1980) 1829;
   E.S. Fradkin and O.K. Kalashnikov, J. Phys. **A8** (1975) 1814; Phys. Lett. **59B** (1975) 159; **64B** (1976) 177;
   E. Ma, Phys. Rev. **D11** (1975) 322; **D17** (1978) 623; **D31** (1985) 1143; Prog. Theor. Phys. **54** (1975) 1828; Phys. Lett. **62B** (1976) 347; Nucl. Phys. **B116** (1976) 195;
   D.I. Kazakov and D.V. Shirkov, *Singular Solutions of Renormalization Group Equations and Symmetry of the Lagrangian*, in Proc. of the 1975 Smolence Conference on *High Energy Particle Interactions*, eds. D. Krupa and J. Pišút (VEDA, Publishing House of the Slovak Academy of Sciences, Bratislava 1976).
5. H. Georgi and A. Pais, Phys. Rev. **D10** (1974) 539.
6. R. Oehme, *Reduction in Coupling Parameter Space*, in Proc. of *Anomalies, Geometry and Topology*, ed. A White (World Scientific, Singapore, 1985) pp. 443;
   Reduction of Coupling Parameters, hep-th/9511006 in Proc. of *the XVIIIth Int. Workshop on High Energy Physics and Field Theory*, June 1995, Moscow-Protvino;
   W. Zimmermann, *Renormalization Group and Symmetries in Quantum Field Theory*, in Proc. of *the 14th ICGTMP*, ed. Y.M. Cho (World Scientific, Singapore, 1985) pp. 145;
   K. Sibold, *Reduction of Couplings*, Acta Physica Polonica, **19** (1989) 295;
   J. Kubo, *Is there any relation between dynamical symmetry breaking and reduction of couplings?*, in Proc. of *the 1989 Workshop on Dynamical Symmetry Breaking*, eds. T. Muta and K. Yamawaki, Nagoya 1989, pp. 48.

7. J. Kubo, M. Mondragón and G. Zoupanos, Acta Phys. Polon. **B27** (1997) 3911.
8. J. Kubo, K. Sibold and W. Zimmermann, Nucl. Phys. **B259** (1985) 331; Phys. Lett. **B200** (1989) 185.
9. J. Avan and H.J. De Vega, Nucl. Phys. **B269** (1986) 621;
H. Meyer-Ortmanns, Phys. Lett. **B186** (1987) 195;
G. Grunberg, Phys. Rev. Lett. **58** (1987) 1180;
K. Sibold and W. Zimmermann, Phys. Lett. **B191** (1987) 427;
T.E. Clark and S.T. Love, Mod .Phys. Lett. **A3** (1988) 661;
K.S. Babu and S. Nandi, Oklahoma State University Preprint, OSU-RN-202 (1988);
W.J. Marciano, Phys. Rev. Lett. **62** (1989) 2793;
J. Kubo, K. Sibold and W. Zimmermann, Phys. Lett. **B220** 1989;
F.M. Renard and D. Schildknecht, Phys. Lett. **B219** (1989) 481;
M. Bastero-Gil and J. Perez-Mercader, Phys. Lett. **B247** (1990) 346;
A. Denner, Nucl. Phys. **B347** (1990) 184;
E. Kraus, Nucl. Phys. **B349** (1991) 563; **B354** (1991) 245;
F. Cooper, Phys. Rev. **D43** (1991) 4129, [Erratum] **D45** (1992) 3012;
L.-N. Chang and N.-P. Chang, Phys. Rev. **D45** (1992) 2988;
L.A. Wills Toro, Z. Phys. **C56** (1992) 635;
R.J. Perry and K.G. Wilson, Nucl. Phys. **B403** (1993) 587;
K.G. Wilson, T.S. Walhout, A. Harindranath, W.-M. Zhang, R.J. Perry and S. D. Glazek, Phys. Rev. **D49** (1994) 6720-6766;
A.B. Lahanas and V.C. Spanos, Phys. Lett. B334 (1994) 378;
H. Skarke, Phys. Lett. **B336** (1994) 32;
E. A. Ammons, Phys. Rev. **D50** (1994) 980;
M. Harada, Y. Kikukawa, T. Kugo and H. Nakano, Prog. Theor. Phys. **92** (1994) 1161;
B. Schrempp and F. Schrempp, Phys. Lett. **B299** (1993) 321;
B. Schrempp, Phys. Lett. **B344** (1995) 193;
A.A. Andrianov and N.V. Romanenko, Phys. Lett. **B343** (1995) 295;
N. Krasnikov,
G. Kreyerhoff and R. Rodenberg, Nuovo Cim. **108A** (1995) 565;
A.A. Andrianov and R. Rodenberg, Nuovo Cim. **108A** (1995) 577;
J. Kubo, Phys. Rev. **D52** (1995) 6475;
M. Atance and J.L. Cortes, Phys. Lett. **B387** (1996) 697; Phys. Rev. **D54** (1996) 4973; **D56** (1997) 3611;
M.V. Chizhov, hep-ph/9610220;
R. Oehme, Phys. Lett. **B399** (1997) 67; hep-th/9808054;
A.Karch and D. Lust and G.Zoupanos, Nucl. Phys. **B529** (1998) 96;
E. Umezawa, Prog. Theor. Phys. **100** (1998) 375;
J. Erdmenger, C. Rupp and K. Sibold, Nucl. Phys. **B530** (1998) 501;
Y. Kawamura, T. Kobayashi and H. Shimabukuro, Phys. Lett. **B436** (1998) 108;
A. Karch, T. Kobayashi, J. Kubo and G. Zoupanos, Phys. Lett. **B441** (1998) 235;
R. J. Perry, To be published in the proceedings of APCTP - RCNP Joint International School on Physics of Hadrons and QCD, Osaka, Japan, 12-13 Oct 1998, nucl-th/9901080, and referencs therein.
10. B. Pendleton and G.G Ross, Phys. Lett. **B98** (1981) 291.
11. W. Zimmermann, Lett. Math. Phys. **30** (1993) 61.
12. W. Zimmermann, Phys. Lett. **B308** (1993) 117.
13. W.A. Bardeen, C.T. Hill and M. Lindner, Phys. Rev. **D41** (1990) 1647.

14. Particle Data Group, C. Caso *et al.*, Eur. Phys. J. **C3** (1998) 1.
15. G. 't Hooft, "Naturalness, chiral symmetry, and spontaneous chiral symmetry breaking", in *Recent developments in gauge theories*, Cargèse, 1979.
16. J. Wess and B. Zumino, Phys. Phys. **B49** 52.
17. J. Iliopoulos and B. Zumino, Nucl. Phys. **B76** (1974) 310;
S. Ferrara, J. Iliopoulos and B. Zumino, Nucl. Phys. **B77** (1974) 413;
K. Fujikawa and W. Lang, Nucl. Phys. **B88** (1975) 61.
18. P. Fayet and J. Iliopoulos, Phys. Lett. **B51** (1974) 461.
19. S. Ferrara, L. Girardello and F. Palumbo, Phys. Rev. **D20** (1979) 403;
K. Harada and N. Sakai, Prog. Theor. Phys. **67** (1982) 1887;
L. Girardello ans M.T. Grisaru, Nucl. Phys. **B194** (1982) 65.
20. H.P. Nilles, Phys. Rep. **110** (1984) 1;
H.E Haber and G.L. Kane, Phys. Rep. **117** (1985) 75.
21. L.J. Dixon, V.S, Kaplunovsky and J. Louis, Nucl. Phys. **B382** (1992) 305;
L. Ibáñez and D. Lüst, Nucl. Phys. **B382** (1992) 305;
Y. Kawamura, H. Murayama and M. Yamaguchi, Phys. Rev. **D51** (1995) 1337.
22. W. Zimmermann, "Reduction of Couplings in Massive Models of Quantum Field Theory", talks given at Kanazawa university, March 1998, at the 12th Max Born Symposium, Wrocł aw, September 1998, Max-Planck-Institute preprint MPI/PhT/98-97, and at the Ringberg Conference on *Trends in Theoretical Particle Physics*, Tegernsee October, 1998.
23. D. Kapetanakis, M. Mondragón and G. Zoupanos, Zeit. f. Phys. **C60** (1993) 181;
M. Mondragón and G. Zoupanos, Nucl.Phys. B (Proc. Suppl) **37C** (1995) 98.
24. S. Mandelstam, Nucl. Phys. **B182** (1981) 125.
25. A. Parkes and P. West, Nucl. Phys. **B222** (1983) 269;
D.R.T. Jones, L. Mezincescu and Y.-P. Yao, Phys. Lett. **B148** (1984) 317;
A.J. Parkes and P.C. West, Phys. Lett. **B138** (1984) 99; Nucl. Phys. **B256** (1985) 340;
P. West, Phys. Lett. **B137** (1984) 371;
D.R.T. Jones and A.J. Parkes, Phys. Lett. **B160** (1985) 267;
D.R.T. Jones and L. Mezinescu, Phys. Lett. **B136** (1984) 242; **B138** (1984) 293;
A.J. Parkes, Phys. Lett. **B156** (1985) 73;
S. Hamidi, J. Patera and J.H. Schwarz, Phys. Lett. **B141** (1984) 349;
X.D. Jiang and X.J. Zhou, Phys. Lett. **B197** (1987) 156; **B216** (1985) 160;
S. Hamidi and J.H. Schwarz, Phys. Lett. **B147** (1984) 301;
D.R.T. Jones and S. Raby, Phys. Lett. **B143** (1984) 137;
J.E. Bjorkman, D.R.T. Jones and S. Raby, Nucl. Phys. **B259** (1985) 503;
J. León *et al.*, Phys. Lett. **B156** (1985) 66;
A.V. Ermushev, D.I. Kazakov and O.V. Tarasov, Nucl. Phys. **B281** (1987) 72;
D.I. Kazakov, Mod. Phys. Let. **A2** (1987) 663; Phys. Lett. **B179** (1986) 352;
D.I. Kazakov and I.N. Kondrashuk, Int. J. Mod. Phys. **A7** (1992) 3869;
K. Yoshioka, Kyoti University preprint KUNS-1444, hep-ph/9705449;
L.E. Ibáñez, hep-ph/9804236;
S. Kachru and E. Silverstein, Phys. Rev. Lett. **80** (1998) 4855;
A. Hanany, M.J. Strassler and A. Uranga, Princeton University preprint IASSNS-HEP-23; hep-ph/9803086;
A. Hanany and Y.-H. He, MIT preprint MIT-CTP-2803.
26. T. Kobayashi, J. Kubo, M. Mondragón and G. Zoupanos, Nucl. Phys. **B511** (1998) 45.

27. C.Lucchesi, O. Piquet and K. Sibold, Helv. Phys. Acta **61** (1988)321; O. Piquet and K. Sibold, Int. J. Mod. Phys. **A1** (1986) 913; Phys. Lett. **B177** (1986) 373.
28. S.L Adler and W.A. Bardeen, Phys. Rev. **182** (1969) 1517.
29. R.G. Leigh and M.J. Strassler, Nucl. Phys. **B447** (1995) 95; M.J. Strassler; Prog. Theor. Suppl. **123** (1996) 373.
30. J. Kubo, M. Mondragón and G. Zoupanos, Nucl. Phys. **B424** (1994) 291.
31. A. Pilaftsis, Phys. Lett. **B435** (1998) 88.
32. K. Inoue, A. Kakuto, H. Komastu and S. Takeshita, Prog. Theor. Phys. **67** (1982) 1889; **68** (1983) 927.
33. L. Hall, R. Rattazzi and U. Sarid, Phys. Rev. **D50** (1994) 7048; M. Carena, M. Olechowski, S. Pokorski and C.E.M. Wagner, Nucl. Phys. **B426** (1994) 269.
34. B.D. Wright, *Yukawa Coupling Thresholds: Application to the MSSM and the Minimal Supersymmetric SU(5) GUT*, University of Wisconsin-Madison report, MAD/PH/812, hep-ph/9404217.
35. H. Georgi, H. Quinn, S. Weinberg, Phys. Rev. Lett. **33** (1974) 451.
36. J. Kubo, M. Mondragón, N.D. Tracas and G. Zoupanos, Phys. Lett. **B342** (1995) 155;
J. Kubo, M. Mondragón, S. Shoda and G. Zoupanos, Nucl. Phys. **B469** (1996) 3.
37. See J. Kubo, M. Mondragón, M. Olechowski and G. Zoupanos, Nucl. Phys. **B479** (1996) 25.
38. O. Piguet and K. Sibold, Phys. Lett. **229B** (1989) 83.
39. I. Jack and D.R.T. Jones, Phys. Lett. **B349** (1995) 294;
I. Jack, D.R.T. Jones and K.L. Roberts, Nucl. Phys. **B455** (1995) 83;
D.I. Kazakov, M.Yu. Kalmykov, I.N. Kondrashuk and A.V. Gladyshev, Nucl. Phys. **B471** (1996) 387.
40. J. Kubo, M. Mondragón and G. Zoupanos, Phys. Lett. **B389** (1996) 523.
41. H.E. Haber, R. Hempfling and A. Hoang, Z. Phys. **C75** (1997) 539, and references therein.
42. Y. Kawamura, T. Kobayashi and J. Kubo, Phys. Lett. **B405** (1997) 64.
43. Y. Yamada, Phys.Rev. **D50** (1994) 3537;
J. Hisano and M. Shifman, Phys. Rev. **D56** (1997) 5475;
I. Jack and D.R.T. Jones, Phys. Lett. **B415** (1997) 383.;
L.V. Avdeev, D.I. Kazakov and I.N. Kondrashuk, Nucl. Phys. **B510** (1998) 289.
44. D.I. Kazakov, Phys. Lett. **B421** (1998) 211.
45. I. Jack, D.R.T. Jones and A. Pickering, Phys. Lett. **B426** (1998) 73.
46. T. Kobayashi, J. Kubo and G. Zoupanos, Phys. Lett. **B427** (1998) 291.
47. I. Jack, D.R.T. Jones and A. Pickering, Phys. Lett. **B432** (1998) 114.
48. V. Novikov, M. Shifman, A. Vainstein and V. Zakharov, Nucl. Phys. **B229** (1983) 381; Phys. Lett. **B166** (1986) 329;
M. Shifman, Int.J. Mod. Phys. **A11** (1996) 5761 and references therein.
49. A. Brignole, L.E. Ibáñez, C. Muñoz and C. Scheich, Z. f. Phys. **C74** (1997) 157;
A. Brignole, L.E. Ibáñez and C. Muñoz, *Soft supersymmetry breaking terms from supergravity and superstring models*, hep-ph/9707209.
50. L.E. Ibáñez, *A chiral $D = 4$, $N = 1$ string vacuum with a finite low-energy effective field theory*, hep-th/9802103;
New perspectives in string phenomenology from dualities, hep-ph/9804236.
51. K. Kikkawa and M. Yamasaki, Phys. Lett. **B149** (1984) 357;
N. Sakai and I. Senda, Prog. Theor. Phys. **75** (1998) 692.

52. J. Polchinski, "String Theory", Vol. I and II, Cambridge University Press (1998).
53. T. Kawamura, T. Kobayashi and J. Kubo, Phys. Lett. **B432** (1998) 108.

The Rigorous Analyticity-Unitarity Program and Its Success

André Martin

Theoretical Physics Division, CERN,
CH-1211 Geneva 23, Switzerland
and
LAPP, F-74941 Annecy le Vieux, France

Abstract. We show how the combination of analyticity properties derived from local field theory and the unitarity condition (in particular positivity) leads to non-trivial physical results, including the proof of the "Froissart bound" from first principles and the existence of absolute bounds on the pion-pion scattering amplitude.

I would like to begin by wishing a very happy birthday to Wolfhart Zimmermann. I have chosen a topic which is close to the interests of Wolfhart and as you will see soon, in which Wolfhart has made a crucial contribution which makes all the work made in the "pre-quark" era still valid now.

My task would have been much easier if the scheduled first speaker of this conference, Harry Lehmann, had been present. Unfortunately he was ill, and, at the time of writing this talk, we know that he left us. As we shall see all through what follows, the contributions of Harry Lehmann to that domain are many and all of them are fundamental.

In 1954, Gell-Mann, Goldberger and Thirring [1] proved that dispersion relation, previously developed in optics could be established for Compton Scattering: $\gamma P \to \gamma P$, from the existence of local fields satisfying the causality property

$$[A(x), A(y)] = 0 \quad \text{for} \quad (x-y)^2 < 0 \ ,$$

i.e., spacelike. This made it possible to express the real part of the forward scattering amplitude as an integral over the imaginary part of the forward scattering amplitude, i.e., by the "optical theorem", an integral over the total cross-section for Compton Scattering. At the same time a general formulation of quantum field theory incorporating causality giving in particular general expression for scattering amplitude was developed by Lehmann, Zimmermann and Symanzik (LSZ) in their pioneering paper (in german!) in Nuovo Cimento [2].

On this basis, dispersion relations were "proposed" for massive particles in the work of Goldberger on the pion-nucleon scattering amplitude [3]. Soon, his "heuristic proof" was turned into a real proof by various authors using the LSZ formalism [4]. One of these proofs is due again to Harry Lehmann!

Before going on, I would like to explain that if these results, even after the discovery that protons and pions are not elementary but made of quarks, are still valid, it is thanks to a fundamental contribution of Wolfart Zimmermann

P. Breitenlohner and D. Maison (Eds.): Proceedings 1998, LNP 558, pp. 127–135, 2000.

entitled "On the bound state problem in quantum field theory" [5], in which it is proved that to a bound state we can associate a local operator. This constitutes an excellent answer to sceptics like Volodia Gribov [6] or Klaus Hepp [7] (qui brûle ce qu'il a adoré!).

Now I believe that it is necessary to give some technical details, even if most of you know about it.

In 3+1 dimensions (3 space, 1 time) the scattering amplitude depends on two variables energy and angle. For a reaction $A + B \to A + B$

$$E_{c.m.} = \sqrt{M_A^2 + k^2} + \sqrt{M_B^2 + k^2} \ , \tag{1}$$

k being the centre-of-mass momentum. The angle is designated by θ. There are alternative variables:

$$s = (E_{CM})^2 \ , \quad t = 2k^2(\cos\theta - 1) \tag{2}$$

(Notice that physical t is NEGATIVE).

We shall need later an auxiliary variable u, defined by

$$s + t + u = 2M_A^2 + 2M_B^2 \tag{3}$$

The Scattering amplitude (scalar case) can be written as a partial wave expansion, the convergence of which will be justified in a moment:

$$F(s, \cos\theta) = \frac{\sqrt{s}}{k} \sum (2\ell + 1) f_\ell(s) P_\ell(\cos\theta) \tag{4}$$

$f_\ell(s)$ is a partial wave amplitude.

The Absorptive part, which coincides for $\cos\theta$ real (i.e., physical) with the imaginary part of F, is defined as

$$A_s(s, \cos\theta) = \frac{\sqrt{s}}{k} \sum (2\ell + 1) \ \mathrm{Im} \ f_\ell(s)(\cos\theta) \tag{5}$$

The Unitarity condition, implies, with the normalization we have chosen

$$\mathrm{Im} \ f_\ell(s) \geq |f_\ell(s)|^2 \tag{6}$$

which has, as a consequence

$$\mathrm{Im} \ f_\ell(s) > 0 \ , \quad |f_\ell| < 1 \ . \tag{7}$$

The differential cross-section is given by

$$\frac{d\sigma}{d\Omega} = \frac{1}{s} \ |F|^2 \ ,$$

and the total cross-section is given by the "optical theorem"

$$\sigma_{total} = \frac{4\pi}{k\sqrt{s}} \ A_s(s, \cos\theta = 1) \ . \tag{8}$$

With these definitions, a dispersion relation can be written as:

$$F(s,t,u) = \frac{1}{\pi}\int \frac{A_s(s',t)ds'}{s'-s} + \frac{1}{\pi}\int \frac{A_u(u',t)du'}{u'-u} \tag{9}$$

with possible subractions, i.e., for instance the replacement of $1/(s'-s)$ by $s^N/s'^N(s'-s)$ and the addition of a polynomial in s, with coefficients depending on t.

The scattering amplitude in the s channel $A+B \to A+B$ is the boundary value of F for $s+i\epsilon$, $\epsilon > 0 \to 0$, $s > (M_A+M_B)^2$. In the same way the amplitude for $A+\bar{B} \to A+\bar{B}$, $\bar{B}$ being the antiparticle of B is given by the boundary value of F for $u+i\epsilon,\ \ \epsilon \to 0\ \ u > (M_A+M_B)^2$. Here we understand the need for the auxiliary variable u.

The dispersion relation implies that, for fixed t the scattering amplitude can be continued in the s complex plane with two cuts. The scattering amplitude possesses the reality property, i.e., for t real it is real between the cuts and takes complex conjugate values above and below the cuts.

In the most favourable cases, dispersion relations have been established for $-T < t \le 0$ $T > 0$. A list of these cases have been given in 1958 by Goldberger [8] and has not been enlarged since then. It is given in the Table.

In the general case, even if dispersion relations are not proved, the crossing property of Bros, Epstein and Glaser states that the scattering amplitude is analytic in a twice cut plane, minus a finite region, for any negative t [9]. So it is possible to continue the amplitude directly from $A+B \to A+B$ to the complex conjugate of $A+\bar{B} \to A+\bar{B}$. By a more subtle argument, using a path with fixed u and fixed s it is possible to continue directly from $A+B \to A+B$ to $A+\bar{B} \to A+\bar{B}$

At this point, we see already that one cannot dissociate analyticity, i.e., dispersion relations, and unitarity, since the discontinuity in the dispersion relations is given by the absorptive part. In the simple case of $t=0$, the absorptive part is given by the total cross-section and the forward amplitude is given, as we said already for the case of Compton Scattering, by an integral over physical quantities.

It was recognized very early that the combination of analyticity and unitarity might lead to very interesting consequences and might give some hope to fulfill at least partially the S matrix Heisenberg program. This was very clearly stated already in 1956 by Murray Gell-Mann [10] at the Rochester conference. Later this idea was taken over by many people, in particular by Geff Chew. To make this program as successful as possible it seemed necessary to have an analyticity domain as large as possible. Dispersion relations are fixed t analyticity properties, in the other variable s, or u as one likes.

Another property derived from local field theory was the existence of the Lehmann ellipse [11], which states that for fixed s, physical, the scattering amplitude is analytic in $\cos\theta$ in an ellipse with foci at $\cos\theta = \pm 1$. $\cos\theta = 1$ corresponds to $t=0$ the ellipse therefore contains a circle

$$|t| < T_1(s) \tag{10}$$

Table 1. Dispersion Relations

a) Proved Relations

Process $k+p \to k'+p'$	Limitation in invariant momentum transfer	Continuation of absorptive part into the unphysical region by convergent partial wave expansion
$\pi+N \to \pi+N$	$T_{\max} = \frac{32m_\pi^2}{3} \frac{2m_p+m_\pi}{2m_p-m_\pi}$	$0 \le T < T_{\max}$
$\pi+\pi \to \pi+\pi$	$T_{\max} = 28m_\pi^2$	$0 \le T < T_{\max}$
$\gamma+N \to \gamma+N^{(*)}$	$T_{\max} = 4m_\pi^2 \left\{ \frac{(2m_p+m_\pi)^2}{4(m_p+m_\pi)^2} + \frac{2m_p+m_\pi}{m_p} \right\}$	$0 \le T < T_{\max}$
$\gamma+N \to \pi+N^{(*)}$ $e+N \to e+\pi+N^{(*)}$	$T_{\max} = 4F(0)^{(**)} \sim 12m_\pi^2$ $T_{\max} = 4F^{(**)}(\gamma)$; $\gamma \equiv k_0^2 - k^2$ $F(-9m_\pi^2) \sim 6m_\pi^2$	$T_{\text{th}} \le T < T_{\max}$ $T_{\text{th}} = \frac{m_p}{m_p+m_\pi} \times (m_\pi - \gamma)$

b) Some unproved relations

	Mass restrictions appearing in proof based upon causality and spectrum; $T=0$	Perturbation theory (every finite order)
$N+N \to N+N$	$m_\pi > (\sqrt[6]{2}-1)m_p$	proved for $T < m_\pi^2$
$K+N \to K+N$	complicated; not fulfilled by narrow margin	
$\pi+D \to \pi+D$	$\epsilon > \frac{m_\pi}{3}$; $m_D = 2m_p - \epsilon$	

In the Lehmann derivation $T_1(s) \to 0$ for $s \to (M_A+M_B)^2$ and $s \to \infty$.

The absorptive part is analytic in the larger ellipse, the "large" Lehmann ellipse, containing the circle

$$|t| < T_2(s) \tag{11}$$

with $T_2(s) \to c > 0$ for $s \to (M_A+M_B)^2$, $T_2(s) \to 0$ for $s \to \infty$.

It was thought by Mandelstam that these two analyticity properties, dispersion relations and Lehmann ellipses, were insufficient to carry very far the analyticity-unitarity program. he proposed the Mandelstam representation [12] which can be written schematically as

$$F = \frac{1}{\pi^2}\int \frac{\rho(s',t')ds'dt'}{(s'-s)\,(t'-t)}$$
$$+\text{circular permutations in } s,t,u$$
$$+\text{one dimensional dispersion integrals}$$
$$+\text{subtractions} \tag{12}$$

This representation is nice. It gives back the ordinary dispersion relations and the Lehmann ellipse when one variable is fixed, but it was never proved nor disproved for all mass cases, even in perturbation theory. One contributor, Jean Lascoux, refused to co-sign a "proof", which, in the end, turned out to be imperfect.

One very impressive consequence of Mandelstam representation was the proof, by Marcel Froissart, that the total cross-section cannot increase faster than (log $s)^2$, the so-called "Froissart Bound" [13].

My own way to obtain the Froissart bound [14] was to use the fact that the Mandelstam representation implies the existence of an ellipse of analyticity in $\cos\theta$ qualitatively larger than the Lehmann ellipse, i.e., such that it contains a circle $|t| < R$, R fixed, independent of the energy. This has a consequence that Im $f_\ell(s)$ decreases with ℓ at a certain exponential rate because of the convergence of the Legendre polynomial expansion and of the polynomial boundedness, but on the other hand the Im $f_\ell(s)$'s are bounded by unity because of unitarity [Eq. (7)]. taking the best bound for each ℓ gives the Froissart bound.

To prove the Froissart bound without using the Mandelstam representation one must find a way to enlarge the "small" and the "large" Lehmann ellipses. In the autumn of 1965, I had very stimulating discussions with Harry Lehmann at the "Institut des Hautes Etudes Scientifiques" about an attempt made in this direction by Nakanishi in which he combined in a not very consistent way positivity and some analyticity properties derived from perturbation theory. He was using a domain shrinking to zero when the energy became physical and this lead nowhere. Finally, in December 1965 [15], I found the way out. The positivity of Im f_ℓ implies, by using expansion (5),

$$\left|\left(\frac{d}{dt}\right)^n A_S(s,t)\right|_{-4k^2\le t\le 0} \le \left|\left(\frac{d}{dt}\right)^n A_S(s,t)\right|_{t=0} \tag{13}$$

To calculate

$$F(s,t) = \frac{1}{\pi}\int_{s_0} \frac{A_s(s't)ds'}{s'-s}$$

(forget the left-hand cut and subtractions!), for s real $< s_0$ one can expand $F(s,t)$ around $t = 0$. From the property (13) one can prove that the successive derivatives can be obtained by differentiating under the integral. When one resums the series one discovers that this can be done not only for s real $< s_0$, but for any s and that the expansion has a domain of convergence in t independent of s. This means that the large Lehmann ellipse must contain a circle $|t| < R$. This is exactly what is needed to get the Froissart bound. In fact, in favourable cases, $R = 4m_\pi^2$, m_π being the pion mass. A recipe to get a lower bound for R was found by Sommer [16]

$$R \geq \sup_{s_0<s<\infty} T_1(s) \tag{14}$$

It was already known that for $|t| < 4m_\pi^2$ the number of subtractions in the dispersion relations was at most two [17], and is lead to the more accurate bound [18]

$$\sigma_T < \frac{\pi}{m_\pi^2} \ (\log s)^2 \tag{15}$$

Notice that this is only a bound, not an asymptotic estimate.

In spite of many efforts the Froissart bound was never qualitatively improved, and it was shown by Kupsch [19] that if one uses only Im $f_\ell \geq |f_\ell|^2$ and full crossing symmetry one cannot do better than Froissart.

Before 1972, rising cross-sections were a pure curiosity. Almost everybody believed that the proton-proton cross-section was approaching 40 millibarns at infinite energy. Only Cheng and Wu [20] had a QED inspired model in which cross-sections were rising and behaving like $(\log s)^2$ at extremely high energy. Yet, Khuri and Kinoshita [21] took seriously very early the possibility that cross-sections rise and proved, in particular, that if the scattering amplitude is dominantly crossing even, and if $\sigma_t \sim (\log s)^2$ then

$$\rho = \frac{ReF}{ImF} \sim \frac{\pi}{\log s} ,$$

where ReF and ImF are the real and imaginary part of the forward scattering amplitude.

In 1972, it was discovered at the ISR, at CERN, that the $p-p$ cross-section was rising by 3 millibarns from 30 GeV c.m. energy to 60 GeV c.m. energy [22]. I suggested to the experimentalists that they should measure ρ and test the Khuri-Kinoshita predictions. They did it [23] and this kind of combined measurements of σ_T and ReF are still going on. In σ_T we have now more than a 50 % increase with respect to low energy values. For an up to date review I refer to the article of Matthiae [24]. it is my strong conviction that this activity should be continued with the future LHC. A breakdown of dispersion relation might be a sign of new physics due to the presence of extra compact dimensions of space according to N.N. Khuri [25]. Future experiments, especially for ρ, will be difficult because of the necessity to go to very small angles, but not impossible [26].

Before leaving the domain of high-energy scattering I would like to indicate the new version of the Pomeranchuk theorem. When it was believed that cross-sections were approaching finite limits, the Pomeranchuk theorem [27] stated that, under a certain assumption on the real part

$$\sigma_T(AB) - \sigma_T(A\bar{B}) \to 0$$

If cross-sections are rising to infinity, one can actually prove, according to Eden [28] and Kinoshita [29] that

$$\sigma_T(AB)/\sigma_T(A\bar{B}) \to 1 .$$

Now I would like to turn to another aspect of analyticity-unitarity. A consequence of the enlargment of the Lehmann ellipse is that, in the special case of $\pi\pi \to \pi\pi$ scattering, one can, by using crossing symmetry, obtain a very large analyticity domain [30], but one can prove that the domain is smaller than the Mandelstam domain [31]. By playing with crossing symmetry and unitarity in a clever way (with years enormous progress has been made according to the Figure), one gets a bound on the scattering amplitude at the "symmetry point" which is [32]

$$|F(s = t = u = 4m_\pi^2/3| < 4 ,$$

where F is normalized in such a way that $F(s = u, t = 0, u = 0)$ is the $\pi_0\pi_0$

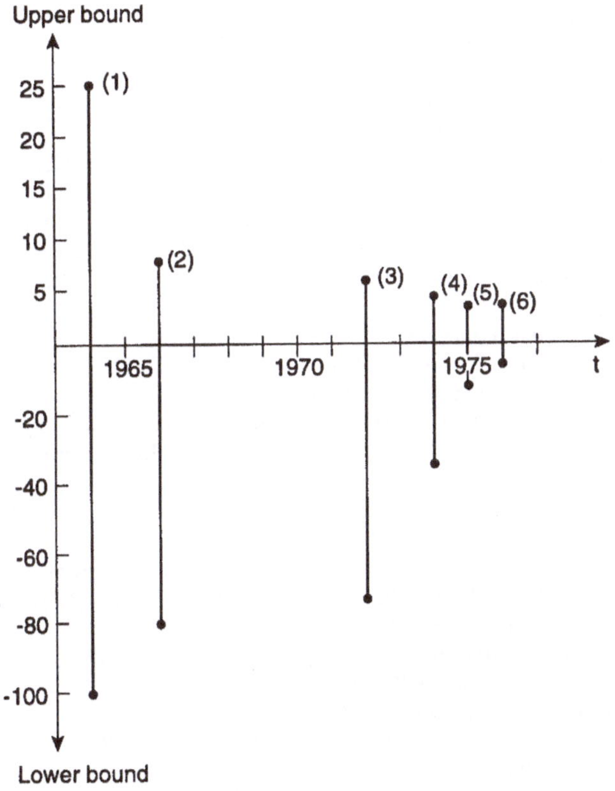

Fig. 1. Bounds on the scattering amplitude at the symmetry point $s = t = u = 4/3m_\pi^2$ as a function of time. Normalization: $F(s = 4m_\pi^2, 0, 0) =$ scattering length.

scattering length, a_{00}. One can also obtain a lower bound on the scattering length, the bound value being [33]

$$a_{00} > -1.75 \ \ (m_\pi)^{-1} ,$$

a number which is off the model predictions only by a factor 10.

Though these latter results may seem "useless", they are remarkable, since they prove that the combination of analyticity and unitarity have a dynamical content.

References

1. M. Gell-Mann, M.L. Goldberger and W. Thirring, *Phys.Rev.* **95** (1954) 1612.
2. H. Lehmann, K. Symanzik and W. Zimmermann, *Nuovo Cimento* (Serie 10) **1** (1955) 205.
3. M.L. Goldberger, *Phys.Rev.* **99** (1975) 979.
4. N.N. Bogoliubov, B.V. Medvedev and M.K. Polivanov, Voprossy Teorii Dispersionnyk Sootnoshenii, V. Shirkov et al. Eds., Moscow 1958;
   K. Symanzik, *Phys.Rev.* **105** (1957) 743;
   H. Lehmann, *Suppl. Nuovo Cimento* **14** (1959) 153.
5. W. Zimmermann, *Nuovo Cimento* **10** (1958) 597.
6. V. Gribov, private communication (1997).
7. K. Hepp, private communication, Zürich, 1996.
8. M.L. Goldberger, in *Proc. Intl. Conf. on High Energy Physics, CERN, Geneva, 1958*, B. Ferretti ed., CERN Scientific Information Service, 1958, p. 208.
9. J. Bros, H. Epstein and V. Glaser, *Commun.Math.Phys.* **1** (1965) 240.
10. M. Gell-Mann, in *Proc. 6th Annual Rochester Conf.*, J. Ballam, V.L.Fitch, T. Fulton, K. Huang, R.R. Rau and S.B. Treiman eds., Interscience Publishers, New York 1956, p. 30.
11. H. Lehmann, *Nuovo Cimento* **10** (1958) 579.
12. S. Mandelstam, *Phys.Rev.* **112** (1958) 1344.
13. M. Froissart, *Phys.Rev.* **123** (1961) 1053.
14. A. Martin, *Phys.Rev.* **129** (1963) 1432; and in *Proc. 1962 Conf. on High energy Physics at CERN*, J. Prentki ed., CERN Scientific Information Service, 1962, p. 567.
15. A. Martin, *Nuovo Cimento* **42** (1966) 901.
16. G. Sommer, *Nuovo Cimento* **A48** (1967) 92.
   In the special case of pion-nucleon scattering a special argument gives $R = 4m_\pi^2$.
   See D. Bessis and V. Glaser, *Nuovo Cimento* (Serie X) **50** (1967) 568.
17. Y.S. Jin and A. Martin, *Phys.Rev.* **B135** (1964) 1375.
18. L. Lukaszuk and A. Martin, *Nuovo Cimento* **52** (1967) 122.
19. J. Kupsch, *Nuovo Cimento* **B70** (1982) 85.
20. H. Cheng and T.T. Wu, *Phys.Rev.Lett.* **24** (1970) 1456.
21. N.N. Khuri and T. Kinoshita, *Phys.Rev.* **B137** (1965) 720.
22. U. Amaldi et al., *Phys.Lett.* **B44** (1973) 112;
   S.R. Amendolia et al., *Phys.Lett.* **B44** (1973) 119.
23. V. Bartenev et al., *Phys.Rev.Lett.* **31** (1973) 1367;
   U. Amaldi et al., *Phys.Lett.* **66B** (1977) 390.
24. G. Matthiae, *Rep.Progr.Phys.* **57** (1994) 743.
25. N.N. Khuri, Rencontres de Physique de la vallée d'Aoste, 1994, M. Greco ed., Editions Frontières 1994, p. 771;
   see also: N.N. Khuri and T.T. Wu, *Phys.Rev.* **D56** (1997) 6779 and 6785.
26. Angela Faus-Golfe, private communication.
27. Y.Ya. Pomeranchuk, *Soviet Phys. JETP* **7** (1958) 499.
28. R.J. Eden, *Phys.Rev.Lett.* **16** (1966) 39.

29. T. Kinoshita, in Perspectives in Modern Physics, R.E. Marshak ed., New York 1966), p. 211;
see also G. Grunberg and T.N. Truong, *Phys.Rev.Lett.* **B31** (1973) 63.
30. A. Martin, *Nuovo Cimento* **44** (1966) 1219.
31. A. Martin, Proceedings of the 1967 International Conference on Particles and Fields, C.!Hagen, G. Guralnik and V.A. Mathur, eds., John Wiley and Sons, New York 1967, p. 255.
32. A. Martin, Preprint, Institute of Theoretical Physics, Stanford University ITP-134 (1964), unpublished;
A. Martin in "High Energy Physics and Elementary particles", ICTP Trieste 1965, International Atomic Energy Agency Vienna (1965), p. 155;
L. Lukaszuk and A. Martin, *Nuovo Cimento* **A47** (1967) 265;
J.B. Healy, *Phys.Rev.* **D8** (1973) 1907;
G. Auberson, L. Epele, G. Mahoux and R.F.A. Simaõ, *Nucl.Phys.* **B94** (1975) 311;
C. Lopez and G. Mennessier, *Phys.Lett.* **B58** (1975) 437;
B. Bonnier, C. Lopez and G. Mennessier, *Phys.Lett.* **B60** (1975) 63;
C. Lopez and G. Mennessier, *Nucl.Phys.* **B118** (1977) 426.
33. I. Caprini and P. Dita, Preprint, Institute of Physics and Engineering, P.O. Box 5206, Bucharest (1978), unpublished.
The initial work on this lower bound was:
B. Bonnier and R. Vinh Mau, *Phys.Rev.* **165** (1968) 1923.

Reduction of Coupling Parameters and Duality

Reinhard Oehme

Enrico Fermi Institute and Department of Physics,
University of Chicago, Chicago, Illinois, 60637, USA **
and
Max-Planck-Institut für Physik, Werner-Heisenberg-Institut,
Föhringer Ring 6, D-80805 München, Germany

Abstract. The general method of the reduction in the number of coupling parameters is discussed. Using renormalization group invariance, theories with several independent couplings are related to a set of theories with a single coupling parameter. The reduced theories may have particular symmetries, or they may not be related to any known symmetry. The method is more general than the imposition of invariance properties. Usually, there are only a few reduced theories with an asymptotic power series expansion corresponding to a renormalizable Lagrangian. There also exist 'general' solutions containing non-integer powers and sometimes logarithmic factors. As an example for the use of the reduction method, the dual magnetic theories associated with certain supersymmetric gauge theories are discussed. They have a superpotential with a Yukawa coupling parameter. This parameter is expressed as a function of the gauge coupling. Given some standard conditions, a unique, isolated power series solution of the reduction equations is obtained. After reparametrization, the Yukawa coupling is proportional to the square of the gauge coupling parameter. The coefficient is given explicitly in terms of the numbers of colors and flavors. 'General' solutions with non-integer powers are also discussed. A brief list is given of other applications of the reduction method.

1 Introduction

The method of reduction in the number of coupling parameters [1–5], [6–10], [11] has found many theoretical and phenomenological applications. It is a very general method, based essentially upon the requirement of renormalization group invariance of the original multi-parameter theory, as well as the related reduced theories with fewer couplings. Combining the renormalization group equations of original and reduced theories, we obtain a set of *reduction equations*. These are differential equations for the removed couplings considered as functions of the remaining parameters. They are necessary and sufficient for the independence of the reduced theories from the normalization mass. We consider massless theories, or mass independent renormalization schemes, so that no mass parameters appear in the coefficient functions of the renormalization group equations. This can be arranged, provided the original coefficient functions have a well defined zero-mass limit [12].

** Permanent Address

P. Breitenlohner and D. Maison (Eds.): Proceedings 1998, LNP 558, pp. 136–156, 2000.

In this paper, we discuss only reductions to a single coupling, which covers most cases of interest. Usually, we can choose one of the original couplings as the remaining parameter. The multi- parameter theory is assumed to be renormalizable with an asymptotic power series expansion in the weak coupling limit. However the reduced theories, as obtained from the reduction equations, may well not all have such expansions in the remaining coupling. Non- integer powers and logarithms can appear, often with undetermined coefficients. Such *general solutions* do not correspond to conventional renomalized power series expansions associated with a Lagrangian. But they are still well defined in view of their embedding in the renormalized multi-parameter theory. Nevertheless, it is the relatively small number of uniquely determined power series solutions of the reduction equations, which is of primary interest. Depending upon the character of the system considered, there may be additional requirements which further reduce the number of these solutions. Although we consider renormalizable theories, with appropriate assumptions, the reduction method can also be applied in cases where the original theory is non-renormalizable.

Regular reparametrization is a very useful tool in connection with the reduction method. For theories with two or more coupling parameters, it is not possible to reduce the β-function expansions to polynomials. However, in the reductions to one coupling, we can usually remove all but the first term in the power series solutions of the reduction equation with determined coefficients. The β-functions of the corresponding reduced theories remain however infinite series. As seen from many examples, these reparametrizations lead to frames which are very natural for the reduced theories.

The imposition of a symmetry on the multi-parameter theory is a conventional way of relating the coupling parameters. If there appear no anomalies, we get a renormalizable theory with fewer parameters so as to implement the symmetry. These situations are all included in the reduction scheme, but our method is more general, leading also to unique power series solutions which exhibit no particular symmetry. This situation is illustrated by an example we have included. An $SU(2)$ gauge theory with matter fields in the adjoint representation. Besides the gauge coupling, there are three additional couplings. With only the gauge coupling remaining after the reduction, we get two acceptable power solutions. One of the reduced theories is an $N = 2$ supersymmetric gauge theory, while the other solution leads to a theory with no particular symmetry.

The main example presented in this article is connected with *duality* [13–16]. We consider $N = 1$ supersymmetric QCD (SQCD) and the corresponding dual theory, magnetic SQCD. The primary interest is in the phase structure of the physical system described by these theories. Essential aspects of this phase structure were first obtained on the basis of supercovergence relations and BRST methods [17–19], and more recently with the help of duality. [13,15]. We exhibit the quantitative agreement of both approaches [20,21]. While duality is formulated only in connection with supersymmetry, the supercovergence arguments can be used also for QCD and similar theories [18,19], [22]. Of particular interest is the transition point at $N_F = \frac{3}{2}N_C$ for SQCD [17,15], where N_F and

N_C are the numbers of flavors and colors respectively. It is the lower end of the conformal window. For smaller values of N_F, the quanta of free, electric SQCD are confined, the system is described by free magnetic excitations of the dual theory (for $N_C > 4$), and eventually by mesons and baryons. (The corresponding transition point for QCD is given by $N_F = \frac{13}{4} N_C$).

As the original theory, SQCD has only the gauge coupling g_e. The dual theory is constructed on the basis of the anomaly matching conditions [13,15]. It involves the two coupling parameters g_m and λ_1, where λ_1 is a Yukawa coupling associated with a superpotential. This potential is required by duality, mainly since theories, which are dual to each other, must have the same global symmetries.

At first, we apply the reduction method to the magnetic theory in the conformal window $\frac{2}{3} N_C < N_F < 3N_C$ [21,23]. We find two power series solutions. After reparametrization, one solution is given by $\lambda_1(g_m^2) = g_m^2 f(N_C, N_F)$, with f being a known function of the numbers of colors and flavors for SQCD. The other solution is $\lambda_1(g_m^2) \equiv 0$. Since the latter removes the superpotential, it is excluded, and we are left with a unique single power solution. This solution implies a theory with a single gauge coupling g_m, and renormalized perturbation expansions which are power series in g_m^2. It is the appropriate dual of SQCD. There are 'general' solutions, but they all approach the excluded power solution $\lambda_1(g^2) \equiv 0$. With one exception, they involve non-integer powers of g_m^2. The reduction can be extended to the 'free electric region' $N_F > 3N_C$, and to the 'free magnetic region' $N_C + 2 < N_F < \frac{2}{3} N_C$, $(N_C > 4)$. The results are similar, and discussed in detail in [23]. In the free magnetic case, we deal however with the approach to a trivial infrared fixed-point.
Possible connections of the reduction results with features of brane dynamics remain to be considered. Internal fluctuations of branes may be of relevance for the field theory properties obtained here.

2 Reduction Equations

We consider renormalizable quantum field theories with several coupling parameters. It is assumed that there is a mass-independent renormalization scheme, so that no mass parameters occur in the coefficient functions of the renormalization group equations. Let $\lambda, \lambda_1, \ldots, \lambda_n$ be $n+1$ dimensionless coupling parameters of the theory. One can reduce this system in various ways, but we want to consider the parameter λ as the primary coupling, and express the remaining n couplings as functions of λ:

$$\lambda_k = \lambda_k(\lambda), \quad k = 1, \ldots, n \ . \tag{1}$$

It is assumed, that these functions $\lambda_k(\lambda)$ are independent of the renormalization mass κ, which can always be arranged.

The Green's functions $G\left(k_i, \kappa^2, \lambda, \lambda_1, \ldots, \lambda_n\right)$ of the original multi-parameter version of the theory satisfy the usual renormalization group equations with the coefficient functions β, β_k, and the anomalous dimension γ_G, which depend upon the $n+1$ coupling parameters. The corresponding Green's functions of the reduced theory are given by

$$G(k_i, \kappa^2, \lambda) = G\left(k_i, \kappa^2, \lambda, \lambda_1(\lambda), \ldots, \lambda_n(\lambda)\right) . \tag{2}$$

Renormalization group invariance requires that they satisfy the equation

$$\left(\kappa^2 \frac{\partial}{\partial \kappa^2} + \beta(\lambda)\frac{\partial}{\partial \lambda} + \gamma_G(\lambda)\right) G(k_i, \kappa_2, \lambda) \; = \; 0 \;, \tag{3}$$

where $\beta(\lambda)$ and $\gamma_G(\lambda)$ are given by the corresponding original coefficients with the insertions $\lambda_k = \lambda_k(\lambda), \quad k = 1, \ldots, n$. Comparison of Eq.(3) with the original multi-parameter renormalization group equation implies then

$$\beta(\lambda)\frac{d\lambda_k(\lambda)}{d\lambda} = \beta_k(\lambda) \;, \quad k = 1, \ldots, n \tag{4}$$

These are the *Reduction Equations*, which are necessary and sufficient for the validity of Eq.(3).

It is of interest to briefly consider the relationship between the reduction method as described above, and the equations for the effective coupling functions $\overline{\lambda}(u), \overline{\lambda}_k(u)$, where u is the dimensionless scaling parameter $u = k^2/\kappa^2$. These functions satisfy the equations

$$\begin{aligned} u\frac{d\overline{\lambda}}{du} &= \beta(\overline{\lambda}, \overline{\lambda}_1, \ldots, \overline{\lambda}_n) \;, \\ u\frac{d\overline{\lambda}_k}{du} &= \beta_k(\overline{\lambda}, \overline{\lambda}_1, \ldots, \overline{\lambda}_n) \;. \end{aligned} \tag{5}$$

With $\overline{\lambda}(u)$ being an analytic function, we can choose a point where $(d\overline{\lambda}(u)/du)$ $\neq 0$ and introduce $\overline{\lambda}(u)$ as a new variable Eqs.(5,6). The result is again the reduction equations (4).

With effective couplings, we study the multi-parameter theory at different mass scales. In the reduction method, we consider the set of different field theories with one coupling parameter (or a reduced number), which can be obtained from a given multi-parameter theory as solutions of the reduction equations. The elements of this set are labeled by the free parameters of the solution, and all are considered at the same fixed mass scale. With some natural assumptions the number of theories in this set is usually smaller than the number of original coupling parameters, and the different theories have characteristic physical and mathematical features. This is best seen in examples, some of which we discuss below. It must be remembered, that the origin of the coupling parameter space is a singular point, so that the Picard-Lindeloef theorem about the uniqueness of solutions at regular points does not apply.

As described so far, the reduction scheme is very general, but in practice we usually know the β-functions only as asymptotic expansions in the small coupling limit. Within the framework of renormalized perturbation theory, we restrict ourselves here to expansions of the form

$$\begin{aligned}\beta(\lambda,\lambda_1,\ldots,\lambda_n) &= \beta_0\lambda^2 + (\ \beta_1\lambda^3 + \beta_{1k}\lambda_k\lambda^2 + \beta_{1kk'}\lambda_k\lambda_{k'}\lambda\ ) \\ &\quad + \sum_{n=4}^{\infty}\sum_{m=0}^{n-1} \beta_{n-2,k_1,\ldots,k_m}\lambda_{k_1}\cdots\lambda_{k_m}\lambda^{n-m}\ , \\ \beta_k(\lambda,\lambda_1,\ldots,\lambda_n) &= (c_k^{(0)}\lambda^2 + c_{k,k'}^{(0)}\lambda_{k'}\lambda + c_{k,k'k''}^{(0)}\lambda_{k'}\lambda_{k''}\ )\ , \\ &\quad + \sum_{n=3}^{\infty}\sum_{m=0}^{n} c_{k,k_1,\ldots,k_m}^{(n-2)}\lambda_{k_1}\cdots\lambda_{k_m}\lambda^{n-m}\ . \end{aligned} \tag{6}$$

In writing the expansions (6), we have assumed that the primary coupling λ is chosen such that $\beta(0,\lambda_1,\ldots,\lambda_n) = 0$.

With the original β-functions given as asymptotic power series expansions, we will consider in the following solutions $\lambda_k(\lambda)$ of the reduction equations, which are also of the form of asymptotic expansions. Of special interest are solutions which are power series expansions. But in general, non-integer powers as well as logarithmic terms are possible.

3 Power Series Solutions

Let us first consider solutions of the reduction equations (4) which are asymptotic power series expansions. Then the Green's functions $G(k_i,\kappa^2;\lambda)$ of the reduced theory have power series expansions in λ and are associated with a corresponding renormalizable Lagrangian. It is reasonable to write

$$\lambda_k(\lambda) = \lambda f_k(\lambda), \quad k = 1,\ldots,n\ , \tag{7}$$

where the functions $f_k(\lambda)$ are bounded for $\lambda \to 0$ so that $\lambda_k(0) = 0$. According to the reduction equations, if we had $\lambda_k(0) \neq 0$, the vanishing of $\beta(0,\lambda_1(0),\ldots,\lambda_n(0))$ would imply that also $\beta_k(0,\lambda_1(0),\ldots,\lambda_n(0))$ vanishes, which is too strong a restriction and not fulfilled by Eq.(6). In terms of the functions $f_k(\lambda)$ the reduction equations are of the form

$$\beta\left(\lambda\frac{df_k}{d\lambda} + f_k\right) = \beta_k\ , \tag{8}$$

where we have introduced the β-functions

$$\beta(\lambda) = \beta(\lambda,\lambda f_1,\ldots,\lambda f_n) = \sum_{n=0}^{\infty} \beta_n(f)\lambda^{n+2}\ , \tag{9}$$

$$\beta_k(\lambda) = \beta_k(\lambda,\lambda f_1,\ldots,\lambda f_n) = \sum_{n=0}^{\infty} \beta_k^{(n)}(f)\lambda^{n+2}\ . \tag{10}$$

Here the argument f stands for $f_1(\lambda),\ldots,f_n(\lambda)$. The coefficients are easily obtained from Eqs.(6). For example, the one-loop terms are given by

$$\beta_0(f) = \beta_0, \quad \beta_k^{(0)}(f) = c_k^{(0)} + c_{kk'}^{(0)}f_{k'} + c_{kk'k''}^{(0)}f_{k'}f_{k''}\ . \tag{11}$$

For the functions $f_k(\lambda)$, we write the expansions

$$f_k(\lambda) = f_k^0 + \sum_{m=1}^{\infty} \chi_k^{(m)} \lambda^m , \tag{12}$$

and insert them, together with the series (9) and (10), into the reduction equations. At the one-loop level, there result then the relations

$$\beta_k(f^0) - f_k^0 \beta_0) = 0 , \tag{13}$$

or in explicit form using Eq.(11),

$$c_k^{(0)} + (c_{kk'}^{(0)} - \beta_0 \delta_{kk'}) f_{k'}^0 + c_{kk'k''}^{(0)} f_{k'}^0 f_{k''}^0 = 0 . \tag{14}$$

These are the fundamental formulae for the reduction.

Given a solution f_k^0 of the quadratic equations (14), we obtain for the expansion coefficients $\chi_k^{(m)}$ the relations

$$\left(M_{kk'}(f^0) - m\beta_0 \delta_{kk'}\right) \chi_{k'}^{(m)} = \left(\beta_m(f^0) f_k^0 - \beta_k^{(m)}(f^0)\right) + X_k^{(m)} , \tag{15}$$

where $m = 1, 2, \ldots, \quad k = 1, \ldots, n$. The matrix $M(f^0)$ is given by

$$M_{kk'}(f^0) = c_{k,k'}^{(0)} + 2c_{k,k'k''}^{(0)} f_{k''}^0 - \delta_{kk'} \beta_0 . \tag{16}$$

The rest term $X^{(m)}$ depends only upon the coefficients $\chi^{(1)}, \ldots, \chi^{(m-1)}$, and upon the β–function coefficients in (9) and (10), evaluated at $f_k = f_k^0$, for order $m-1$ and lower. They vanish for $\chi^{(1)} = \ldots = \chi^{(m-1)} = 0$.

We see that the *one–loop* criteria

$$det\left(M_{kk'}(f^0) - m\beta_0 \delta_{kk'}\right) \neq 0 \;\; for \;\; m = 1, 2, \ldots \tag{17}$$

are sufficient to insure that all coefficients $\chi^{(m)}$ in the expansion (12) are determined. Then the reduced theory has a renormalized power series expansion in λ. All possible solutions of this kind are determined by the one–loop equation (13) for f_k^0.

With the coefficients $\chi^{(m)}$ fixed, we can use regular *reparametrization transformations* in order to remove all but the first term in the expansion (12) of the functions $f_k(\lambda)$. These reparametrization transformations are of the form

$$\begin{aligned} \lambda' &= \lambda'(\lambda, \lambda_1, \ldots, \lambda_n) = \lambda + a^{(20)}\lambda^2 + a_k^{(11)} \lambda_k \lambda + \cdots \;, \\ \lambda_k' &= \lambda_k'(\lambda, \lambda_1, \ldots, \lambda_n) = \lambda_k + b_{kk'k''}^{(20)} \lambda_{k'} \lambda_{k''} + b_{kk'}^{(11)} \lambda_{k'} \lambda + \cdots . \end{aligned} \tag{18}$$

They leave invariant the one-loop quantities

$$f_k^0, \; \beta_0(f^0), \; \beta_k^{(0)}(f^0), \; M_{kk'}(f^0) . \tag{19}$$

Given the condition (17), we then have a frame where

$$\lambda_k(\lambda) = \lambda f_k^0 . \tag{20}$$

This result is valid to all orders of the asymptotic expansion and determined by one-loop information. With the expressions (20), the β-function expansions (9) and (10) of the reduced theory have constant coefficients $\beta_m(f^0)$, $\beta_k^{(m)}(f^0)$, but they are generally not polynomials. They satisfy the relations

$$\beta_k^{(m)}(f^0) - f_k^0 \beta_m(f^0) = 0. \tag{21}$$

for all values of m. Only the relations for $m = 0$ are reparametrization invariant. They are the fundamental formulae (13).

So far, we have implicitly assumed that $f_k^0 \neq 0$. But it is straightforward to include the cases where $f_k^0 = 0$. They are of particular interest for supersymmetric theories. Suppose we have a solution of the reduction equations with the asymptotic expansion

$$f_k(\lambda) = \chi_k^{(N)} \lambda^N + \sum_{m=N+1}^{\infty} \chi_k^{(m)} \lambda^m , \tag{22}$$

where $N \geq 1$ and $\chi_k^{(N)} \neq 1$. Then coefficients appearing in this equation are again determined except for the first one, which is invariant. Hence, using regular reparametrization, there is a frame where

$$f_k(\lambda) = \chi_k^{(N)} \lambda^N. \tag{23}$$

We have considered here only expansions at the origin in the space of coupling parameters. However, one can use the method also in connection with any non-trivial fixed point of the theory.

4 General Solutions

At first, let us briefly consider the case where the determinant appearing in Eq.(17) vanishes. Suppose there is a positive eigenvalue of the matrix $\beta_0^{-1} M(f^0)$ for some $m = N \leq 1$, $\beta_0 \neq 0$. Then the asymptotic power series must be supplemented by terms of the form $\lambda^m (lg\lambda)^p$, with $m \leq N$ and $1 < p < \sigma(N)$. After reparametrization, we obtain then an expansion of the form

$$f_k(\lambda) = f_k^0 + \chi_k^{(N,1)} \lambda^N \lg \lambda + \chi_k^{(N)} \lambda^N + \dots , \tag{24}$$

All parameters in Eq.(24) are determined except the vector $\chi_k^{(N)}$, which contains as many free parameters as the degeneracy of the eigenvalue. Even though the theory considered here can have logarithmic terms in the asymptotic expansion, it is 'renormalized' in view of it's embedding into the original, renormalized multi-parameter theory. In special cases it may happen that the coefficients of the logarithmic terms vanish, as in the example of the massless Wess-Zumino model.

We now return to systems with non-vanishing determinant for all values of m. In addition to the power series solutions described before, there can be *general* solutions of the reduction equations, which approach the latter asymptotically. In order to describe a characteristic case, we assume that $\beta_0 \neq 0$ and that the matrix $\beta_0^{-1} M(f^0)$ has one positive eigenvalue η which is non-integer, with all others being negative. Then the reduction equations (4) have solutions of the form

$$f_k(\lambda) = f_k^0 + \sum_{a,b} \chi_k^{(a\eta+b)} \lambda^{a\eta+b} + \sum_m \chi_k^{(m)} \lambda^m \tag{25}$$

with $a = 1, 2, \ldots$, $b = 0, 1, \ldots$, $a\eta + b =$ non-integer. After reparametrization, powers with $m < \eta$ are removed, and we have

$$f_k(\lambda) = f_k^0 + \chi_k^{(\eta)} \lambda^\eta + \ldots . \tag{26}$$

In this expansion all coefficients are determined except $\chi_k^{(\eta)}$, which may contain up to r arbitrary parameters if the eigenvalue η is r-fold degenerate:

$$\chi_k^{(\eta)} = C_1 \xi_k^{(1)} + \ldots + C_r \xi_k^{(r)} \ , \tag{27}$$

where the $\xi_k^{(i)}$ are the eigenvectors.

The results described above can be generalized to situations with several positive, non-integer eigenvalues. In special cases, where the matrix also has a zero eigenvalue, logarithmic factors may appear.

So far, we have assumed that $\beta_0 \neq 0$, and obtained general solutions which approach the power series solution (20) asymptotically with a power law as indicated in Eq.(26). The situation is quite different if $\beta_0 = 0$. Then th Matrix M is given by

$$M_{kk'}(f^0) = \left(\frac{\partial \beta_k^{(0)}(f)}{\partial f_{k'}} \right)_0 \tag{28}$$

and we find that the general solutions and the power series solutions differ asymptotically by terms which vanish exponentially. We refer to [2] for more details.

Besides the general solutions, which approach the power series solutions asymptotically, there can be others which move away in the limit $\lambda \to 0$. These are not calculable unless the β-functions are known more explicitly. However, we can get information about the existence or non-existence of such solutions on the basis of the linear part of the reduction differential equations (4). We find that the theorems of Lyapunov [24] , with generalizations by Malkin [25], are applicable here [26]. We refer to [5] for some more discussion, and to [27] for an application. Generally, it turns out that a power series solution (20) is asymptotically stable if there are no negative eigenvalues of the matrix $\beta_0^{-1} M(f^0)$ (or the matrix $\beta_N^{-1} M(f^0)$ in the case of the solution (23)). A solution is unstable if there is at least one negative eigenvalue.

5 Gauge Theory

It should be most helpful to discuss briefly an example. We use a gauge theory with one Dirac field, one scalar and one pseudoscalar field, all in the adjoint representation of SU(2) [4]. Besides the usual gauge couplings, the direct interaction part of the Lagrangian is given by

$$\mathcal{L}_{dir.int.} = i\sqrt{\lambda_1}\,\epsilon^{abc}\overline{\psi}^a(A^b + i\gamma_5 B^b)\psi^c \\ - \frac{1}{4}\lambda_2(A^aA^a + B^aB^a)^2 + \frac{1}{4}\lambda_3(A^aA^b + B^aB^b)^2 \; . \tag{29}$$

Writing $\lambda = g^2$, where g is the gauge coupling, and $\lambda_k = \lambda f_k$, with k=1,2,3 , the one-loop β-function coefficients of this theory are given by

$$\begin{aligned} (16\pi^2)\beta_{g0} &= -4 \\ (16\pi^2)\beta_1^0 &= 8f_1^2 - 12f_1 \\ (16\pi^2)\beta_2^0 &= 3f_3^2 - 12f_3f_2 + 14f_2^2 + 8f_1f_2 - 8f_1^2 - 12f_2 + 3 \\ (16\pi^2)\beta_3^0 &= -9f_3^2 + 12f_3f_2 + 8f_3f_1 - 12f_3 - 3. \end{aligned} \tag{30}$$

The algebraic reduction equations (14) have four real solutions, which are given by

$$\begin{aligned} &f_1^0 = 1, \quad f_2^0 = 1, \quad f_3^0 = 1 \\ &f_1^0 = 1, \quad f_2^0 = \frac{9}{\sqrt{105}}, \quad f_3^0 = \frac{7}{\sqrt{105}}, \end{aligned} \tag{31}$$

and two others with reversed signs of f_2^0 and f_3^0, so that the classical potential approaches negative infinity with increasing magnitude of the scalar fields. These latter solutions will not be considered further. We note that the Yukawa coupling is required for the consistency of the reduction.

The eigenvalues of the matrix $\beta_{g0}^{-1}M(f^0)$ are respectively

$$\left(-2, -3, +\frac{1}{2}\right) \tag{32}$$

and

$$\left(-2, -\frac{3}{4}\,\frac{25+\sqrt{343}}{\sqrt{105}}, -\frac{3}{4}\,\frac{25-\sqrt{343}}{\sqrt{105}}\right) = (-2, -3.189\ldots, -0.470\ldots) \, . \tag{33}$$

There are no positive integers appearing in the equations (32) or (33). Hence the coefficients of the power series solutions are determined and can be removed by reparametrization, except for the invariant first term. With $\lambda = g^2$ as the primary coupling, g being the gauge coupling, these solutions are

$$(a) \quad \lambda_1 = \lambda_2 = \lambda_3 = g^2 \; , \tag{34}$$

which corresponds to an $N = 2$ extended SUSY Yang-Mills theory, and

$$(b)\quad \lambda_1 = g^2,\ \ \lambda_2 = \frac{9}{\sqrt{105}}\, g^2,\ \ \lambda_3 = \frac{7}{\sqrt{105}}\, g^2\ , \tag{35}$$

which is not associated with any known symmetry, at least in four dimensions. Both theories are 'minimally' coupled gauge theories with matter fields. The eigenvalues of the matrix $\beta_{g0}^{-1} M(f^0)$, given in Eqs.(32),(33), are all negative with the exception of the third one for the N=2 supersymmetric theory. In this case we have a general solution corresponding to Eq.(26) with $\eta = +\frac{1}{2}$, and with the coefficient given by $\chi^{(\frac{1}{2})} = (0, C, 3C)$, where C is an arbitrary parameter. The theory with $C \neq 0$ corresponds to one with hard breaking of SUSY. It has an asymptotic power series in g and not in g^2, as is the case for the invariant theory.

As we see from Eqs.(32) and (33), both power series solutions have some negative eigenvalues of the matrix $\beta_{g0}^{-1} M(f^0)$, and are therefore unstable. Not all nearby solutions approach them asymptotically.

From the present example, and many others, we realize that the special frame, where the power series solutions of the reduction equations are of the simple form (20), is a natural frame as far as the reduced one-parameter theories are concerned. The β- functions of the reduced theories are still power series and are not reduced to polynomials.

6 Dual SQCD

As the main application of the reduction method, we consider here the reduction of multi-parameter theories appearing in connection with duality. As a particular example, we discuss the dual magnetic theory associated with SQCD [13,16]. While SQCD, as the 'electric' theory, has the gauge coupling g_e as the only coupling parameter, the dual 'magnetic' theory has two parameters: the magnetic gauge coupling g_m and a Yukawa coupling λ_1, which measures the strength of the interaction of color-singlet superfields with the magnetic quark superfields. It is our aim to discuss the reduced theories where the Yukawa coupling is expressed in terms of the gauge coupling.

For SQCD the gauge group is $SU(N_C)$ with N=1 supersymmetry. There are N_F quark superfields Q_i and their antifields $\tilde{Q}^i$, $i = 1, 2, \ldots, N_F$ in the fundamental representation. For completeness and later reference, we give here the β-function coefficients for the electric SQCD theory:

$$\beta_e(g_e^2) \;=\; \beta_{e0}\, g_e^4 \;+\; \beta_{e1}\, g_e^6 \;+\; \cdots, \tag{36}$$

with

$$\begin{aligned} \beta_{e0} \;&=\; (16\pi^2)^{-1}(-3N_C \;+\; N_F) \\ \beta_{e1} \;&=\; (16\pi^2)^{-2}\left(2N_C(-3N_C + N_F) \;+\; 4N_F\frac{N_C^2 - 1}{2N_C}\right)\ . \end{aligned} \tag{37}$$

The corresponding dual magnetic theory is constructed mainly on the basis of the anomaly matching conditions [13,15,28]. It involves the gauge group $G^d = SU(N_C^d)$ with $N_C^d = N_F - N_C$. Here N_F is the number of quark superfields q_i, $\tilde{q}^i$, $i = 1, 2, \ldots, N_F$ in the fundamental representation of G^d. Because both theories must have the same global symmetries, the number of flavors N_F should be the same for SQCD and it's dual. As we have mentioned, duality requires a non-vanishing Yukawa coupling in the form of a superpotential

$$W = \sqrt{\lambda_1} M_j^i q_i \tilde{q}^j . \tag{38}$$

The N_F^2 gauge singlet superfields M_j^i are independent and cannot be constructed from q and $\tilde{q}$. The superpotential not only provides for the coupling of the M superfield, but also removes a global $U(1)$ symmetry acting on M, which would have no counterpart in the electric theory.

In the following, we will be dealing essentially only with the magnetic theory. For convenience, we therefore write g in place of g_m for the corresponding gauge coupling. We also omit the subscript m for the β-function coefficients. Then the β-function expansions of the magnetic theory are

$$\begin{aligned} \beta(g^2, \lambda_1) &= \beta_0\, g^4 + (\beta_1\, g^6 + \beta_{11}\, g^4 \lambda_1) + \cdots \\ \beta_1(g^2, \lambda_1) &= c_1^{(0)} g^2 \lambda_1 + c_{11}^{(0)} \lambda_1^2 + \cdots . \end{aligned} \tag{39}$$

The coefficients are given by [21,23,29], [30]

$$\begin{aligned} \beta_0 &= (16\pi^2)^{-1}(3N_C - 2N_F) \\ \beta_1 &= (16\pi^2)^{-2}\left(2(N_F - N_C)(3N_C - 2N_F) + 4N_F \frac{(N_F - N_C)^2 - 1}{2(N_F - N_C)}\right) \\ \beta_{11} &= (16\pi^2)^{-2}\left(-2N_F^2\right) \\ c_1^{(0)} &= (16\pi^2)^{-1}\left(-4\frac{(N_F - N_C)^2 - 1}{2(N_F - N_C)}\right) \\ c_{11}^{(0)} &= (16\pi^2)^{-1}\left(3N_F - N_C\right)) . \end{aligned} \tag{40}$$

Already at the one-loop level, we see some important features from Eqs.(37), (40). In the interval

$$\frac{3}{2} N_C < N_F < 3N_C, \tag{41}$$

both theories are asymptotically free at large momenta, in particular the magnetic theory for $N_F > \frac{3}{2} N_C$. For $N_F > 3N_C$, the electric theory is not asymptotically free in the UV but in the IR, where the magnetic version remains strongly coupled. Hence we expect that the original electric excitations are present in the 'physical' state space. The situation is reversed for $N_F < \frac{3}{2} N_C$, where the electric quanta are confined, and the elementary magnetic excitations describe the system, at least for $N_F > N_C + 2$ where the dual theory exists which is the 'free magnetic region'. This is the duality picture as proposed by Seiberg, with both theories describing the same physical system.

In the *conformal window* given in Eq.(41), the electric as well as the magnetic theory are in an interacting non-Abelian Coulomb phase, and it is indicated that they both have non-trivial conformal fixed points at zeroes of the exact β-functions. At these fixed points the theories are actually equivalent. Near an endpoint of the window, in the infrared limit, one theory may be in a weak coupling situation, and the other, dual theory in a strong coupling regime. Since both theories represent the same system, we can describe the strongly coupled field theory by the weakly coupled dual. The free excitations of the latter may be considered as composites of those of the former theory.

Within the framework of this duality picture, the system undergoes an important phase transition at the point $N_F = \frac{3}{2}N_C$. As has already been mentioned above, below this point the elementary electric quanta are confined in the sense that they are not elements of the physical state space. In the original electric theory, the transition at $N_F = \frac{3}{2}N_C$ is not apparent from the β-function coefficients, in contrast to the phase change at $N_F = 3N_C$, where $\beta_{e0} = 0$. But in the duality picture, we have $\beta_0 = 0$ at $N_F = \frac{3}{2}N_C$ for the magnetic theory, and this is the indication for the phase transition of the system.

Many years ago, we have obtained the phase transition of SQCD at $N_F = \frac{3}{2}N_C$ by using a rather different method [17]. It involves analyticity and superconvergence of the gauge field propagator, as well as the BRST-cohomology in order to define the physical state space of the theory [18,21]. The superconvergence relations, where they exist, are exact. They connect long and short distance information, and are not valid in perturbation theory [31–33].

The asymptotic form of the gauge field propagator is governed by the ratio γ_{00}/β_0, where γ_{00} is the anomalous dimension of the gauge field (not the superfield) at the fixed point $\alpha = 0$. Here $\alpha \geq 0$ is the conventional gauge parameter. Because this parameter is effectively a function of the momentum scale, it tends to a fixed point asymptotically. For example, in general covariant gauges, the discontinuity of the structure function has the asymptotic form

$$-k^2\rho(k^2,\kappa^2,g,\alpha) \;\simeq\; C(g^2,\alpha)\left(-\beta_0\, ln\frac{k^2}{\kappa^2}\right)^{-\gamma_{00}/\beta_0}, \tag{42}$$

which is independent of α with the possible exception of the coefficient.

For the discussion of confinement using the BRST cohomology or the quark-antiquark potential, it is most convenient to work in the Landau gauge, where the superconvergence relation is of the form

$$\int_{-0}^{\infty} dk^2\rho(k^2,\kappa^2,g,0) \;=\; 0, \tag{43}$$

provided $\gamma_{00}/\beta_0 > 0$. For general gauges $\alpha \geq 0$, the relation is the same except that the right hand side is given by α/α_0, were $\alpha_0 = -\frac{\gamma_{00}}{\gamma_{01}}$, with $\gamma_0(\alpha) = \gamma_{00} + \alpha\gamma_{01}$ [33].

In contrast to the duality arguments, the superconvergence method is applicable to non-supersymmetric theories like QCD, where the interval corresponding to the window (53) for SQCD is given by

$$\frac{13}{4}N_C < N_F < \frac{22}{4}N_C, \tag{44}$$

For $N_F < \frac{13}{4}N_C$ for QCD and for $N_F < \frac{3}{2}N_C$ for SQCD, our arguments show that the transverse gauge field excitations are not elements of the physical state space and hence confined. With some further arguments one can extend this result to quark fields.

For SQCD and similar theories, the connection between duality and superconvergence results is *quantitative*. For electric and magnetic SQCD, we have the anomalous dimensions

$$\begin{aligned}\gamma_{e00} &= (16\pi^2)^{-1}(-\frac{3}{2}N_C + N_F)\\ \gamma_{m00} &= (16\pi^2)^{-1}\frac{-1}{2}(-3N_C + N_F)\ ,\end{aligned} \tag{45}$$

and with the β-function coefficients from Eqs.(37),(40), we obtain the relations [20,21]

$$\begin{aligned}\beta_{m0}(N_F) &= -2\gamma_{e00}(N_F)\\ \beta_{e0}(N_F) &= -2\gamma_{m00}(N_F)\ ,\end{aligned} \tag{46}$$

where the argument N_F on both sides refers to matter fields with different quantum numbers corresponding to electric and magnetic gauge groups. We have restored the subscript m for these duality relations. We see that $\gamma_{e00}(N_F)$ changes sign at the same point $N_F = \frac{3}{2}N_C$ as $\beta_{m0}(N_F)$, and the ratio $\gamma_{e00}(N_F)/\beta_{e0}(N_F)$ is positive below this point, indicating superconvergence and confinement as discussed before.

The exact relations (46) are an indication, that the anomalous dimension coefficients of the gauge fields at the fixed point $\alpha = 0$ may have a more fundamental significance, similar to the one-loop β-function coefficients.

Our discussion about the relation of superconvergence and duality results can be extended to similar supersymmetric gauge theories with other gauge groups [21,34,35]. The results are analogous. However, in the presence of matter superfields in the adjoint representation [36], the problem is more complicated. There the construction of dual theories requires a superpotential already for the original electric theory, and a corresponding reduction of couplings would be called for. Also the application of the superconvergence arguments is not straight forward. These cases deserve further study.

Duality in general superconformal theories has been discussed in [37], and for softly broken SQCD in [38].

7 Reduced Dual SQCD

The magnetic theory dual to SQCD contains two parameters, the gauge coupling g and the Yukawa coupling λ_1. We now want to apply the reduction method described in the previous sections and express the coupling parameter λ_1 as a function of g^2. With Eq.(7) we write

$$\lambda_1(g^2) = g^2 f_1(g^2) \ , \ \text{with} \ \ f_1(g^2) = f^0 + \sum_{l=1}^{\infty} \chi^{(l)} g^{2l} \ . \tag{47}$$

The essential one-loop reduction equation is then

$$\beta_0 f^0 \ = \ \left(c_{11}^{(0)} f^0 \ + \ c_1^{(0)} \right) f^0 \ . \tag{48}$$

There are two solutions:

$$f^0 = f_{01} = \frac{\beta_0 - c_1^{(0)}}{c_{11}^{(0)}} \quad \text{and} \quad f^0 = f_{00} = 0 \ , \tag{49}$$

where f_{01} is a function of N_C and N_F, and is given by

$$f_{01}(N_C, N_F) \ = \ \frac{N_C\,(N_F - N_C - 2/N_C)}{(N_F - N_C)(3N_F - N_C)} \ , \tag{50}$$

using the explicit expressions (40) for the coefficients. Here and in the following, we do not consider possible additional terms which vanish exponentially or faster [5]. The criteria for the unique definition of the coefficients $\chi^{(l)}$ in the expansion (47) are given by

$$\left(M(f^0) - l\beta_0 \right) \neq 0 \ \ for \ \ l = 1, 2, \ldots \tag{51}$$

with

$$M(f^0) = c_1^{(0)} + 2c_{11}^{(0)} f^0 - \beta_0 \ . \tag{52}$$

Upon substitution of the solutions (49) and the explicit form of the coefficients from Eqs.(40), wie find

$$\begin{aligned} M(f_{01}) - l\beta_0 &= -\beta_0(\xi + l) \\ M(f_{00}) - l\beta_0 &= +\beta_0(\xi - l) \ , \end{aligned} \tag{53}$$

with β_0 from Eq.(40) and ξ as a function of N_C and N_F given by

$$\xi(N_C, N_F) \ = \ \frac{N_C\,(N_F - N_C - 2/N_C)}{(N_F - N_C)(2N_F - 3N_C)} \ . \tag{54}$$

The equations for the coefficients $\chi^{(l)}$ are of the general form given in Eqs.(15). For $l+1$ loops, they are simply

$$\left(M(f^0) - l\beta_0 \right) \chi^{(l)} = \left(\beta_l(f^0) f^0 - \beta^{(l)}(f^0) \right) + X^{(l)}, \tag{55}$$

where $l = 1, 2, \ldots$, and where f^0 is to be replaced by the solutions f_{01} or f_{00} respectively. The β-function coefficients are as in Eq.(11) with appropriate substitutions.

In the following , we consider characteristic intervals in N_F separately, and concentrate on the *conformal window.*

We have already discussed the window $\frac{3}{2}N_C < N_F < 3N_C$, where both SQCD and dual SQCD are asymptotically free at small distances. Considering first the solution $f_{01}(N_C, N_F)$ as given in Eq.(50), we see that it is positive in the window, as is the function $\xi(N_C, N_F)$. Since also $\beta_0 < 0$, the coefficients in Eq.(55) do not vanish. Consequently the expansion coefficients $\chi^{(l)}$ are uniquely determined and can be removed by a regular reparametrization transformation. We are left with the explicit solution

$$\lambda_1(g^2) \;=\; g^2\, f_{01}(N_C, N_F)\;, \tag{56}$$

with f_{01} given by Eq.(50). The β-functions of the reduced theory, as defined by the solution (56), are now simply given by Eqs.(9) and (10)) with the argument f of the coefficient functions replaced by $f_{01}(N_C N_F)$, so that they are constants:

$$\beta(g^2) = \beta(g^2, g^2 f_{01}) = \sum_{l=0}^{\infty} \beta_l(f_{01})(g^2)^{l+2}\;,\quad \beta_1(g^2) = f_{01}\beta(g^2)\;. \tag{57}$$

The second relation follows from the reduction equation (4) with Eq.(56). The coefficient β_0 is as given in Eq.(40), and for $\beta_1(f_{01})$ we obtain explicitly [21,23]

$$\begin{aligned}(16\pi^2)^2\beta_1(f_{01}) = 2(N_F - N_C)(3N_C - 2N_F) \;+\; 4N_F\frac{(N_F - N_C)^2 - 1}{2(N_F - N_C)} \\ -4N_F^2\,\frac{N_C(N_F - N_C - 2/N_C)}{2(N_F - N_C)(3N_F - N_C)}\;.\end{aligned} \tag{58}$$

These relations are used later in connection with the infrared fixed point of dual SQCD in the conformal window near $N_F = \frac{3}{2}N_C$. We must note here, that for the expansion (57), in addition to β_0, the two-loop coefficient $\beta_1(f_{01})$ is *reparametrization invariant.* This result follows because f_{01} satisfies the reduction equation (48).

It remains to consider the second solution presented in Eq.(49), with $f^0 = f_{00} = 0$. In this case the second expression in Eq.(53) is relevant for the determination of the higher coefficients in the expansion of $f_1(g^2)$. There could be a zero if $\xi(N_C, N_F)$ is a positive integer in the window. Generally however, this is not the case (at least for $N_C < 16$), with the characteristic exception of $N_C = 3, N_F = 5$, where $\xi(3,5) = 2$ and the magnetic gauge group is $SU(2)$. Ignoring this case, we have again the situation that all coefficients $\chi^{(l)}$ are determined and can be removed by regular reparametrization. Then the second power series solution of the reduction equations is given by

$$\lambda_1(g^2) \;\equiv\; 0\;, \tag{59}$$

and leads to a theory without superpotential. As we have discussed earlier, this situation is not acceptable for the dual magnetic theory.

Returning to the exceptional case with the magnetic gauge group $SU(2)$, we find that, after reparametrization, it leads to a solution of the form

$$\lambda_1(g^2) = Ag^6 + \chi^{(3)}g^8 + \cdots , \tag{60}$$

where the coefficient A is undetermined, and the higher ones are fixed once A is given. They vanish if $A = 0$. We do not discuss this case here any further.

Finally, we briefly consider possible 'general' solutions of the reduction equations. It turns out that for dual SQCD there are no such solutions which asymptotically approach the relevant polynomial solution $\lambda_1(g^2) = g^2 f_{01}$ given in Eq.(56). The only general solution we obtain is associated with the excluded polynomial solution $\lambda_1(g^2) \equiv 0$. It is given by

$$\lambda_1(g^2) = A(g^2)^{1+\xi} + \cdots, \tag{61}$$

where A is again an undetermined parameter with properties analogous to those discussed above for Eq.(60). As we have pointed out, the exponent ξ, as given in Eq.(54), is positive and generally non-integer in the limit. The only exception is for $N_C = 3, N_F = 5$, in which case we are back to the exceptional solution (60) discussed above.

We see that, within the set of solutions of the reduction equations for magnetic SQCD, the power series $\lambda_1(g^2) = g^2 f_{01}$ is the unique choice for duality. Ignoring the isolated $SU(2)$ case, the second power series solution $\lambda_1(g^2) \equiv 0$ is excluded. The general solution (61), which is associated with it, leads to asymptotic expansions of Green's functions involving non- integer powers. This is not consistent with a conventional, renormalizable Lagrangian formulation. Since there are no general solutions approaching the power solution $\lambda_1(g^2) = g^2 f_{01}$, the latter is isolated or unstable.

From the one- and two-loop expressions for the β-functions of the electric and the reduced magnetic theories given in Eqs.(37) and (40), we can obtain some information about non-trivial infrared fixed points in the conformal window [39,40]. These expansions are useful as long as the fixed points occur for values of N_C, N_F near the appropriate endpoint of the window. We find [23]

$$\beta_m(g^{*2}) = 0 \quad \text{for} \quad \frac{g^{*2}}{16\pi^2} = \frac{7}{3}\,\frac{N_F - \frac{3}{2}N_C}{\frac{N_C^2}{4} - 1} + \cdots , \tag{62}$$

and

$$\beta_e(g_e^{*2}) = 0 \quad \text{for} \quad \frac{g_e^{*2}}{16\pi^2} = \frac{3N_C - N_F}{6(N_C^2 - 1)} + \cdots , \tag{63}$$

for sufficiently small and positive values of $3N_C - N_F$ and $N_F - \frac{3}{2}N_C$ respectively. Larger values of N_C may be needed in order to have a useful approximation. Higher order terms have been calculated and may be found in [29].

With the reduced dual theory depending only upon the magnetic gauge coupling, it is straightforward to obtain the critical exponent $\gamma_m(N_C, N_F)$ near the lower end of the window at $N_F = \frac{3}{2}N_C$ [29]. This exponent is relevant for describing the rate at which a given charge approaches the infrared fixed point. With Eqs.(40), (58) and (62), the lowest order term is given by

$$\gamma_m = \left(\frac{d\beta_m(g^2)}{dg^2}\right)_{g^2=g^{*2}} = \frac{14}{3}\frac{(N_F - \frac{3}{2}N_C)^2}{\frac{N_C^2}{4} - 1} + \cdots , \tag{64}$$

where we have written g in place of g_m as before. For the electric theory in the window near $N_F = 3N_C$, the corresponding expression is

$$\gamma_e = \left(\frac{d\beta_e(g_e^2)}{dg_e^2}\right)_{g_e^2=g_e^{*2}} = \frac{1}{6}\frac{(3N_C - N_F)^2}{N_C^2 - 1} + \cdots . \tag{65}$$

In both cases we refer to [29] for the next order.

In this report we consider mainly the reduction of dual magnetic SQCD in the conformal window. A detailed discussion of the situation in the *free magnetic phase* $N_C + 2 \leq N_F < \frac{3}{2}N_C$ may be found in [23]. This interval is non-empty for $N_F > 4$. The electric theory is UV-free and the magnetic theory IR-free. At low energies, it is the latter which describes the spectrum. Because of the lack of UV-asymptotic freedom, one may be concerned that the magnetic theory may not exist as a strictly local field theory. However, it can be considered as a long distance limit of an appropriate brane construction in superstring theory, which can also confirm duality [41]. Except for special cases involving again $SU(2)$ as the magnetic gauge group, the unique power series solution (56) remains the appropriate choice also for this phase. It correspond here to the approach to the *trivial infrared fixed point*. Below $N_F = N_C + 2$ there is no dual magnetic gauge theory, and the spectrum should contain massless baryons and mesons associated with gauge invariant fields.

In the free electric phase for $N_F > 3N_C$, the magnetic theory remains UV-free, and the results of the reduction method are the same as in the conformal window.

8 Conclusions

In the application to Duality, we see that the reduction method is most helpful in bringing out characteristic features of theories with superpotentials. In the case of the dual of SQCD, we get an essentially unique solution of the reduction equations, which corresponds to a renormalizable Lagrangian theory with an asymptotic power series expansion in the remaining gauge coupling. This dual magnetic theory is asymptotically free. It is UV-free in the conformal window and above, and IR-free in the free magnetic region below the window. In this latter region, it describes the low energy excitations. These can be considered as composites of the free quanta of the electric theory, which is strongly coupled there.

As we have mentioned before, dual theories can be obtained as appropriate limits of brane systems [41]. In these brane constructions, duality corresponds essentially to a reparametrization of the quantum moduli space of vacua of a given brane structure. It is of interest to find out how the reduction solutions are related to special features of these constructions, in particular as far as the unique power solution (56) is concerned.

Besides the use of the reduction method in connection with duality, which we have described in this article, there are many other theoretical as well as phenomenological applications. Examples of applications in more phenomenological situations are discussed in this volume by J. Kubo [42].

Without detailed discussions, we mention here only a few applications:

- Construction of gauge theories with "minimal" coupling of Yang-Mills and matter fields [4].
- Proof of conformal invariance (finiteness) for $N = 1$ SUSY gauge theories with vanishing lowest order β-function on the basis of one-loop information [43,44].
- Reduction of the infinite number of coupling parameters appearing in the light-cone quantization method [45].
- Reduction in an effective field theory formulation of quantum gravity and in effective scalar field theory [46].
  We see that the reduction method can be used also within the framework of non-renormalizable theories, where the number of couplings is infinite a priori.
- Applications of reduction to the standard model (non-SUSY) give values for the top-quark mass which are too small, indicating the need for more matter fields [47].
- Gauge-Yukawa unifications within the framework of SUSY GUT's. Successful calculations of top-quark and bottom-quark masses within the framework of finite and non-finite theories [9,48,42].
- Reduction and soft symmetry breaking parameters. In softly broken $N = 1$ SUSY theories with gauge-Yukawa reduction, one finds all order renormalization group invariant sum rules for soft scalar masses [49,50,42]. There are interesting agreements with results from superstring based models.

There are other problems where the reduction scheme is a helpful and often an important tool [51].

Acknowledgements

It was a pleasure to have participated in the Ringberg Symposium in June 1998, where I have presented this talk. I would like to thank Peter Freund, Einan Gardi, Jisuke Kubo, Klaus Sibold, Wolfhart Zimmermann and George Zoupanos for very helpful discussions. I am grateful to Jisuke Kubo for correspondence concerning our respective contributions to the Festschrift. Thanks are also due to Wolfhart Zimmermann, and the theory group of the Max Planck Institut für Physik - Werner Heisenberg Institut - for their kind hospitality in München. This work has been supported in part by the National Science Foundation, grant PHY 9600697.

References

1. R. Oehme and W. Zimmermann, Max Planck Institut Report MPT-PAE/Pth 60/82 (1982); Commun. Math. Phys. **97**, 569 (1985).
2. R. Oehme, K. Sibold and W. Zimmermann, Phys. Lett. **B147**, 115 (1984).
3. W. Zimmermann, Commun. Math. Phys. **97**, 211 (1985), (Symanzik Memorial Volume).
4. R. Oehme, K. Sibold and W. Zimmermann, Phys. Lett. **B153**, 142 (1985).
5. R. Oehme, Prog. Theor. Phys. Suppl. **86**, 215 (1986), (Nambu Festschrift); CERN-Report TH.42-45/1985 .
6. R. Oehme, "Reduction in Coupling Parameter Space", in *Anomalies, Geometry and Topology*, Proc. Symposium, Argonne, Illinois, March 1985, ed. A. White (World Scientific, Singapore, 1985) pp. 443-458.
7. W. Zimmermann, "Renormalization Group and Symmetries in Quantum Field Theory", in *Proc. 14th ICGTMP*, ed. Y. M. Cho (World Scientific, Singapore, 1985) pp. 145-154.
8. K. Sibold, "Reduction of Couplings", Acta Physica Polonica, **19**, 295 (1988).
9. J. Kubo, M. Mondragón, M. Olechowski and G. Zoupanos, "Unification of Gauge and Yukawa Couplings without Symmetry", Proc. 5th Hellenic Summerschool, Corfu, Sept. 1995, hep-ph/9606434.
10. R. Oehme, "Reduction of Coupling Parameters", in *Proc XVIIIth Intl. Workshop on High Energy Physics and Field Theory, June 1995, Moscow-Protvino, Russia*, eds. V.A. Petrov, A.P. Samokhin and R.N. Rogalyov, pp. 251-270; hep-th/9511006.
11. N.-P. Chang, Phys. Rev. **D10**, 2706 (1974);
N.-P. Chang, A. Das and J. Perez-Mercader, Phys. Rev. **D22**, 1829 (1980);
E.S. Fradkin and O.K. Kalashnikov, J. Phys. **A8** (1975) 1814; Phys. Lett. **59B** (1975) 159; **64B** (1976) 177;
E. Ma, Phys. Rev. **D11** (1975) 322; **D17** (1978) 623; **D31** (1985) 1143; Prog. Theor. Phys. **54** (1975) 1828; Phys. Lett. **62B** (1976) 347; Nucl. Phys. **B116** (1976) 195;
D.I. Kazakov and D.V. Shirkov, "Singular Solutions of Renormalization Group Equations and Symmetry of the Lagrangian", in *Proc. 1975 Smolence Conference High Energy Particle Interactions*, eds. D. Krupa and J. Pišút (VEDA, Publishing House of the Slovak Academy of Sciences, Bratislava 1976).
12. O. Piguet and K. Sibold, Phys. Lett. **B229**, 83 (1989);
D. Maison, (unpublished);
W. Zimmermann, "Reduction of Couplings in Massive Models of Quantum Field Theory", MPI/PhT/98-97, (To appear in the Proceedings of the 12th Max Born Symposium, Wrocklow, September 1998).
13. N. Seiberg, Phys. Rev. **D49**, 6857 (1994).
14. N. Seiberg and E. Witten, Nucl. Phys. **B426**, 19 (1994); **B431**, 484 (1994).
15. N. Seiberg, Nucl. Phys. **B435**, 129 (1995).
16. P.C. Argyres, M.R. Plesser and N. Seiberg, Nucl.Phys. **B471**, 159 (1996); **B483**, 172 (1996).
17. R. Oehme, 'Superconvergence, Supersymmetry and Conformal Invariance',
in *Leite Lopes Festschrift*, eds. N. Fleury, S. Joffily, J. A. M. Simões, and A. Troper (World Scientific, Singapore, 1988) pp. 443 - 457;
University of Tokyo Report UT 527 (1988).
18. R. Oehme, Phys. Rev. **D42**, 4209 (1990); Phys. Lett. **B195**, 60 (1987).
19. R. Oehme, Phys. Lett. **B232**, 489 (1989).

20. R. Oehme, 'Supercovergence, Confinement and Duality', in *Proc. Intl. Workshop on High Energy Physics, Novy Svit, Crimea, September 1995*; eds. G.V. Bugrij and L. Jenkovsky, (Bogoliubov Institute, Kiev, 1995) pp. 107-116; hep-th/9511014.
21. R. Oehme, Phys. Lett. **B399**, 67 (1997).
22. K. Nishijima, Prog. Theor. Phys. **75**, 22 (1986); Nucl. Phys. **B238**, 601 (1984); Prog. Theor. Phys. **77**, 1053 (1987); K. Nishijima in *Symmetry in Nature*, Festschrift for Luigi A. Radicati di Brozolo (Scuola Normale Superiore, Pisa, 1989) pp. 627-655; Int.J.Mod.Phys. **A9**, 3799 (1994).
23. R. Oehme, "Reduction of Dual Theories", Phys. Rev. **D59**, 0850XX (1999) , EFI 98-27, MPI-Ph/98-52, hep-th /9808054.
24. A.M. Ljapunov, 'General Problems of the Stability of Motion' , (in Russian) (Charkov 1892); French translation in Annals of Mathematical Studies, No.17 (Princeton University Press, Princeton, 1947).
25. I.G. Malkin, 'Theory of the Stability of Motion' , (in Russian) (Ghostekhizdat, Moscow, 1952); German translation (Oldenbourg, München, 1959).
26. R. Oehme, K. Sibold, and W. Zimmermann, (unpublished).
27. E. Kraus, Nucl. Phys. **B354**, 281 (1991).
28. K. Intriligator and N. Seiberg, Nucl. Phys. Proc. Suppl. **BC45**, 1 (1996); Nucl. Phys. **B444**, 125 (1995).
29. E. Gardi and G. Grunberg, "Confomel Window in QCD and Supersymmetric QCD", CPTH-S645-0998, hep-th/9810192.
30. D.R.T. Jones, Nucl. Phys. **B87**, 127 (1975);
R. Barbieri et al., Phys. Lett. **B115**, 212 (1982);
A.J. Parkes and P.C. West, Nucl. Phys. **B256**, 340 (1985);
P. West, Phys. Lett. **B137**, 371 (1984);
D.R.T. Jones and L. Mezincescu, Phys. Lett. **B136**, 293 (1984);
I.I. Kogan, M. Shifman, and A. Vainstein, Phys. Rev. **D53**, 4526 (1996).
31. R. Oehme and W. Zimmermann, Phys. Rev. **D21**, 471, 1661 (1980).
32. R. Oehme, Phys. Lett. **B252**, 641 (1990).
33. R. Oehme and W. Xu, Phys. Lett. **B333**, 172 (1994); **B384**, 269 (1996).
34. R. Oehme, (unpublished).
35. M. Tachibana, Phys. Rev. **D58**, 045015 (1998).
36. D. Kutasov, Phys. Lett. **B351**, 230 (1995);
D. Kutasov and A Schwimmer, Phys. Lett. **B354**, 315 (1995);
K. Intriligator, R. Leigh and M. Strassler, Nucl. Phys. **B456**, 567 (1995).
37. A. Karsch, D. Lüst and G. Zoupanos, Phys. Lett. **B430**, 254 (1998);
A. Karsch, D. Lüst and G. Zoupanos, Nucl.Phys. **B529**, 96 (1998).
38. A. Karsch, T. Kobayashi, J. Kubo and G. Zoupanos, Phys. Lett. **B441**, 235 (1998).
39. T. Banks and A. Zaks, Nucl. Phys. **B196**, 189 (1982);
J. Kubo, Phys. Rev. **D52**, 6475 (1995).
40. M.J. Strassler, Progr. Theor. Phys. Suppl. **123**, 373 (1996).
41. S. Elizur, A. Giveon, D. Kutasov, E. Rabinovici and A. Schwimmer, Nucl. Phys. **B505**, 202 (1997);
H. Ooguri and C. Vafa, Nucl. Phys. **B500**, 62 (1997);
A. Brandhuber, J. Sonnenschein, S, Theisen and S. Yankielowicz, Nucl. Phys. B502, 125 (1997); A. Giveon and D. Kutasov, hep-th/9802067, Rev. Mod. Phys. (to be published), [this paper contains further references].
42. J. Kubo, "Applications of Reduction of Couplings", in *Recent Developments in Quantum Field Theory*, edited by P. Breitenlohner, D. Maison and J. Wess (Springer Verlag, Heidelberg, New York, 1999), (to be published).
We refer to this article for detailed references to phenomenological applications.

43. C. Lucchesi, O. Piguet and K. Sibold, Phys. Lett. **B201**, 241 (1988); Helv. Physica Acta **61**, 321 (1988);
O. Piguet, "Supersymmetry, Ultraviolet Finiteness and Grannd Unification", Alushta 1996, hep-th/9606045; "Supersymmetry, Supercurrent and Scale Invariance", WHEP96, Rio de Janeiro, hep-th/9611003.
44. C. Lucchesi and G. Zoupanos, Fortsch. Phys. **45**, 129 (1997).
45. K. G. Wilson, T. S. Walhout, A. Harianas, W-M. Zhang and R. J. Perry, Phys. Rev. **D49**, 6720 (1994);
R. J. Perry and K. G. Wilson, Nucl. Phys. **B403**, 587 (1993);
R. J. Perry, Ann. of Physics (N.Y.), **232**, 116 (1994); "Light Front QCD: A Constituent Picture of Hadrons", Cambridge 1997, pp 263-306, hep-th/9710175; "Light Front Quantum Chromodynamics", nucl-th/9901080.
46. M. Atance and J. L. Cortés, Phys.Lett. **B387**, 697, (1996); Phys. Rev. **D54**, 4973 (1996); **D56**, 3611 (1997);
L. A. W. Toro, Z. Phys. **C56**, 635 (1992).
47. J. Kubo, K. Sibold and W. Zimmermann, Nucl. Phys. **B259**, 331 (1985); Phys. Lett. **B220**, 185 (1989);
K. Sibold and W. Zimmermann, Phys. Lett. **B191**, 427 (1987);
J. Kubo, Phys. Lett. **B262**, 472 (1991);
W. Zimmermann, Phys. Lett. **B311**, 249 (1993);
M. Atance, J. L. Cortés and I. G. Irastorza, Phys. Lett. **B403**, 80 (1997).
48. D. Kapetanakis, M. Mondragón and G. Zoupanos, Z.Physik **C60**, 181 (1993);
J. Kubo, M. Mondragón and G. Zoupanos, Nucl. Phys. **B424**, 291 (1994);
J. Kubo, M. Mondragón, M. Olechowsky and G. Zoupanos, Nucl. Phys. **B479**, 25 (1996);
T. Kobayashi, J. Kubo, M. Mondragón, and G. Zoupanos, Nucl. Phys. **B511**, 45 (1998).
49. I. Jack and D.R.T. Jones, Phys. Lett. **B333**, 372 (1994) ;
Phys. Lett. **B349**, 294 (1995);
I. Jack, D.R.T. Jones and A. Pickering, Phys. Lett. **B426**, 73 (1998).
50. J. Kubo, M. Mondragón and G. Zoupanos, Phys. Lett. **B389**, 523 (1996);
Y. Kawamura, T. Kobayashi und J. Kubo, Phys. Lett. **B405**, 64 (1997);
T. Kobayashi, J. Kubo and G Zoupanos, Phys.Lett **B427**, 291 (1998).
51. See the HEP Data Base.

The Bogoliubov Renormalization Group in Theoretical and Mathematical Physics

Dmitrij V. Shirkov

N.N. Bogoliubov Laboratory of Theoretical Physics,
JINR, Dubna, Russia

Abstract. This text follows the line of a talk on Ringberg symposium dedicated to Wolfhart Zimmermann 70th birthday. The historical overview (Part 1) partially overlaps with corresponding text of my previous commemorative paper – see Ref. [61] in the list. At the same time second part includes some recent results in QFT (Sect. 2.1) and summarize (Sect. 2.4) an impressive progress of the "QFT renormalization group" application in mathematical physics.

1 Early History of the RG in the QFT

1.1 The birth of Bogoliubov's renormalization group

In the spring of 1955 a small conference on "Quantum Electrodynamics and Elementary Particle Theory" was organized in Moscow. It took place at the Lebedev Institute in the first half of April. Among the participants there were several foreigners, including Hu Ning and Gunnar Källén. Landau's survey lecture "Fundamental Problems in QFT", in which the issue of ultraviolet (UV) behaviour in the QFT was discussed, constituted the central event of the conference. Not long before, the problem of short-distance behaviour in QED was advanced substantially in a series of articles [1] by Landau, Abrikosov, and Khalatnikov. They succeeded in constructing a closed approximation of the Schwinger–Dyson equations, which admitted an explicit solution in the massless limit and, in modern language, it resulted in the summation of the leading UV logarithms.

The most remarkable fact was that this solution turned out to be self-contradictory from the physical point of view because it contained a "ghost pole" in the renormalized amplitude of the photon propagator or, in terms of bare notions, the difficulty of "zero physical charge".

At that time our meetings with Nicolai Nicolaevich Bogoliubov (N.N. in what follows) were regular and intensive because we were tightly involved in the writing of final text[1] of our big book. N.N. was very interested in the results of Landau's group and proposed me to consider the general problem of evaluating their reliability by constructing, e.g., the second approximation (including *next-to-leading UV logs*) to the Schwinger–Dyson equations, to verify the stability of the UV asymptotics and the very existence of a ghost pole.

[1] Just at that time the first draft of a central part of the book has been published [2] in the form of two extensive papers.

P. Breitenlohner and D. Maison (Eds.): Proceedings 1998, LNP 558, pp. 157–176, 2000.

Shortly after the meeting at the Lebedev Institute, Alesha Abrikosov told me about Gell–Mann and Low's article[3] which had just appeared. The same physical problem was treated in this paper, but, as he put it, it was hard to understand and to combine it with the results obtained by the Landau group.

I looked through the article and presented N.N. with a brief report on the methods and results, which included some general assertions on the scaling properties of the electron charge distribution at short distances and rather cumbersome functional equations – see, below, Section 1.3.

N.N. immediate comment was that Gell–Mann and Low's approach is very important: it is closely related to the *la groupe de normalisation* discovered a couple of years earlier by Stueckelberg and Petermann [4] in the course of discussing the structure of the finite arbitrariness in the scattering matrix elements arising upon removal of the divergences. This group is an example of the continuous groups studied by Sophus Lie. This implied that functional group equations similar to those of paper [3] should take place not only in the UV limit but also in the general case as well.

Within the next few days I succeeded in recasting Dyson's finite transformations and obtaining the desired functional equations for the QED propagator amplitudes, which have group properties, as well as the group differential equations, that is, the Sophus Lie equations of the renormalization group (RG). Each of these resulting equations — see, below Eqs.(3 — contained a specific object, the product of the squared electron charge $\alpha = e^2$ and the transverse photon propagator amplitude $d(Q^2)$. We named this product, $e^2(Q^2) = e^2 d(Q^2)$, the *invariant charge*. From the physical point of view this function is an analogue of the so–called *effective charge* of an electron, first discussed by Dirac in 1933 [5], which describes the effect of the electron charge screening due to quantum vacuum polarization. Also, the term "renormalization group" was first introduced in our Doklady Akademii Nauk SSSR publication [6] in 1955 (and in the English language paper [7]).

At the above–mentioned Lebedev meeting Gunnar Källén presented a paper written with Pauli on the so–called "Lee model", the exact solution of which contained a *ghost pole* (which, in contrast to the physical one corresponding to a bound state, had negative residue) in the nucleon propagator. Källén–Pauli's analysis led to the conclusion that the Lee model is physically void.

In view of the argument on the presence of a similar pole in the QED photon propagator (which follows from the abovementioned solution of Landau's group as well as from an independent analysis by Fradkin [8]) obtained in Moscow, Källén's report resulted in a heated discussion on the possible inconsistency of QED. In the discussion Källén argued that no rigorous conclusion about the properties of sum of an infinite nonconvergent series can be drawn from the analysis of a finite number of terms.

Nevertheless, before long a publication by Landau and Pomeranchuk (see, e.g., the review paper[9]) appeared arguing that not only QED but also local QFT were self–contradictory.

Without going into details, remind that our analysis of this problem carried out [10] with the aid of the RG formalism just appeared led to the conclusion that such a claim cannot have the status of a ***rigorous result, independent of perturbation theory.***

1.2 Renormalization and renormalization invariance

As is known, the regular formalism for eliminating ultraviolet divergences in quantum field theory (QFT) was developed on the basis of covariant perturbation theory in the late 40s. This breakthrough is connected with the names of Tomonaga, Feynman, Schwinger and some others. In particular, Dyson and Abdus Salam carried out the general analysis of the structure of divergences in arbitrarily high orders of perturbation theory. Nevertheless, a number of subtle questions concerning so-called overlapping divergences remained unclear.

An important contribution in this direction based on a thorough analysis of the mathematical nature of UV divergences was made by Bogoliubov. This was achieved on the basis of a branch of mathematics which was new at that time, namely, the Sobolev–Schwartz ***theory of distributions***. The point is that propagators in local QFT are distributions (similar to the Dirac delta–function) and their products appearing in the coefficients of the scattering matrix expansion require supplementary definition in the case when their arguments coincide and lie on the light cone. In view of this the UV divergences reflect the ambiguity in the definition of these products.

In the mid 50ies on the basis of this approach Bogoliubov and his disciples developed a technique of supplementing the definition of the products of singular Stueckelberg–Feynman propagators [2] and proved a theorem [11] on the finiteness and uniqueness (for renormalizable theories) of the scattering matrix in any order of perturbation theory. The prescription part of this theorem, namely, ***Bogoliubov's R-operation***, still remains a practical means of obtaining finite and unique results in perturbative calculations in QFT.

The Bogoliubov algorithm works, essentially, as follows:

- To remove the UV divergences of one-loop diagrams, instead of introducing some regularization, for example, the momentum cutoff, and handling (quasi) infinite counterterms, it suffices to complete the definition of divergent Feynman integral by subtracting from it certain polynomial in the external momenta which in the simplest case is reduced to the first few terms of the Taylor series of the integral.
- For multi-loop diagrams (including ones with overlapping divergencies) one should first subtract all divergent subdiagrams in a hierarchical order regulated by the R–operator.

The uniqueness of computational results is ensured by special conditions imposed on them. These conditions contain specific degrees of freedom (related to different renormalization schemes and momentum scales) that can be used to establish the relationships between the Lagrangian parameters (masses, coupling constants) and the corresponding physical quantities. The fact that physical predictions are independent of the arbitrariness in the renormalization conditions, that is, they are *renorm–invariant*, constitutes the conceptual foundation of the renormalization group.

An attractive feature of this approach is that it is free from any auxiliary nonphysical attributes such as bare masses, bare coupling constants, and regularization parameters which turn out to be unnecessary in computations employing Bogoliubov's approach. As a whole, this method can be regarded as *renormalization without regularization and counterterms.*

1.3 The discovery of the renormalization group

The renormalization group was discovered by Stueckelberg and Petermann [4] in 1952-1953 as a group of infinitesimal transformations related to a finite arbitrariness arising in the elements of the scattering S-matrix upon elimination of the UV divergences. This arbitrariness can be fixed by means of certain parameters c_i:

> "... *we must expect that a group of infinitesimal operators* $\mathbf{P}_i = (\partial/\partial c_i)_{c=0}$, *exists, satisfying*
>
> $$\mathbf{P}_i S = h_i(m,e)\partial S(m,e,\ldots)/\partial e \ ,$$
>
> *admitting thus a renormalization of e.*"

These authors introduced the *normalization group* generated (as a Lie group) by the infinitesimal operators $\mathbf{P}_i$ connected with renormalization of the coupling constant e.

In the following year, on the basis of Dyson's transformations written in the regularized form, Gell-Mann and Low [3] derived functional equations for QED propagators in the UV limit. For example, for the renormalized transverse part d of the photon propagator they obtained an equation of the form

$$d\left(\frac{k^2}{\lambda^2}, e_2^2\right) = \frac{d_C(k^2/m^2, e_1^2)}{d_C(\lambda^2/m^2, e_1^2)} \ , \quad e_2^2 = e_1^2 d_C(\lambda^2/m^2, e_1^2) \ , \tag{1}$$

where λ is the cutoff momentum and e_2 is the physical electron charge. The appendix to this article contains the general solution (obtained by T.D.Lee) of this functional equation for the photon amplitude $d(x, e^2)$ written in two equivalent forms:

$$e^2 d\left(x, e^2\right) = F\left(x F^{-1}\left(e^2\right)\right) \ , \quad \ln x = \int_{e^2}^{e^2 d} \frac{dy}{\psi(y)} \ , \tag{2}$$

with

$$\psi(e^2) = \frac{\partial(e^2 d)}{\partial \ln x} \quad \text{at} \quad x = 1 \ .$$

A qualitative analysis of the behaviour of the electromagnetic interaction at small distances was carried out with the aid of (2). Two possibilities, namely, infinite and finite charge renormalizations were pointed out:

> *Our conclusion is that the* **shape** *of the charge distribution surrounding a test charge in the vacuum does not, at small distances, depend on the coupling constant except through the scale factor. The behavior of the propagator functions for large momenta is related to the magnitude of the renormalization constants in the theory. Thus it is shown that the unrenormalized coupling constant $e_0^2/4\pi\hbar c$, which appears in perturbation theory as a power series in the renormalized coupling constant $e_1^2/4\pi\hbar c$ with divergent coefficients, many behave either in two ways:*
> *It may really be infinite as perturbation theory indicates;*
> *It may be a finite number independent of $e_1^2/4\pi\hbar c$.*

Note, that the latter possibility corresponds to the case when ψ vanishes at a finite point: $\psi(\alpha_\infty) = 0$. Here, α_∞ is known now as a fixed point of the renormalization group transformations.

The paper [3] paid no attention to the group character of the analysis and the results obtained there. The authors failed to establish a connection between their results and the standard perturbation theory and did not discuss the possibility that a ghost pole might exist.

The final step was taken by Bogoliubov and Shirkov [6,12] – see also the survey [7] published in English in 1956. Using the group properties of finite Dyson transformations for the coupling constant and the fields, these authors derived functional group equations for the propagators and vertices in QED in the general case (that is, with the electron mass taken into account). For example, the equation for the transverse amplitude of the photon propagator and electron propagator amplitude were obtained in the form

$$\begin{aligned} d(x,y;e^2) &= d(t,y;e^2)d\left(\frac{x}{t},\frac{y}{t};e^2 d(t,y;e^2)\right) , \\ s(x,y;e^2) &= s(t,y;e^2)s\left(\frac{x}{t},\frac{y}{t};e^2 d(t,y;e^2)\right) , \end{aligned} \tag{3}$$

in which the dependence not only on momentum transfer $x = k^2/\mu^2$ (where μ is a certain normalizing scale factor), but also on the mass variable $y = m^2/\mu^2$ is taken into account.

As can be seen, the product $e^2 d$ of electron charge squared and photon propagator amplitude enters in both functional equations. This product is invariant with respect to Dyson transfermation. We called this function – *invariant charge*.

In the modern notation, the first equation (which in the massless case $y = 0$ is equivalent to (1)) is an equation for the invariant charge (now widely known as an effective or running coupling) $\bar{\alpha} = \alpha d(x,y;\alpha = e^2)$:

$$\bar{\alpha}(x,y;\alpha) = \bar{\alpha}\left(x/t, y/t; \bar{\alpha}(t,y;\alpha)\right) . \tag{4}$$

Let us emphasize that, unlike in the Ref.[3] approach, in our case there are no simplifications due to the massless nature of the UV asymptotics. Here the homogeneity of the transfer momentum scale is violated explicitly by the mass m. Nevertheless, the symmetry (even though a bit more complex one) underlying the renormalization group, as before, can be stated as an *exact symmetry* of the solutions of the quantum field problem – see eq. (11) below. This is what we mean when using the term *Bogoliubov's renormalization group* or *renorm-group* for short.

The differential group equations (DGEs) for $\bar{\alpha}$ and for the electron propagator:

$$\frac{\partial\bar{\alpha}(x,y;\alpha)}{\partial\ln x}=\beta\left(\frac{y}{x},\bar{\alpha}(x,y;\alpha)\right)\ ;\ \frac{\partial s(x,y;\alpha)}{\partial\ln x}=\gamma\left(\frac{y}{x},\bar{\alpha}(x,y;\alpha)\right)s(x,y;\alpha)\ , \tag{5}$$

with

$$\beta(y,\alpha)=\frac{\partial\bar{\alpha}(\xi,y;\alpha)}{\partial\xi}\ ,\quad \gamma(y,\alpha)=\frac{\partial s(\xi,y;\alpha)}{\partial\xi}\quad \text{at}\ \ \xi=1\ . \tag{6}$$

were first derived in [6] by differentiating the functional equations. In this way an explicit realization of the DGEs mentioned in the citation from [4] was obtained. These results established a conceptual link with the Stueckelberg–Petermann and Gell-Mann – Low approaches.

1.4 Creation of the RG method

Another important achievement of paper [6] consisted in formulating a simple algorithm for improving an approximate perturbative solution by combining it with the Lie differential equations (modern notation is used in this quotation from [6]):

> *Formulae (5) show that to obtain expressions for $\bar{\alpha}$ and s valid for all values of their arguments one has only to define $\bar{\alpha}(\xi,y,\alpha)$ and $s(\xi,y,\alpha)$ in the vicinity of $\xi=1$. This can be done by means of the usual perturbation theory.*

In our adjacent publication [12] this algorithm was effectively used to analyse the UV and infrared (IR) asymptotic behaviour in QED. The one-loop and two-loop UV asymptotics

$$\bar{\alpha}^{(1)}_{RG}(x;\alpha)\equiv\bar{\alpha}^{(1)}_{RG}(x,0,\alpha)=\frac{\alpha}{1-\frac{\alpha}{3\pi}\cdot\ln x}\ , \tag{7}$$

$$\bar{\alpha}^{(2)}_{RG}(x;\alpha)=\frac{\alpha}{1-\frac{\alpha}{3\pi}\ln x+\frac{3\alpha}{4\pi}\ln(1-\frac{\alpha}{3\pi}\ln x)} \tag{8}$$

of the photon propagator as well as the IR asymptotics

$$s(x,y;\alpha)\approx(x/y-1)^{-3\alpha/2\pi}=(p^2/m^2-1)^{-3\alpha/2\pi}$$

of the electron propagator in transverse gauge were obtained. At that time these expressions had already been known only at the one–loop level. It should be noted that in the mid 50s the problem of the UV behaviour in local QFT was quite urgent. As it has been mentioned already a substantial progress in the analysis of QED at small distances was made by Landau and his collaborators [1]. However, Landau's approach did not provide a prescription for constructing subsequent approximations.

An answer to this question was found only within the new renorm–group method. The simplest UV asymptotics of QED propagators obtained in our paper [12], for example, expression (7), agreed precisely with the results of Landau's group.

Within the RG approach these results can be obtained in just a few lines of argumentation. To this end, the massless one-loop approximation

$$\bar{\alpha}^{(1)}_{PTh}(x;\alpha) = \alpha + \frac{\alpha^2}{3\pi}\ell + \dots \ , \qquad \ell = \ln x$$

of perturbation theory should be substituted into the right-hand side of the first equation in (6) to compute the generator $\beta(0,\alpha) = \psi(\alpha) = \alpha^2/3\pi$, followed by an elementary integration of the first of Eqs.(5).

Moreover, starting from the two-loop expression $\bar{\alpha}^{(2)}_{PTh}(x,;\alpha)$ containing the $\alpha^2\ell/4\pi^2$ term we arrive at the second renormalization group approximation (8) performing summation of the next-to-leading UV logs. Comparing solution (8) with (7) one can conclude that two-loop correction is extremely essential just in the vicinity of the ghost pole singularity at $x_1 = \exp(3\pi/\alpha)$. This demonstrates that the RG method is a regular procedure, within which it is quite easy to estimate the range of applicability of the results.

The second order renorm–group solution (8) for the invariant coupling first obtained in [12] contains the nontrivial log–of–log dependence which is now widely known of the two–loop approximation for the running coupling in quantum chromodynamics (QCD).

Quite soon [13] this approach was formulated for the case of QFT with two coupling constants g and h, namely, for a model of pion–nucleon interactions with self-interaction of pions. To the system of functional equations for two invariant couplings

$$\bar{g}^2\left(x,y;g^2,h\right) = \bar{g}^2\left(\frac{x}{t},\frac{y}{t},\bar{g}^2(t,y;g^2,h),\bar{h}\left(t,y;g^2,h\right)\right) \ ,$$
$$\bar{h}\left(x,y;g^2,h\right) = \bar{h}\left(\frac{x}{t},\frac{y}{t},\bar{g}^2\left(t,y;g^2,h\right),\bar{h}\left(t,y;g^2,h\right)\right)$$

there corresponds a coupled system of nonlinear differential equations. It was analysed [14] in one-loop appriximation to carry out the UV analysis of the renormalizable model of pion-nucleon interaction.

In Refs. [6,12,13] and [14] the RG was thus directly connected with practical computations of the UV and IR asymptotics. Since then this technique, known as the *renormalization group method* (RGM), has become the sole means of asymptotic analysis in local QFT.

1.5 Other early RG applications

Another important general theoretical application of the RG method was made in the summer of 1955 in connection with the (then topical) so-called ghost pole problem. This effect, first discovered in quantum electrodynamics [8,15], was at first thought [15] to indicate a possible difficulty in QED, and then [9,16] as a proof of the inconsistency of the whole local QFT.

However, the RG analysis of the problem carried out in [10] on the basis of massless solution (2) demonstrated that no conclusion obtained with the aid of finite–order computations within perturbation theory can be regarded as a complete proof. This corresponds precisely to the impression, one can get when comparing (7) and (8). In the mid 50s this result was very significant, for it restored the reputation of local QFT. Nevertheless, in the course of the following decade the applicability of QFT in elementary particle physics remained doubtful in the eyes of many theoreticians.

In the general case of arbitrary covariant gauge the renormalization group analysis in QED was carried out in [17]. Here, the point was that the charge renormalization is connected only with the transverse part of the photon propagator. Therefore, under nontransverse (for example, Feynman) gauge the Dyson transformation has a more complex form. This issue has been resolved by considering the treating the gauge parameter as another coupling constant.

Ovsyannikov [18] found the general solution to the functional RG equations taking mass into account:

$$\Phi(y,\alpha) = \Phi\left(y/x, \bar{\alpha}(x,y;\alpha)\right)$$

in terms of an arbitrary function Φ of two arguments, reversible in its second argument. To solve the equations, he used the differential group equations represented as linear partial differential equations of the form (which are now widely known as the Callan—Symanzik equations):

$$\left\{x\frac{\partial}{\partial x} + y\frac{\partial}{\partial y} - \beta(y,\alpha)\frac{\partial}{\partial \alpha}\right\}\bar{\alpha}(x,y,\alpha) = 0\ .$$

The results of this "period of pioneers" were collected in the chapter "Renormalization group" in the monograph [19], the first edition of which appeared in 1957 (shortly after that translated into English and French [20]) and very quickly acquired the status of the "QFT folklore".

2 Further Bogoliubov's RG Development

2.1 Quantum field theory

The next decade and a half brought a calm period, during which there was practically no substantial progress in the renorm–group method.

1. New possibilities for applying the RG method were discovered when the technique of operator expansion at small distances (on the light cone) appeared

[21]. The idea of this approach stems from the fact that the RG transform, regarded as a Dyson transformation of the renormalized vertex function, involves the simultaneous scaling of all its invariant arguments (normally, the squares of the momenta) of this function. The expansion on the light cone, so to say, "separates the arguments", as a result of which it becomes possible to study the physical UV asymptotic behaviour by means of the expansion coefficients (when some momenta are fixed on the mass shell). As an important example we can mention the evolution equations for moments of QCD structure functions [22].

2. In the early 70ies S. Weinberg [23] proposed the notion of the *running mass* of a fermion. If considered from the viewpoint of [17], this idea can be formulated as follows:
any parameter of the Lagrangian can be treated as a (generalized) coupling constant, and its effective counterpart should be included into the renorm-group formalism.

However, the results obtained in the framework of this approach turned out to be, practically, the same as before. For example, the most familiar expression for the fermion running mass

$$\bar{m}(x,\alpha) = m_\mu \left(\frac{\alpha}{\bar{\alpha}(x,\alpha)}\right)^{\nu} ,$$

in which the leading UV logarithms are summed, was known for the electron mass in QED (with $\nu = 9/4$) since the mid 50s (see [1], [12]).

3. The end of the calm period can be marked well enough by the year 1971, when the renormalization group method was applied in the quantum theory of non-Abelian gauge fields, in which the famous effect of ***asymptotic freedom*** has been discovered [24].

The one-loop renorm-group expression

$$\bar{\alpha}_s^{(1)}(x;\alpha_s) = \frac{\alpha_s}{1+\alpha_s\beta_1 \ln x} ,$$

for the QCD effective coupling $\bar{\alpha}_s$ exhibits a remarkable UV asymptotic behaviour thanks to β_1 being positive. This expression implies, in contrast to Eq.(7), that the effective QCD coupling decreases as x increases and tends to zero in the UV limit. This discovery, which has become technically possible only because of the RG method use, is the most important physical result obtained with the aid of the renorm–group approach in particle physics.

4. One more interesting application of the RG method in the multicoupling case, ascending back in 50ies [14], refers to special solutions, so called separatrices in a phase space of several invariant couplings. These solutions relate effective couplings and represent a scale invariant trajectories, like, e.g., $g_i = g_i(g_1)$ in the phase space which are straight lines at the one-loop case.

Some of them, that are "attractive" (or stable) in the UV limit, are related to symmetries that reveal themselves in the high-energy domain. It has been conjectured that these trajectories may be connected to ***hidden symmetries of a***

Lagrangian and even could serve as a tool to find them. On this basis the method has been developed [25] for finding out these symmetries. It was shown that in the phase space of the invariant charges the internal symmetry corresponds to a singular solution that remain straight-line when taking into account the higher order corrections. Such solutions corresponding to supersymmetry have been found for some combinations of Yukawa and quartic interactions.

Generally, these singular solutions obey the relations

$$\frac{dg_i}{dt} = \frac{dg_i}{dg_1}\frac{dg_1}{dt}\,, \quad t = \ln x$$

which are known since Zimmermann's paper [26] as *the reduction equations.* In the 80ies they have been used [27] (see also review paper [28] and references therein) in the UV analyzis of asymptotically free models. Just for these cases the one-loop reduction relations are adequate to physics.

Quite recently some other application of this technique has been found in a supersymmetrical generalizations of Grand Unification scenario in the Standard Model. It has been shown [29–31] that it is possible to achieve complete UV finiteness of a theory if Yukawa couplings are related to the gauge ones in a way corresponding to these special solutions,that is to reduction relations.

5. A general method of approximate *solution of the massive RG equations* has been developed [32]. Analytic expressions of high level of accuracy for an effective coupling and one-argument function have been obtained up to four- and three-loop order [33]. For example, the two-loop massive expression for the invariant coupling

$$\begin{aligned}&\bar{\alpha}_s(Q^2,m^2)_{\mathrm{rg},2} = \\ &\quad \alpha_s \left\{1 + \alpha_s A_1(Q^2,m^2) + \alpha_s \frac{A_2(Q^2,m^2)}{A_1(Q^2,m^2)} \ln\left(1 + \alpha_s A_1(Q^2,m^2)\right)\right\}^{-1}\end{aligned} \tag{9}$$

at small α_s values corresponds to adequate perturbation expansion

$$\begin{aligned}&\bar{\alpha}_s(Q^2,m^2)_{\mathrm{pert},2} = \\ &\quad \alpha_s \left\{1 - \alpha_s A_1(Q^2,m^2) + \alpha_s^2 A_1^2(Q^2,m^2) - \alpha_s^2\, A_2(Q^2,m^2) + \ldots\right\}\,.\end{aligned} \tag{10}$$

At the same time, it smoothly interpolates between two massless limits (with $A_\ell \simeq \beta_\ell \ln Q^2 + c_\ell$) at $Q^2 \ll m^2$ and $Q^2 \gg m^2$ described by equation analogous to Eq.(8). In the latter case it can be represented in the form usual for the QCD practice:

$$\bar{\alpha}_s^{-1}(Q^2/\Lambda^2)_{\mathrm{rg},2} \to \beta_1 \left\{\ln\frac{Q^2}{\Lambda^2} + b_1 \ln\left(\ln\frac{Q^2}{\Lambda^2}\right)\right\}\;;\; b_1 = \frac{\beta_2}{\beta_1^2}\,.$$

The solution (9) demonstrate, in particular, that the threshold crossing generally changes the subtraction scheme [34].

Our investigation [32,33,35] was prompted by the problem of explicitly taking into account heavy quark masses in QCD. However, the results obtained are important from a more general point of view for a discussion of the scheme dependence problem in QFT. The method used could also be of interest for RG applications in other fields within the situation with disturbed homogeneity, such as, e.g., intermediate asymptotics in hydrodynamics, finite-size scaling in critical phenomena and the excluded volume problem in polymer theory.

In the paper [35] this method was used for the effective couplings evolution in Standard Model (SM). Here, new analytic solution of a coupled system of three mass-dependent two-loop RG evolution equations for three SM invariant gauge couplings has been obtained.

6. One more recent QFT development relevant to renorm-group is the *Analytic approach* to perturbative QCD (pQCD). It is based upon the procedure of *Invariant Analytization* [36] ascending to the end of 50ies.

The approach consists in a combining of two ideas: the RG summation of leading UV logs with analyticity in the Q^2 variable, imposed by spectral representation of the Källén–Lehmann type which implements general properties of local QFT including the Bogoliubov condition of microscopic causality. This combination was first devised [38] to get rid of the ghost pole in QED about forty years ago.

Here, the pQCD invariant coupling $\bar{\alpha}_s(Q^2)$ is transformed into an "analytic coupling" $\alpha_{\rm an}(Q^2/\Lambda^2) \equiv \mathcal{A}(x)$, which, by constuction, is free of ghost singularities due to incorporating some nonperturbative structures.

This analytic coupling $\mathcal{A}(x)$ has no unphysical singularities in the complex Q^2-plane; its conventional perturbative expansion precisely coincides with the usual perturbation one for $\bar{\alpha}_s(Q^2)$; it has no extra parameters; it obeys an universal IR limiting value $\mathcal{A}(0) = 4\pi/\beta_0$ that is independent of the scale parameter Λ; it turns out to be remarkably stable with respect to higher loop corrections and, in turn, to scheme dependence.

Meanwhile, the "analytized" perturbation expansion [39] for an observable F, in contrast with the usual case, may contain specific functions $\mathcal{A}_n(x)$, instead of powers $(\mathcal{A}(x))^n$. In other words, the pertubation series for $F(x)$, due to analyticity imperative, may change its form [40] turning into an asymptotic expansion à la Erdélyi over a nonpower set $\{\mathcal{A}_n(x)\}$.

2.2 Ways of the RG expanding

As is known, in the early 70ies Wilson [41] succeeded in transplanting the RG philosophy from relativistic QFT to a quite another branch of modern theoretical physics, namely, the theory of phase transitions in spin lattice systems. This new version of the RG was based on Kadanoff's idea[42] of joninig in "blocks" of few neighbouring spins with appropriate change (renormalization) of the coupling constant.

To realize this idea, it is necessary to average spins in each block. This operation reducing the number of degrees of freedom and simplifying the system under consideration, preserves all its long-range properties under a suitable renormalization of the coupling constant. Along with this, the above procedure gives rise to a new theoretical model of the original physical system.

In order that the system obtained by averaging be similar to the original one, one must also discard those terms of a new effective Hamiltonian which turns out to be irrelevant in the description of infrared properties. As a result of this *Kadanoff–Wilson decimation*, we arrive at a new model system characterized by new values of the elementary scale (spacing between blocks) and coupling constant (of blocks interaction). By iterating this operation, one can construct a discrete ordered set of models. From the physical point of view the passage from one model to some other one is an irreversible approximate procedure. Two passages of that sort applied in sequence should be equivalent to one, which gives rise to a group structure in the set of transitions between models. However, in this case the RG is an approximative and is realized as a semigroup.

This construction, obviously in no way connected with UV properties, was much clearer from the general physical point of view and could therefore be readily understood by many theoreticians. Because of this, in the seventies the RG concept and its algorithmic structure were successfully carried over to diverse branches of theoretical physics such as polymer physics [43], the theory of noncoherent transfer [44], and so on.

Apart from constructions analogous to that of Kadanoff–Wilson, in a number of cases the connection with the original quantum field renorm–group was established. This has been done with help of the functional integral representation. For example, the classic Kolmogorov–type turbulence problem was connected with the RG approach by the following steps [45]:

1. Define the generating functional for correlation functions.
2. Write for this functional the path integral representation.
3. By a change of functional integration variable establish an equivalence of the given classical statistical system with some QFT model.
4. Construct the Schwinger–Dyson equations for this equivalent QFT.
5. Use the Feynman diagram technique and perform a finite renormalization.
6. Write down the standard RG equations and use them to find fixed point and scaling behavior.

The physics of renormalization transformation in the turbulence problem is related to a change of UV cutoff in the wave-number variable.

Hence, in different branches of physics the RG evolved in two directions:

- The construction of a set of models for the physical problem at hand by direct analogy with the Kadanoff–Wilson approach (by averaging over certain degrees of freedom) — in polymer physics, noncoherent transfer theory, percolation theory, and others;
- The search for an exact RG symmetry directly or by proving its equivalence to some QFT: for example, in turbulence theory [45,46] and turbulence in plasma [48].

What is the nature of the symmetry underlying the renormalization group?

a) In QFT the renorm–group symmetry is an exact symmetry of a solution described in terms of the notions of the equation(s) and some boundary condition(s).
b) In turbulence and some other branches of physics it is a symmetry of a solution of an equivalent QFT model.
c) In spin lattice theory, polymer theory, noncoherent transfer theory, percolation theory, and so on (in which the Kadanoff–Wilson blocking concept is used) the RG transformation involves transitions inside a set of auxiliary models (constucted especially for this purpose). To formulate RG, one should define an ordered set $\mathcal{M}$ of models M_i. The RG transformation connecting various models has the form $R(n)M_i = M_{ni}$. Here, the symmetry can be formulated only in the terms of whole set $\mathcal{M}$.

There is also a purely mathematical difference between the aforesaid RG realizations. In QFT the RG is a *continuous symmetry group.* On the contrary, in the theory of critical phenomena, polymers, and other cases (when an averaging operation is necessary) we have an *approximate discrete semigroup.* It must be pointed out that in dynamical chaos theory, in which RG ideas and terminology can sometimes be applied too, functional iterations do not constitute a group at all, in general. An entirely different terminology is sometimes adopted in the above–mentioned domains of theoretical physics. Terms like "the real–space RG", "the Wilson RG", "the Monte–Carlo RG", or "the chaos RG" are in use.

Nevertheless, the affirmative answer to the question *"Are there distinct renormalization groups?"* implies no more than what has just been said about the differences between cases a) and b) on the one hand and c) on the other.

For this reason, we shall use notation of the "Bogoliubov Renormalization Group" for the exact Lie group, as it emerged from the QFT original papers [4,6,7] (see also chapter "Renormalization Group" in the monograph [19,20]) of mid-fifties. This is to make clear distinction between exact group and the Wilson construction for which the term "Renormalization Group" is widely used in the current literature.

2.3 Functional self–similarity

An attempt to analyse the relationship between these formulations on a simple common basis was undertaken about fifteen years ago [49]. In this paper (see also our surveys [50–52]) it was demonstrated that all the above–mentioned realizations of the RG could be considered in a unified manner by using only some common notions of mathematical physics.

In the general case it proves convenient to discuss the symmetry underlying the renorm–group with the aid of a continuous one–parameter transformation of two variables x and g

$$R_t : \{x \to x' = x/t,\ g \to g' = \bar{g}(t, g)\} \ . \tag{11}$$

Here, x is the basic variable subject to a scaling transformation, while g is a physical quantity undergoing a more complicated functional transformation. To form a group, the transform R_t must satisfy the composition law

$$R_t R_\tau = R_{t\tau} \ ,$$

which yields the functional equation for $\bar{g}$:

$$\bar{g}(x,g) = \bar{g}\left(x/t, \bar{g}(t,g)\right) \ . \tag{12}$$

This equation has the same form as the functional equation (4) for the effective coupling in QFT in the massless case, that is, at $y = 0$. It is therefore clear that the contents of RG equation can be reduced to the group composition law.

In physical problems the second argument g of the transformation usually is related to the boundary value of a solution of the problem under investigation. This means that the symmetry underlying the RG approach is a symmetry of a solution (not of equation) describing the physical system at hand, involving a transformation of the parameters entering the boundary conditions.

Therefore, in the simplest case the renorm–group can be defined as a continuous one–parameter group of transformations of a solution of a problem fixed by a boundary condition. The RG transformation affects the parameters of a boundary condition and corresponds to changing the way in which this condition is introduced for *one and the same solution.*

Special cases of such transformations have been known for a long time. If we assume that $F = \bar{g}$ is a factored function of its arguments, then from Eq.(12) it follows that $F(z,f) = fz^k$, with k being a number. In this particular case the group transform takes the form

$$P_t \ : \ \{ \ z \to z' = z/t \ , \quad f \to f' = ft^k \ \} \ ,$$

which is known in mathematical physics long since as a power *self-similarity transformation.* More general case R_t with functional transformation law (11) can be characterized [49] as a *functional self–similarity* (FSS) transformation.

2.4 Recent application in mathematical physics

We can now answer the question concerning the physical meaning of the symmetry that underlies FSS and the Bogoliubov renorm–group. As we have already mentioned, it is not a symmetry of the physical system or the equations of the problem at hand, but a symmetry of a solution considered as a function of the essential physical variables and suitable boundary conditions. A symmetry like that can be related, in particular, to the invariance of a physical quantity described by this solution with respect to the way in which the boundary conditions are imposed. The changing of this way constitutes a group operation in the sense that the group property can be considered as the transitivity property of such changes.

Homogeneity is an important feature of the physical systems under consideration. However, homogeneity can be violated in a discrete manner. Imagine that such a discrete inhomogeneity is connected with a certain value of x, say,

$x = y$. In this case the RG transformation with canonical parameter t will have the form:

$$R_t \ : \ \{ \ x' = x/t \ , \quad y' = y/t \ , \quad g' = \bar{g}(t, y; g) \ \} \ . \tag{13}$$

The group composition law yields precisely the functional equation (4).

The symmetry connected with FSS is a very simple and frequently encountered property of physical phenomena. It can easily be "discovered" in various problems of theoretical physics: in classical mechanics, transfer theory, classical hydrodynamics, and so on [51–54].

Recently, some interesting attempts have been made to *use the RG concept in classical mathematical physics*, in particular, to study strong nonlinear regimes and to investigate asymptotic behavior of physical systems described by nonlinear partial differential equations (DEs).

About a decade ago the RG ideas were applied by late Veniamin Pustovalov with co-authors [56] to analyze a problem of generating of higher harmonics in plasma. This problem, after some simplification, was reduced to a couple of partial DEs with the boundary parameter – solution "characteristic" – explicitly included. It was proved that these DEs admit an exact symmetry group, that takes into account transformations of this boundary parameter, which is related to the amplitude of the magnetic field at a critical density point. The solution symmetry obtained was then used to evaluate the efficiency of harmonics generation in cold and hot plasma. The advantageous use of the RG-approach in solving the above particular problem gave promise that it may work in other cases and this was illustrated in [57] by a series of examples for various boundary value problems.

Moreover, in Refs. [51,57] the possibility of devising a regular method for finding a special class of symmetries of the equations in mathematical physics, namely, RG-type symmetries, was discussed. The latter are defined as solution symmetries with respect to transformations involving parameters that enter into the solutions through the equations as well as through the boundary conditions in addition to (or even rather than) the natural variables of the problem present in the equations.

As it is well known, the aim of modern group analysis [58,59], which goes back to works of Sophus Lie [60], is to find symmetries of DEs. This approach does not include a similar problem of studying the symmetries of solutions of these equations. Beside the main direction of both the classical and modern analysis, there also remains the study of solution symmetries with respect to transformations involving not only the variables present in the equations, but also parameters entering into the solutions from boundary conditions.

From the aforesaid it is clear that the symmetries which attracted attention in the 50s in connection with the discovery of the RG in QFT were those involving the parameters of the system in the group transformations. It is natural to refer to these symmetries related to FSS as the *RG-type symmetries.*

It should be noted that the procedure of revealing the RG symmetry (RGS), or some group feature, similar to RG regularity, in any partial case (QFT, spin lattice, polymers, turbulence and so on) up to now is not a regular one. In practice, it needs some imagination and atypical manipulation "invented" for every particular case — see discussion in [61]. By this reason, the possibility to find a regular approach to constructing RGS is of principal interest.

Recently a possible scheme of this kind was presented in application to mathematical model of physical system that is described by DEs. The leading idea [54,57,62] in this case is based on the fact that solution symmetry for such system can be found in a regular manner by using the well-developed methods of modern group analysis. The scheme that describes devising of RGS is then formulated [63] as follows.

Firstly, a specific RG-manifold should be constructed. Secondly, some auxiliary symmetry, i.e., the most general symmetry group admitted by this manifold is to be found. Thirdly, this symmetry should be restricted on a particular solution to get the RGS. Fourthly, the RGS allows one to improve an approximate solution or, in some cases, to get an exact solution.

Depending on both a mathematical model and boundary conditions, the first step of this procedure can be realized in different ways. In some cases, the desired RG-manifold is obtained by including parameters, entering into a solution via equation(s) and boundary condition, in the list of independent variables. The extension of the space of variables involved in group transformations, e.g., by taking into account the dependence of coordinates of renorm–group operator upon differential and/or non-local variables (which leads to the Lie-Bäcklund and non-local transformation groups [59]) can also be used for constructing the RG-manifold. The use of the Ambartsumian's invariant embedding method [64] and of differential constraints sometimes allows reformulations of a boundary condition in a form of additional DE(s) and enables one to construct the RG-manifold as a combination of original equations and embedding equations (or differential constraints) which are compatible with these equations. At last, of particular interest is the perturbation method of constructing the RG-manifold which is based on the presence of a small parameter.

The second step, the calculating of a most general group $\mathcal{G}$ admitted by the RG-manifold, is a standard procedure in the group analysis and has been described in detail in many texts and monographs – see, for example, [58,65,66].

The symmetry group $\mathcal{G}$ thus constructed can not as yet be referred to as a renorm–group. In order to obtain this, the next, third step should be done which consists in restricting $\mathcal{G}$ on a solution of a boundary value problem. This procedure utilizes the invariance condition and mathematically appears as a "combining" of different coordinates of group generators admitted by the RG-manifold.

The final step, i.e., constructing analytic expression for solution of boundary value problem on the basis of the RGS, usually presents no specific problems. A review of results, that were obtained on the basis of the formulated scheme can be found, for example, in [63,67,68].

Up to now the described regular method is feasible for systems that can be described by DEs and is based on the formalism of modern group analysis. However, it seems also possible to extend our approach on physical systems that are not described just by differential equations. A chance of such extension is based on recent advances in group analysis of systems of integro-differential equations [69,70] that allow transformations of both dynamical variables and functionals of a solution to be formulated [71]. More intriguing is the issue of a possibility of constructing a regular approach for more complicated systems, in particular to that ones having an infinite number of degrees of freedom. The formers can be represented in a compact form by functional (or path) integrals.

Acknowledgements

The author is indebted to D.V. Kazakov, V.F.Kovalev, B.V.Medvedev and to late V.V. Pustovalov for useful discussion and comments. This work was partially supported by grants of Russian Foundation for Fundamental Research (projects No 96-01-00195 and 96-15-96030) as well as by the Heisenberg–Landau program (JINR, Dubna) and the Werner Heisenberg MPG Institute at Munich.

References

1. L.D. Landau, A.A. Abrikosov and I.M. Khalatnikov, *Doklady AN SSSR*, **95** (1954) 497-499; 773-776; 1117-1120; **96** (1954) 261-264 (in Russian).; *Nuovo Cim.* Supp.3, 80-104.
2. N.N. Bogoliubov and D.V. Shirkov, Problems in quantum field theory. I and II, *Uspekhi Fiz. Nauk* **55**, (1955), 149-214 and **57**, (1955), 3-92 – in Russian; for German translation see *Fortschr. d. Physik* **3** (1955) 439-495 and **4** (1956) 438-517.
3. M. Gell-Mann and F. Low, Quantum Electrodynamics at Small Distances, *Phys. Rev.* **95** (1954) 1300-1312.
4. E.C.G. Stueckelberg and A. Petermann, La normalisation des constantes dans la theorie des quanta" *Helv. Phys. Acta*, **26** (1953) 499-520.
5. P.A.M. Dirac in Theorie du Positron (7-eme Conseil du Physique Solvay: Structure et propriete de noyaux atomiques, Octobre 1933), Gauthier-Villars, Paris,1934, pp.203-230.
6. N.N. Bogoliubov and D.V. Shirkov, On the renormalization group in QED, *Doklady AN SSSR* **103** (1955) 203-206 (in Russian).
7. N.N. Bogoliubov and D.V. Shirkov, Charge renormalization group in QFT, *Nuovo Cim.* **3** (1956) 845-637.
8. E.S. Fradkin, *Soviet Phys. JETP* **1** (1955).
9. L.D. Landau, On the Quantum Theory of Fields, In: Niels Bohr and the development of physics, Eds. W. Pauli et al., Pergamon, London, 1955, pp 52-69.
10. N.N. Bogoliubov and D.V. Shirkov, Lee type model in QED, *Doklady AN SSSR* **105** (1955) 685-688 (in Russian).
11. N.N. Bogoliubov and O.S. Parasyuk, *Doklady AN SSSR*, **100** (1955) 25–28, 429–432 (in Russian); *Acta Mathematica*, **97** (1957), 227–266.
12. N.N. Bogoliubov and D.V. Shirkov, Application of the renormalization group to improve the formulae of perturbation theory, *Doklady AN SSSR*, **103** (1955) 391-394 (in Russian).
13. D.V. Shirkov, Renormalization group with two coupling constants in the theory of pseudoscalar mesons, *Doklady AN SSSR* **105** (1955) 972-975 (in Russian).

14. I.F. Ginzburg, *Doklady AN SSSR*, **110** (1956) 535–538 (in Russian).
15. L.D. Landau and I.Ya. Pomeranchuk, *Doklady AN SSSR* **102** (1955) 489-492 (in Russian).
16. I.Ya. Pomeranchuk, *Doklady AN SSSR*, **103** (1955) 1005–1008; **105** (1955) 461–464 (in Russian); *Nuovo Cim.* **10** (1956) 1186-1203.
17. A.A. Logunov, *Soviet Phys. JETP* **3** (1956).
18. L.V. Ovsyannikov, *Doklady AN SSSR* **109** (1956) 1112-1115 (in Russian); English translation in In the Intermissions ... Ed. Yu.A.Trutnev, WS, 1998, 76-79.
19. N.N. Bogoliubov and D.V. Shirkov, Introduction to the theory of quantized fields, Nauka, Moscow 1957, 1973, 1976 and 1984 (in Russian); for English and French translations: see [20].
20. N.N. Bogoliubov and D.V. Shirkov, Introduction to the theory of quantized fields; two American editions 1959 and 1980, Wiley-Interscience, N.Y.; Bogolioubov N.N., Chirkov D.V. Introduction à la Theorie des Champes Quantique, Paris, Dunod, 1960.
21. K. Wilson, *Phys.Rev.* **179** (1969) 1499-1515.
22. G. Altarelli and G. Parisi, *Nucl. Phys.* **B 126** (1977) 298-318.
23. S. Weinberg, *Phys.Rev.* **D 8** (1973) 605-625.
24. D. Gross and P. Wilczek, *Phys. Rev.* **D 8** (1973) 3633-3652; H. Politzer. *Phys. Rev. Lett.* **30** (1973) 1346-1349.
25. D.I. Kazakov and D.V. Shirkov, Singular Solutions of RG Eqs. and the Symmetry of the Lagrangian, JINR Preprint E2-8974, 1975; in High Energy Particle Interaction (Proceed. 1975 Smolenice Conf.) Eds. D.Krupa and Pišút, Veda, Bratislava, 1976, 255-78.
26. W. Zimmermann, *Commun. Math. Phys.* **97** (1985) 211.
27. R. Oehme and W. Zimmermann, *Commun. Math. Phys.* **97** (1985) 569; R. Oehme, K. Sibold and W. Zimmermann, *Phys. Lett.* **B 147** (1984) 115; *Phys. Lett.* **B 153** (1985) 142.
28. W. Zimmermann, in Renormalization Group, Eds. D.V. Shirkov et al, WS, Singapore, 1988, pp 55 - 64.
29. A.V. Ermushev, D.I. Kazakov, O.V. Tarasov, *Nucl. Phys.* **B 281** (1987) 72
30. C. Lucchesi, O. Piguet and K. Sibold, *Phys. Lett.* **B 201** (1988) 241; *Helv. Phys. Acta* **61** (1988) 32.
31. D.I. Kazakov, *Phys. Lett.* **B 421** (1998) 211-216.
32. D.V. Shirkov, Threshold effects in two-loop approximation and parameterization of the real QCD, *Sov. J. Nucl. Phys.* **34(2)** (1981) 300-2; Mass Dependence in Renorm-Group Solutions, *Theor. Math. Fiz.* **49** (1981) 1039-42.
33. D.V. Shirkov, Perturbative analysis of general RG solution in a massive case, *Nucl. Phys.* **B 371** (1992) 467-81.
34. D.V. Shirkov, Mass Effects in Running Coupling Evolution and Hard Processes, in Perspectives in Particle Physics , Eds. D.Klabucar et al., WS, 1995, pp 1–13; On continuous mass-dependent analysis of DIS data, in Proc. EPSHEP95 Conf. (Bruxelles, July 1995), Eds. J.Lemonne et al., WS, pp 141-2.
35. D.V. Shirkov, Mass and Scheme Effects in Coupling Constant Evolution, Munich preprint MPI-Ph/92-94, Oct.1992; Published in English in *Teor. Math. Fizika* (1992) **93** 466-72.
36. D.V. Shirkov and I.L. Solovtsov, JINR Rapid Comm. No. 2[76]-96, 5, e-Print Archive: hep-ph/9604363; *Phys. Rev. Lett.* **79** (1997) 1209.
37. D.V. Shirkov, *Nucl. Phys.* (Proc. Suppl.) **B 64**, (1998) 106.
38. N.N.Bogoliubov, A.A.Logunov and D.V. Shirkov, *Sov. Phys. JETP* **10** (1959) 574-581.

39. K.A. Milton, I.L. Solovtsov and O.P. Solovtsova, *Phys. Lett.* **B 415** (1997) 104.
40. D.V. Shirkov, Renorm-Group, Causality and Non-power Perturbation Expansion in QFT, JINR preprint, E2-98-311, Dubna, 1998; e-Print Archive: hep-th/9810246, to appear in the April issue of *Theor. Math. Fiz.* **119** (1999) No.1.
41. K. Wilson, *Phys.Rev.* **B 4** (1971) 3174-3183.
42. L. Kadanoff, *Physica* **2** (1966) 263.
43. P.G. De Gennes, *Phys. Lett.* **38A** (1972) 339-340;
J. des Cloiseaux, *J. Physique (Paris)* **36** (1975) 281.
44. T.L. Bell et al, *Phys. Rev.* **A17** (1978) 1049-1057;
G.F. Chapline, *Phys. Rev.* **A21** (1980) 1263-1271.
45. C. DeDominicis, *Phys. Rev.* **A19** (1979) 419-422.
46. L. Adjemyan, A. Vasil'ev, and M. Gnatich, *Teoret. Mat. Fiz.* **58** (1984) 72; **65** (1985) 196; see also
A.N. Vasiliev, Quantum Field Renormalization Group in the Theory of Turbulence and in Magnetic Hydrodynamics in [47] , pp 146-159.
A.N. Vasiliev, Quantum Field RG in the Theory of Critical Behavior and Stochastic Dynamics Sankt-Peterburg, PINP Publ. House, 1998.
47. Renormalization Group, (Proceed. 1986 Dubna Conf.), Eds. D.V. Shirkov et al, WS, Singapore, 1988.
48. G. Pelletier, *J. Plasma Phys.* **24** (1980) 421-443.
49. D.V. Shirkov, The renormalization group, the invariance principle, and functional self-similarity, *Soviet Phys. Dokl.* **27** (1982) 197-199.
50. D.V. Shirkov, in Nonlinear and turbulent processes in physics, Ed. R.Z. Sagdeev, Harwood Acad.Publ., N.Y. 1984, v.**3**, pp 1637-1647.
51. D.V. Shirkov, Renormalization group in modern physics, in [47] pp 1-32; also *Int. J. Mod. Phys.* **A3** (1988) 1321-1341.
52. D.V. Shirkov, Renormalization group in different fields of theoretical physics, KEK Report 91-13, Feb. 1992.
53. M.A. Mnatsakanyan, *Soviet Phys. Dokl.* **27** (1982).
54. D.V. Shirkov, Several topics on renorm-group theory. In [55] , pp 1-10.
55. Renormalization Group '91 Proceed. of 1991 Dubna Conf., Eds. D.V. Shirkov, V.B. Priezzhev, WS, Singapore, 1992.
56. V.F. Kovalev, V.V. Pustovalov, Strong nonlinearity and generation of high harmonics in laser plasma, in Proceedings of the Conf. on Plasma Physics (Kiev, USSR, April 6-12, 1987), Ed. A.G. Sitenko (Naukova Dumka, Kiev, 1987), **1**, 271-273; Influence of laser plasma temperature on the high harmonics generation process, *ibid*, **1**, 274-277;
V.F. Kovalev, V.V. Pustovalov, Functional self-similarity in a problem of plasma theory with electron nonlinearity, *Theor. Math. Physics* **81**, No 1, 1060-1071 (1990).
57. V.F. Kovalev, S.V. Krivenko, V.V. Pustovalov, The Renormalization group method based on group analysis, in [55], 300-314.
58. L.V. Ovsyannikov, Group analysis of differential equations, Acad. Press, N.Y., 1982.
59. N.H. Ibragimov, Transformation groups applied to mathematical physics, Reidel Publ., Dordrecht–Lancaster 1985.
60. M. Sophus Lie. Gesammelte Abhandlungen. Leipzig–Oslo, Bd.5, 1924; Bd 6, 1927.
61. D.V. Shirkov, Bogoliubov renormgroup, *Russian Math. Surveys* **49:5**, 155-176 (1994) - with numerous misprints; for corrected version see: "The Bogoliubov Renormalization Group (second English printing), JINR Comm. E2-96-15, also in hep-th/9602024.

This article originally appeared in August 1994 in Russian as a preprint of Joint Institute for

Nuclear Research No. P2-94-310 being simultaneously submitted to "Uspekhi Math. Nauk". Unfortunately, in the second half of published version [*Uspehi Mat. Nauk*, v.49, No.5, (1994) 147-164], due to the editorail office irresposibility, there appeared more than 25 misprints related to the references of papers. In spite of the author's signal for the UMN editorial board, these errors have been reproduced in the English translation. Moreover, this latter publication [*Russian Math. Surveys*, 49:5 (1994) 155-176], due to the translator's poor qualification both in the subject terminology and Russian, contains a lot (ca 50) of additional errors distorting the author's text. By these reasons the author published anew the corrected English text in the form of JINR Communication E2-96-15. It is adequate to the Russian text of preprint P2-94-310.

62. D.V. Shirkov, Renormalization Group Symmetry and Sophus Lie Group Analysis, *Intern. J. Mod. Physics* **C 6** (1995) 503-512.
63. V.F. Kovalev, V.V. Pustovalov, D.V. Shirkov, Group analysis and renormgroup symmetries, *J.Math.Phys.* **39** (1998) 1170-1188.
64. To our knowledge, the embedding method was first introduced by V.A.Ambartzumyan, in *Astr.Journ.* **19** (1942) 30 – in Russian; later on this method enjoyed wide application to different problems – see, e.g. J.Casti, R.Kalaba, Imbedding methods in applied mathematics (Addison-Wesley, Reading, Ma, 1973) and references therein.
65. CRC Handbook of Lie Group Analysis of Differential Equations, Ed. N.H.Ibragimov (CRC Press, Boca Raton, Florida, USA). Vol.1: Symmetries, Exact Solutions and Conservation Laws, 1994; Vol.2: Applications in Engineering and Physical Sciences, 1995; Vol.3: New trends in theoretical developments and computational methods, 1996.
66. Peter J. Olver, Applications of Lie groups to differential equations, Springer, N. Y.,1986.
67. V.F.Kovalev, Group and renormgroup symmetries of boundary value problems, in Modern group analysis VI: developments in theory, computation and application, Eds. N.H.Ibragimov, F.M.Mahomed, 1997, New Age Internatl (P) Ltd Publishers, India, N. Delhi, 225-238.
68. V.F. Kovalev and D.V. Shirkov, Renormalization group in mathematical physics and some problems of laser optics, *J. Nonlin. Opt. Phys. & Mater.* **6** (1997) 443-454.
69. Yu.N. Grigoryev, S.V. Meleshko, *Sov. Phys. Dokl.* **32** (1987) 874-876.
70. V.F. Kovalev, S.V. Krivenko, V.V. Pustovalov, Group analysis of the Vlasov kinetic equation, *Diff.Equations* **29** (1993) No 10, 1568-1578; No 11, 1712-1721.
71. V.F. Kovalev, S.V. Krivenko, V.V. Pustovalov, Symmetry group of Vlasov-Maxwell equations in plasma theory, *J. Nonlin. Math. Phys.* **3** (1996) 175-180.

Algebraic Renormalization, from Supersymmetry to the Standard Model

Klaus Sibold

Institut für Theoretische Physik, Universität Leipzig,
Augustusplatz 10/11, D-04109 Leipzig, Germany

Abstract. It is reviewed how supersymmetry and BRS invariance can be established to all orders of perturbation theory without involving a specific subtraction scheme. The same algebraic technique can also be applied to the electroweak standard model and brings to light which postulates uniquely determine its structure.

1 Introduction

The year 1975 represents a landmark in renormalization theory. In that year the normal product algorithm originally developped by Zimmermann [1] for massive theories had been generalized by him and Lowenstein to include massless fields [2,3]. This opened the way to treat non-abelian gauge theories in a rigorous fashion, in fact independent of a specific renormalization scheme once the mathematical existence of Green functions was ensured. This had been demonstrated by Becchi, Rouet and Stora in the same year [4]. Using the idea of Piguet that a symmetry, more precisely a Ward identity, should characterize a model and not a Lagrangian[1] [5], they based their reasoning on the action principle [6] and consistency conditions which had been formulated before by Wess and Zumino in the context of effective theories [7]. So within one year an extremely powerful machinery was built up and called for its application. And the subject for application also waited around the corner: supersymmetry had just been discovered [8], a few papers on its renormalization had been written, but it became clear very soon that for supersymmetric Yang-Mills-theory an invariant regularization did not exist hence the recourse to the algebraic technique of BRS was unavoidable and the existence of normal products also in massless theories established at the right moment.

In the present note I would like to describe the program of renormalizing supersymmetric theories in this spirit and then go on to the electroweak standard model, where the same technique can be applied so fruitfully. My point of view will be purely subjective: I shall tell the story how I witnessed it.

2 Elements of supersymmetry

Central in supersymmetry is the algebra of its generators

$$\{Q_\alpha, \bar{Q}_{\dot\alpha}\} \quad = 2\sigma^\mu_{\alpha\dot\alpha} P_\mu \tag{1}$$

[1] "A Lagrangian is an opinion." (Becchi); or: "... a letter of intent." (Seiler)

P. Breitenlohner and D. Maison (Eds.): Proceedings 1998, LNP 558, pp. 177–191, 2000.

$$\{Q_\alpha, Q_\beta\} = 0 = \{\bar{Q}_{\dot\alpha}, \bar{Q}_{\dot\beta}\}. \tag{2}$$

Here Q_α is a *spinorial* charge, hence transforms under Lorentz-transformations according to

$$[M_{\mu\nu}, Q_\alpha] = -\tfrac{1}{2}\sigma_{\mu\nu\alpha}{}^{\beta} Q_\beta. \tag{3}$$

$\bar{Q} := (Q)^\dagger$ is its conjugate and we work in Weyl basis, i.e. the range of the indices α, $\dot\alpha$, ... is 1 and 2.

$$\begin{array}{ll} \sigma^\mu \equiv \{\mathbb{1}, \sigma^i\} & \bar\sigma^\mu \equiv \{\mathbb{1}, -\sigma^i\} \\ \sigma_{\mu\nu} \equiv \frac{i}{2}(\sigma_\mu\bar\sigma_\nu - \sigma_\nu\bar\sigma_\mu) & \sigma^i \ : \ \text{Pauli matrices} \end{array} \tag{4}$$

P_μ is the generator of translations. Linear representations of supersymmetry on fields, *superfields*, are to be obtained from the group action. If $G(a,\xi,\bar\xi) = e^{i(a^\mu P_\mu + \xi^\alpha Q_\alpha + \bar\xi_{\dot\alpha}\bar Q^{\dot\alpha})}$ is a group element it operates on a second one according to

$$G(a,\xi,\bar\xi)\, G(x,\theta,\bar\theta) = G(x + a + i\xi\sigma\bar\theta - i\theta\sigma\bar\xi, \theta + \xi, \bar\theta + \bar\xi) \tag{5}$$

and thus gives rise to an infinitesimal motion described by

$$\delta_\alpha = \frac{\partial}{\partial\theta^\alpha} - i\sigma^\mu_{\alpha\dot\alpha}\bar\theta^{\dot\alpha}\frac{\partial}{\partial x^\mu}. \tag{6}$$

It furnishes a linear representation via

$$i[Q_\alpha, \Phi(x,\theta,\bar\theta)] = \delta_\alpha\Phi. \tag{7}$$

Covariant derivatives are defined by

$$\begin{aligned} D_\alpha &\equiv \frac{\partial}{\partial\theta^\alpha} + i\sigma^\mu_{\alpha\dot\alpha}\bar\theta^{\dot\alpha}\frac{\partial}{\partial x^\mu} \\ \bar D_{\dot\alpha} &\equiv -\frac{\partial}{\partial\bar\theta^{\dot\alpha}} - i\theta^\alpha\sigma^\mu_{\alpha\dot\alpha}\frac{\partial}{\partial x^\mu}, \end{aligned} \tag{8}$$

meaning that $D_\alpha\Phi$, $\bar D_{\dot\alpha}\Phi$ are superfields if Φ is a superfield. They permit to impose covariant constraints

$$\begin{array}{lll} \bar D_{\dot\alpha} A = 0 & A \text{ chiral} & A = \mathcal{A} + \theta\psi + \theta^2 F \\ D_\alpha \bar A = 0 & \bar A \text{ antichiral} & \bar A = \bar{\mathcal{A}} + \bar\theta\bar\psi + \bar\theta^2 \bar F \\ DD\Phi = 0 & \Phi \text{ linear} & \\ \bar D\bar D\Phi = 0 & \Phi \text{ linear} & \\ \bar\Phi = \Phi & \Phi \text{ real} & \Phi = C + \theta\chi + \bar\theta\bar\chi + \frac{1}{2}\theta^2 M \\ & & \quad + \frac{1}{2}\bar\theta^2\bar M + \theta\sigma^\mu\bar\theta v_\mu + \frac{1}{2}\bar\theta^2\theta\lambda \\ & & \quad + \frac{1}{2}\theta^2\bar\theta\bar\lambda + \frac{1}{4}\theta^2\bar\theta^2 D \end{array} \tag{9}$$

Invariant actions can be formed out of the highest θ-components of superfields because they transform as total derivatives. I.e. $\int d^4x\,(\quad)_{F,D}$ is invariant. One can covariantly project onto them by:

$$\begin{array}{lll} \int d\bar S \equiv \int d^4x\, \bar D\bar D & & \text{chiral measure} \\ \int dS \equiv \int d^4x\, DD & & \text{antichiral measure} \\ \int dV \equiv \int d^4x\, DD\bar D\bar D & & \text{vector measure} \end{array} \tag{10}$$

One goes on building invariant actions by observing that sums and products of one type of superfield are superfields of the same type. The most notable examples are:
Chiral model (Wess-Zumino model)

$$\begin{aligned}\Gamma_{\mathrm{cl}}^{\mathrm{inv}} = & \tfrac{1}{16}\int \mathrm{dV}\, \left(\mathrm{e}^{i\theta\sigma\bar\theta\partial} A(x,\theta)\right)\left(\mathrm{e}^{-i\theta\sigma\bar\theta\partial}\bar A(x,\bar\theta)\right) \\ & -\tfrac{1}{4}\int \mathrm{dS}\, \left(\tfrac{m}{2}A^2(x,\theta)+\tfrac{g}{12}A^3(x,\theta)\right) + c.c. \\ = & \int \mathrm{d}^4x\, \left(\partial\mathscr{A}\partial\bar{\mathscr{A}} + \tfrac{1}{2}\psi i\sigma\partial\bar\psi + F\bar F\right) \\ + & \int \mathrm{d}^4x\, \left(m\mathscr{A}F - \tfrac{m}{4}\psi\psi + \tfrac{g}{4}\mathscr{A}^2F - \tfrac{3g}{8}\mathscr{A}\psi\psi + c.c.\right)\end{aligned} \tag{11}$$

Supersymmetric QED (SQED)

$$\begin{aligned}\Gamma_{\mathrm{cl}} = & \tfrac{1}{128}\int \mathrm{dV}\,\left(D\bar D\bar D D - \tfrac{1}{2\xi}\{DD,\bar D\bar D\} + 8M^2\right)\Phi \\ & + \tfrac{1}{16}\int \mathrm{dV}\,\left(A_+\mathrm{e}^{g\Phi}\bar A_+ + A_-\mathrm{e}^{-g\Phi}\bar A_-\right) - \tfrac{m}{4}\int \mathrm{dS}\, A_+A_- + c.c. \\ = & \int \mathrm{d}^4x\, v^\mu\left(\eta_{\mu\nu}\Box - \partial_\mu\partial_\nu) - \tfrac{1}{\xi}\partial_\mu\partial_\nu + M^2\eta_{\mu\nu}\right)v^\nu \\ + & \int \mathrm{d}^4x\left(D^\mu\mathscr{A}_\pm D_\mu\bar{\mathscr{A}}_\pm + \psi_\pm i\not{D}\,\psi_\pm + m\mathscr{A}_+F_- - \tfrac{m}{4}\psi_+\psi_- \right.\\ & \left. + \ldots + \ldots + c.c.\right)\end{aligned} \tag{12}$$

Supersymmetric Yang-Mills (SYM)

$$\begin{aligned}\Gamma_{\mathrm{cl}} = & -\tfrac{1}{128g^2}\mathrm{Tr}\int \mathrm{dS}\,\bar D\bar D\left(\mathrm{e}^{-\Phi}D^\alpha\mathrm{e}^{\Phi}\right)\bar D\bar D\left(\mathrm{e}^{-\Phi}D_\alpha\mathrm{e}^{\Phi}\right) + c.c. \\ & + \tfrac{1}{16}\int \mathrm{dV}\, A\mathrm{e}^{T^a\Phi^a}\bar A + \int \mathrm{dS}\,\lambda_{ijk}A_iA_jA_k + c.c. \\ = & -\tfrac{1}{4g^2}\,\mathrm{Tr}\int \mathrm{d}^4x\, F^{\mu\nu}F_{\mu\nu} + \ldots \\ + & \int \mathrm{d}^4x\, D^\mu\mathscr{A}D_\mu\bar{\mathscr{A}} + \ldots \\ + & \int \mathrm{d}^4x\,\left(-4\lambda_{ijk}\mathscr{A}_i\mathscr{A}_jF_k + \ldots\right)\end{aligned} \tag{13}$$

with $\Phi \equiv \Phi^a\tau^a$, $F_{\mu\nu} \equiv F^a_{\mu\nu}\tau^a$ and D^μ, $\not{D}$ gauge covariant derivatives.

3 Renormalization

The aim of renormalization is to construct finite Green functions to all orders of perturbation theory such that relevant symmetries are maintained and axioms can be proved. If the S-matrix exists it should be Lorentz invariant, unitary and causal.

3.1 Supersymmetry

Green functions are constructed with the help of the Gell-Mann-Low formula, amended by a subtraction scheme, rendering the diagrams finite. Since – as we shall indicate below – for massless vector superfields there arises an infrared problem already off-shell, we assume for the time being all fields to be massive. As subtraction scheme one may think of Zimmermann normal products suitably generalized to superspace [9]. Our reasoning will not depend essentially on this ingredient. On the classical level supersymmetry can be expressed by Ward identities (WI).

$$W_\alpha \Gamma_{\rm cl} = 0\,, \qquad \bar{W}_{\dot\alpha} \Gamma_{\rm cl} = 0\,, \tag{14}$$

with

$$W_\alpha \equiv -i \int \mathrm{d}z\, \delta_\alpha \underline{\phi} \frac{\delta}{\delta \underline{\phi}}\,, \qquad \bar{W}_{\dot\alpha} \equiv -i \int \mathrm{d}z\, \bar\delta_{\dot\alpha} \underline{\phi} \frac{\delta}{\delta \underline{\phi}} \tag{15}$$

and $\Gamma_{\rm cl}$ being the classical action. It is important that $\delta_\alpha \phi$ is linear in the fields. (dz denotes the appropriate measure.) The Ward identity operators satisfy an algebra:

$$\begin{aligned} \{W_\alpha, \bar{W}_{\dot\alpha}\} &= 2\sigma^\mu_{\alpha\dot\alpha} W^P_\mu \\ \{W_\alpha, W_\beta\} &= 0 = \{\bar{W}_{\dot\alpha}, \bar{W}_{\dot\beta}\} \end{aligned} \tag{16}$$

i.e. take over to the level of functionals the role of the charges Q_α. Perturbation theory consists in the loopwise expansion

$$\Gamma = \Gamma_{\rm cl} + \hbar \Gamma^{(1)} + \hbar^2 \Gamma^{(2)} + \ldots \tag{17}$$

of the generating functional Γ of one-particle-irreducible Green functions whose zeroth approximation can be identified with the classical action. Γ is translation invariant

$$W^P_\mu \Gamma = 0\,, \tag{18}$$

whereas for supersymmetry we only know from the action principle [6] that

$$W_\alpha \Gamma = \Delta_\alpha \cdot \Gamma\,, \qquad \bar{W}_{\dot\alpha} \Gamma = \bar\Delta_{\dot\alpha} \cdot \Gamma \tag{19}$$

with Δ_α, $\bar\Delta_{\dot\alpha}$ being *local* insertions. The algebra (16) now leads to consistency conditions for these insertions. They have to satisfy

$$\begin{aligned} W_\alpha(\bar\Delta_{\dot\alpha} \cdot \Gamma) + \bar{W}_{\dot\alpha}(\Delta_\alpha \cdot \Gamma) &= 0 \\ W_\alpha(\Delta_\beta \cdot \Gamma) + W_\beta(\Delta_\alpha \cdot \Gamma) &= 0 \\ \bar{W}_{\dot\alpha}(\bar\Delta_{\dot\beta} \cdot \Gamma) + \bar{W}_{\dot\beta}(\Delta_{\dot\alpha} \cdot \Gamma) &= 0\,. \end{aligned} \tag{20}$$

Eqns. (19) and (20) should actually hold even outside of perturbation theory, but within perturbation theory we can make use of the fact that for every local insertion

$$\Delta \cdot \Gamma = \Delta + o(\hbar \Delta)\,, \tag{21}$$

where the first term on the right hand side denotes the trivial contribution in terms of diagrams whereas the second stands for all contributions having at least one loop. The consistency conditions reduce in lowest oder to

$$\begin{aligned} W_\alpha \bar{\Delta}_{\dot\alpha} + \bar{W}_{\dot\alpha} \Delta_\alpha &= 0 \\ W_\alpha \Delta_\beta + W_\beta \Delta_\alpha &= 0 \\ \bar{W}_{\dot\alpha} \bar{\Delta}_{\dot\beta} + \bar{W}_{\dot\beta} \bar{\Delta}_{\dot\alpha} &= 0 \,, \end{aligned} \tag{22}$$

i.e. an algebraic problem for classical integrated field monomials! Its solution is given by a

Theorem:

$$\Delta_\alpha = W_\alpha \hat{\Delta} \,, \qquad \bar{\Delta}_{\dot\alpha} = \bar{W}_{\dot\alpha} \hat{\Delta} \,, \tag{23}$$

i.e. algebraically there is no anomaly for $N = 1$ supersymmetry [10].

Remarks:

1. *As fields ϕ are admitted: chiral, antichiral, vector.*
2. *The off-shell IR problem has been avoided by assuming masses for vector fields.*
3. *Spontaneous breaking of susy (i.e. shifts in W_α) are permitted.*

The theorem implies that susy can be restored if $\hat{\Delta}$ can be absorbed in Γ as a counterterm. This is always possible as far as UV power counting is concerned; it may be forbidden if its IR dimension is too low. An example is provided by the *O'Raifeartaigh model*:

$$\Gamma_{\text{cl}} = \Gamma_{\text{kin}} + \left(\int \text{dS} \, \left(\tfrac{\lambda}{4} A_0 + \tfrac{m}{4} A_1 A_2 + \tfrac{g}{32} A_0 A_1^2 \right) + c.c. \right) \tag{24}$$

λ, m, g are real; the model being defined by parity, $I : A_0 \to A_0$, $A_{1,2} \to -A_{1,2}$, R-symmetry: $\delta_R A = i(n + \theta\partial_\theta)A$, $n(A_0) = n(A_2) = -2$, $n(A_1) = 0$.

Susy turns out to be spontaneously broken: the multiplet A_0 is massless, it contains the Goldstone spinor; in multiplet A_1 the spinor has mass m, the scalar resp. pseudoscalar components have mass^2: $m^2 \pm \frac{1}{4}\lambda g$.

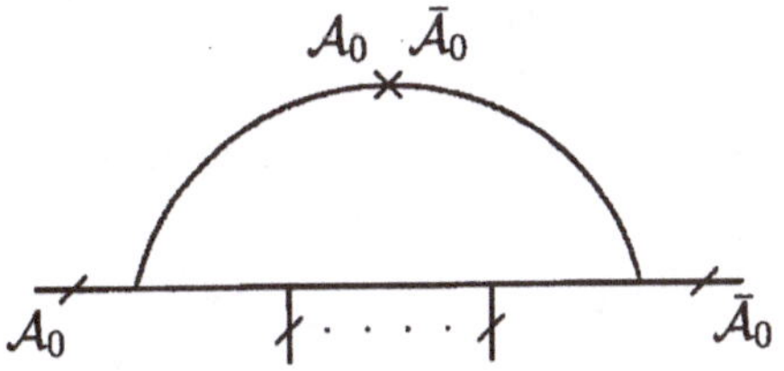

Fig. 1. IR anomaly

In higher orders [4]

$$W_\alpha \Gamma = u^{(1)} W_\alpha \int \text{d}^4x \, \mathscr{A}_0 \bar{\mathscr{A}}_0 + o(\hbar u) \,, \tag{25}$$

i.e. algebraically there is no obstruction, the breaking is the variation of a mass term but this term cannot be absorbed as counterterm because it could cause IR divergencies (see Fig. 1): *IR anomaly.*

Fig. 2. Sum over insertions

There is a (somewhat formal) solution to this problem [12]: just perform the infinite sum over all such insertions to a given $\mathcal{A}_0\bar{\mathcal{A}}_0$-line (see Fig. 2). This should yield a nonzero, computable mass for the field $\mathcal{A}$. The appropriate general setting has been formulated in [13] for purely scalar field models: admit a perturbation series as

$$\Gamma = \sum_{n,k(n)} \sqrt{\hbar^n}(\ln \hbar)^{k(n)} \Gamma^{(n,k)} . \tag{26}$$

For the O'Raifeartaigh model it turned out that the expansion

$$\Gamma = \sum_{n,k(n)} \hbar^n (\ln \hbar)^{k(n)} \Gamma^{(n,k)} \tag{27}$$

is realizable. Susy can then be strictly established:

$$W_\alpha \Gamma = 0 \qquad \bar{W}_{\dot{\alpha}} \Gamma = 0 , \tag{28}$$

the propagator of the field $\mathcal{A}_0$ has a pole (with computable position), this pole is invariant under the renormalization group [14].

3.2 Abelian gauge symmetry

The abelian gauge transformations read

$$\delta \Phi = i(\Lambda - \bar{\Lambda}) \qquad \begin{aligned} \delta A_\pm &= \mp i g \Lambda A_\pm \\ \delta \bar{A}_\pm &= \pm i g \bar{\Lambda} \bar{A}_\pm \end{aligned} \tag{29}$$

The classical action (12) is the solution of the local WI

$$w_\Lambda \Gamma \equiv \left(\bar{D}\bar{D}\frac{\delta}{\delta \Phi} - g A_+ \frac{\delta}{\delta A_+} + g A_- \frac{\delta}{\delta A_-} \right) \Gamma = \tfrac{1}{8\xi}(\Box + \xi M^2)\bar{D}\bar{D}\Phi \tag{30}$$

and its conjugate, some normalization conditions having been imposed. The main problem for establishing higher order Green functions originates from the propagator:

$$\langle T\Phi\Phi \rangle = -\frac{8i}{p^2 - M^2}\frac{\theta_{12}^4}{16} - i\xi \frac{\mathrm{e}^{-\bar{\theta}_1 \gamma \theta_2 p}\left(1 + \frac{\theta_{12}^4}{4}p^2\right)}{(p^2 - M^2)(p^2 - \xi M^2)} \tag{31}$$

$(\theta_{12}^4 \equiv \theta_{12}^2 \bar{\theta}_{12}^2 \equiv (\theta_1 - \theta_2)^2 (\bar{\theta}_1 - \bar{\theta}_2)^2)$
For vanishing M, $\xi \neq 1$ the propagator is no longer locally integrable at $p = 0$. In the abelian theory this problem can be *circumvented*. One constucts the massless theory at $\xi = 1$ which is stable and studies gauge parameter dependence in the massive version. For all quantities of physical interest the massless limit turns out to be controllable. The local WI (30) can be established to all orders. Like in ordinary QED it ensures unitarity ($M \neq 0$) since it says that the longitudinal parts $DD\Phi$, $\bar{D}\bar{D}\Phi$ of the vector field Φ are free. In the massless case the same is true, unitarity, of course, being only formal [15].

3.3 Non-abelian gauge symmetry

The non-abelian gauge transformations of a vector multiplet transform

$$e^{\Phi} \to e^{-i\Lambda} e^{\Phi} e^{i\bar{\Lambda}}, \tag{32}$$

$\Phi \equiv \Phi^a \tau^a$, $\Lambda \equiv \Lambda^a \tau^a$, $\bar{\Lambda} \equiv \bar{\Lambda}^a \tau^a$,
τ^a generates the fundamental representation of the simple gauge group G. The infinitesimal form of (32) reads

$$\delta\Phi = i(\Lambda - \bar{\Lambda}) + \tfrac{i}{2}[\Phi, \Lambda + \bar{\Lambda}] + \tfrac{i}{12}[\Phi, [\Phi, \Lambda - \bar{\Lambda}]] + \ldots \tag{33}$$

and permits the transition to the BRS transformations

$$\begin{aligned} s\Phi &= c_+ - \bar{c}_+ + \tfrac{1}{2}[\Phi, c_+ + \bar{c}_+] + \tfrac{1}{12}[\Phi, [\Phi, c_+ - \bar{c}_+]] + \ldots \\ &\equiv Q_s(\Phi) \end{aligned} \tag{34}$$

(s for special).

An important observation is that the requirement of nilpotency of the transformation law

$$s^2 \Phi = 0 \tag{35}$$

admits in $s\Phi$ an arbitrary parameter a_2,

$$s\Phi = Q_s(\Phi) + a_2\{\Phi, c_+ - \bar{c}_+\}' - \frac{a_2^2}{3}[\Phi, [\Phi, c_+ - \bar{c}_+]] + \ldots . \tag{36}$$

The prime at the anticommutator indicates that trace terms are to be omitted. A closer analysis in fact reveals that at *every* order in the field Φ new parameters may appear. The reason simply is that every new vector field

$$\Phi \to \mathscr{F}(\Phi) = \Phi + a_2\Phi^2 + a_3\Phi^3 + \ldots \tag{37}$$

is as good a dynamical variable as the old one. In $\Phi = C + \theta\chi + \ldots$ the component C has canonical dimension zero hence we have to expect generalized field amplitude renormalizations of the type (37) with a_2, a_3, ... as parameters. What saves the theory and prevents a disaster is the fact that these infinitely many parameters are *gauge* parameters. (The components C, χ, $\bar{\chi}$, M, $\bar{M}$ are *longitudinal* components of the gauge vector superfield Φ.)

Constructing Green functions proceeds in two steps: in step one vector fields are assumed to be massive, all desired WI's are established modulo soft breaking terms; in step two the off-shell infrared problem is solved by singling out which quantities of physical interest really exist.

For step one we impose the rigid WI, the gauge condition and the Slavnov-Taylor identity (ST)

$$\mathcal{W}\Gamma = 0 \tag{38}$$

$$\frac{\delta\Gamma}{\delta B} = \xi \bar{D}\bar{D}\bar{B} + \tfrac{1}{8}\bar{D}\bar{D}DD\Phi \tag{39}$$

$$\mathcal{S}(\Gamma) \sim 0\,. \tag{40}$$

Here $\mathcal{S}(\Gamma)$ is a Γ-bilinear functional

$$\mathcal{S}(\Gamma) \equiv \mathrm{Tr}\int\left(\frac{\delta\Gamma}{\delta\rho}\frac{\delta\Gamma}{\delta\phi} + \left(\frac{\delta\Gamma}{\delta\sigma}\frac{\delta\Gamma}{\delta c_+} + B\frac{\delta\Gamma}{\delta c_-} + \text{c.c.}\right) + \text{matter}\right) = 0 \tag{41}$$

which arises because one organizes the renormalization properties of non-linear field transformations in general by coupling them to external fields

$$\begin{aligned} sc_+ &= -c_+c_+ \longrightarrow \frac{\delta\Gamma}{\delta\sigma} & \qquad sc_- &= B \\ s\Phi &= Q_s(\Phi) \longrightarrow \frac{\delta\Gamma}{\delta\rho} & \qquad sB &= 0 \end{aligned} \tag{42}$$

In order to find consistency conditions one linearizes:

$$\mathcal{S}_\Gamma \equiv \mathrm{Tr}\int\left(\frac{\delta\Gamma}{\delta\rho}\frac{\delta}{\delta\Phi} + \frac{\delta\Gamma}{\delta\Phi}\frac{\delta}{\delta\rho} + \left(\frac{\delta\Gamma}{\delta\sigma}\frac{\delta}{\delta c_+} + \frac{\delta\Gamma}{\delta c_+}\frac{\delta}{\delta\sigma} + B\frac{\delta}{\delta c_-} + \text{c.c.}\right)\right.$$
$$\left. + \text{matter}\right) \tag{43}$$

and finds the crucial relations

$$\mathcal{S}_\Gamma\mathcal{S}(\Gamma) = 0 \qquad \forall\Gamma \tag{44}$$

$$\mathcal{S}_\Gamma\mathcal{S}_\Gamma = 0 \qquad \text{for } \Gamma \text{ with } \mathcal{S}(\Gamma) = 0\,. \tag{45}$$

They can be used to constrain the deviation from BRS invariance in higher orders

$$\mathcal{S}(\Gamma) = \Delta\cdot\Gamma = \Delta + o(\hbar\Delta) \tag{46}$$

$$\begin{aligned} \mathcal{S}_\Gamma\mathcal{S}(\Gamma) &= \mathcal{S}_\Gamma\Delta + o(\hbar\Delta) \\ 0 &= \mathcal{S}_{\Gamma_{\text{cl}}}\Delta\,. \end{aligned} \tag{47}$$

I.e. the local insertion Δ which in (46) was only restricted by power counting has to satisfy (47). Since furthermore

$$\mathcal{S}_{\Gamma_{\text{cl}}}\mathcal{S}_{\Gamma_{\text{cl}}} = 0\,, \tag{48}$$

one can utilize $\mathcal{s}_{\Gamma_{\rm cl}}$ to solve (47). Due to the vanishing dimension of Φ this is a fairly involved problem but it has been solved with the solution

$$\Delta = \mathcal{s}_{\Gamma_{\rm cl}} \hat{\Delta} + r\mathscr{A} \tag{49}$$

$$\mathscr{A} = \mathrm{Tr} \int \mathrm{dS}\, c_+ \bar{D}\bar{D}D\Phi\bar{D}\bar{D}D\Phi - \mathrm{Tr} \int \mathrm{d\bar{S}}\, \bar{c}_+ DD\bar{D}\Phi DD\bar{D}\Phi + o(4)\,. \tag{50}$$

The coefficient r is independent from all gauge parameters

$$\partial_{a_k} r = 0 \tag{51}$$

and has the value

$$r = \frac{1}{3 \cdot 2^{12}\pi^2} \frac{d^{ijk}}{d^2} \mathrm{Tr}\; T^i T^j T^k\,, \tag{52}$$

where T^i generates the representation of the matter fields

$$\begin{aligned} sA &= -c_+^i T^i A \\ s\bar{A} &= \bar{A} T^i \bar{c}_+^i\,. \end{aligned} \tag{53}$$

d^{ijk} is a totally symmetric tensor of the group G (e.g. $SU(n)$) [16]. One can in fact prove a

Non-Renormalization Theorem [17]:
$r \neq 0 \Longrightarrow r = r^{(1)}$.

I.e. if r is non-vanishing it starts with *one* loop.

Crucial for step two, the solution of the off-shell IR problem, is the observation that the redefinition (37) does not violate BRS invariance but rather redefines it. If we then perform the redefinition

$$\Phi \to (1 + \tfrac{1}{2}\mu^2\theta^2\bar{\theta}^2)\Phi\,, \tag{54}$$

we shall also not violate BRS, but – of course – supersymmetry. In $\Gamma_{\rm bil}$ this field transformation means

$$D^2 \to (D + 2\mu^2 C + \Box C)^2 \tag{55}$$

and leads to the new propagator

$$\langle TCC \rangle = \tfrac{-i}{4} \frac{\xi - g^2}{p^2 - \mu^2 + i\epsilon} \tag{56}$$

which is IR regulated! Hence the parameter μ^2 breaks supersymmetry softly, regulates IR-wise and is a *gauge* parameter like $a_2, a_3, \ldots$. The dependence of the theory on μ^2 can be followed by letting it vary under BRS

$$s\mu^2 = \nu^2 \tag{57}$$

like

$$sa_k = \xi_k\,, \tag{58}$$

ν^2 and χ_k being Grassmann parameters. They contribute to the ST identity. Establishing the enlarged ST identity then means that Green functions of BRS invariant operators exist IR-wise because they are gauge parameter independent. On these quantities supersymmetry is linearly realized [18].

These results had been derived prior to 1985. They are complemented by a study of the currents of the theory which form an interesting structure under superconformal transformations. For a review see [19].

In order to facilitate the transition to non-supersymmetric theories Piguet and collaborators reformulated renormalized supersymmetry in the Wess-Zumino gauge. The technical complication arises from the fact that susy is non-linearly realized [20]. Similarly the work on the supercurrent is being pushed forward on the level of multiple insertions of it [21].

4 The electroweak standard model

Since experimentally supersymmetry has not yet been found one could regard the algebraic method as an academic game within an academic field. That – to the contrary – it is an eminently practical and useful tool will follow by considering the most relevant model for physical applications: the electroweak standard model (SM).

4.1 The problems

If one attempts to renormalize the electroweak standard model to all orders of perturbation theory one is faced with the observation that no obvious invariant regularization is known. Dimensional regularization has to cope with the γ_5-problem, whereas BPHZ or analytic regularization spoil BRS invariance. Hence an all order treatment can only be based on the algebraic method which we exemplified above in the context of supersymmetric models. It has to deal with the following peculiarities specific to the SM:

1. The gauge group $SU(2) \times U(1)$ is not semisimple and the position of the unbroken $U(1)$ subgroup has to be determined and fixed in the course of renormalization.
2. The photon has to be kept massless. Its mixing with Z_μ has to be controlled such that a particle interpretation is possible. Off-shell IR problems have to be avoided.
3. $W_\mu^\pm$, Z_μ are unstable: the definition of their mass is non-trivial and gauge parameter dependence is a crucial issue.

In order to solve problem 1) one needs

- the Slavnov-Taylor identity,
- (deformed) rigid Ward identities,
- an abelian local Ward identity.

The solution of problem 2) requires

- careful IR power counting,
- suitable normalization conditions.

For problem 3) a complete solution to all orders is not yet known, but the ST identity and reasonable normalization conditions guarantee unitarity and permit the LSZ asymptotic limit at least in a formal sense.

Under the simplifying assumption that CP is maintained and families are not mixed problems 1) and 2) have been solved by E. Kraus [22]. The remarks that follow are based on this paper.

4.2 The abelian subgroup

It turns out that fixing the abelian subgroup is equivalent to finding equations whose solutions together with normalization conditions characterize uniquely the model. In order not to miss parameters or representations one *sharpens* the algebraic method: one does not give the WI-operators beforehand but prescribes only type (scalar, vector, spinor) and number of fields and the *algebra* of the WI-operators. For the rigid transformations one requires

$$[W_\alpha, W_\beta] = i\hat{\epsilon}_{\alpha\beta\gamma}\tilde{I}_{\gamma\gamma'}W_{\gamma'} \tag{59}$$

for consistency with the ST identity

$$W_\alpha \mathcal{S}(\Gamma) - \mathcal{S}_\Gamma W_\alpha \Gamma = 0 \qquad \forall \Gamma\,. \tag{60}$$

$\hat{\epsilon}_{\alpha\beta\gamma} : \begin{cases} \hat{\epsilon}_{+-3} = i \\ \hat{\epsilon}_{+-4} = 0 \end{cases}$ is totally antisymmetric.

$\tilde{I} = \begin{pmatrix} 0\,1\,0\,0 \\ 1\,0\,0\,0 \\ 0\,0\,0\,0 \\ 0\,0\,0\,0 \end{pmatrix}$ correlates $+,-$ of the electric charge.

We present a sample for the respective ansatz (contribution of some vector fields)

$$W_\alpha = \tilde{I}_{\alpha\dot{\alpha}} \int \mathrm{d}^4x \cdots V_b^\mu \hat{a}^V_{bc,\alpha'} \tilde{I}_{cc'} \frac{\delta}{\delta V^\mu_{c'}} + \cdots \tag{61}$$

$$\begin{aligned} \mathcal{S}(\Gamma) = \int & z_4(\sin\theta_3^g \partial_\mu c_Z + \cos\theta_3^g \partial_\mu c_A) \times \\ & \times \left(\sin\theta_4^V \frac{\delta\Gamma}{\delta Z_\mu} + \cos\theta_4^V \frac{\delta\Gamma}{\delta A_\mu}\right) + \cdots \\ & + \frac{\delta\Gamma}{\delta\rho_3^\mu} z_g \left(\cos\theta_3^V \frac{\delta\Gamma}{\delta Z_\mu} - \sin\theta_3^V \frac{\delta\Gamma}{\delta A_\mu}\right) + \cdots \end{aligned} \tag{62}$$

(Here the first two lines stand for linearly transforming vector pieces in ST, the third line for non-linearly transforming ones. ρ_3^μ is an external field coupled to a part of the BRS transformations of A_μ.)

The parameters $\hat{a}^V_{\dots}$, θ^g_3, $\theta^V_{3,4}$, z_4, z_g are to be chosen such that (59) and (60) is satisfied. In order to make this ansatz conceivable we give their tree approximation values in the conventional parametrizations.

$$\hat{a}^V_\alpha = \mathcal{O}^T(\theta_w)\hat{\epsilon}_\alpha \mathcal{O}(\theta_w) \tag{63}$$

$$\mathcal{O}(\theta_w) = \begin{pmatrix} 1 & 0 & 0 & 0 \\ 0 & 1 & 0 & 0 \\ 0 & 0 & \cos\theta_w & -\sin\theta_w \\ 0 & 0 & \sin\theta_w & \cos\theta_w \end{pmatrix} \tag{64}$$

$$\begin{aligned} \theta^g_3 &= \theta^V_3 = \theta^V_4 = \theta_w \\ z_4 &= z_g = 1 \end{aligned} \tag{65}$$

It is to be noted that electric charge and Faddeev-Popov-charge neutrality is naively maintained by the ansatz. Similarly one chooses the parameter values such that

$$W^*_+ = W_- \qquad W_{\substack{\pm\\3\\4}} \overset{\text{CP}}{\longrightarrow} -W_{\substack{\mp\\3\\4}} \,. \tag{66}$$

This work has to be performed for all sectors (vectors, scalars, fermions, ghosts); then the eqns. (59), (60) have to be solved and the solutions have to be parametrized such that the free parameters can either be fixed naively or via normalization conditions. It is most remarkable that a *non-diagonal* transformation

$$s\bar{c}_a = \hat{g}_{ab} B_b \tag{67}$$

is compatible with the algebra. If one has succeeded with this first step, namely solving the algebra, one can go on and find now in a second step the most general classical solution of the rigid WI

$$W_\alpha \Gamma_{\text{cl}} = 0 \tag{68}$$

and the ST identity

$$\mathcal{S}(\Gamma_{\text{cl}}) = 0 \,. \tag{69}$$

As experience tells one and as one confirms in the present case too, this yields all possible renormalizations in the form of possible redefinitions of fields and parameters. With the help of the action principle and the consistency conditions as inferred from (59), (60) one performs in an analogous manner the search for the solutions of the WI's (68), (69) to all orders. Here contact is made with the work of BBBC [23] because it turns out that this analysis can equivalently be performed in terms of physical fields ($W^\pm_\mu$, Z_μ, A_μ, ...). The absence of unitarity ruining anomalies follows from the structure of the standard multiplets.

In a third and last step one can now indeed proceed to the identitification of the abelian subgroup. First of all one has to note that the naive electromagnetic WI operator

$$W_{\text{em}} = \int d^4x\, w_{\text{em}} = i \int d^4x \sum_a Q^{\text{em}}_a \phi_a \frac{\delta}{\delta\phi_a} \tag{70}$$

is *not* abelian. In particular

$$w_{\mathrm{em}}\Gamma = \Box B_{\mathrm{em}} + Q_{\mathrm{em}} \cdot \Gamma \tag{71}$$

with Q_{em} a non-trivial insertion. On the non-integrated level it is not the electromagnetic direction which is abelian in the sense of having a trivial right hand side (which could then naively be constructed to all orders). It rather turns out $w_4^Q := w_{\mathrm{em}} - w_3$ is a good starting point leading eventually to

$$\hat{w}_4^Q = g_1 w_4^Q - \frac{1}{r_Z^V} \sin\theta^V \partial_\mu \frac{\delta}{\delta Z_\mu} - \frac{1}{r_A^V} \cos\theta^V \partial_\mu \frac{\delta}{\delta A_\mu} \,. \tag{72}$$

This operator is singled out by

$$[\hat{w}_4^Q, W_\alpha] \qquad = 0 \tag{73}$$

$$\mathcal{S}_\Gamma \hat{w}_4^Q \Gamma - \hat{w}_4^Q \mathcal{S}(\Gamma) = 0 \qquad \forall \Gamma \,. \tag{74}$$

It satisfies a local WI

$$\hat{w}_4^Q \Gamma = \frac{\sin\theta^V}{r_Z^V} \Box B_Z + \frac{\cos\theta^V}{r_A^V} \Box B_A \tag{75}$$

which can be postulated and established to all orders of perturbation theory because the right hand side is linear in propagating fields. This WI, which can only be required – due to its characterization by (73) and (74) – after the rigid WI and the ST identity have been established, fixes eventually the instabilities of the abelian subgroup.

The parameter

$$g_1 = \frac{e}{\cos\theta_w} + o(\hbar) \tag{76}$$

is in QED-like normalization conditions fixed on this local WI. The parameters θ^V, $r_{Z,A}^V$ are determined in $W_\pm$.

As far as interpretation is concerned one has to note that algebraically no distinction is possible between gauging the electromagnetic current or lepton- and quark number currents; hence this local WI is needed as additional requirement.

4.3 Photon/Z mixing

In order to keep the photon massless and also control the mixing one imposes as normalization conditions

$$\Gamma_{AA}^T(p^2 = 0) \qquad = 0 \tag{77}$$

$$\Gamma_{ZA}^T(p^2 = 0) \qquad = 0 \tag{78}$$

$$\mathrm{Re}\ \Gamma_{ZZ}^T(p^2 = M_Z^2) = 0 \,. \tag{79}$$

Eqns. (77) and (78) are automatic if one uses the BPHZL scheme [1–3]. The crucial point is now to check that one has indeed enough parameters at ones disposal to satisfy these normalization conditions after having arranged all WI's. Since IR dangerous terms like $\int \bar{c}_A c_Z$ are to be avoided as counterterms this task is non-trivial. It turns out that one can avoid off-shell IR danger by using the freedom left in the transformation law (67) of the antighost fields. This introduces a *ghost angle* θ_G as an important paramter into the theory.

All other masses are similarly introduced via two-point-functions in order to ensure poles. This in turn requires to have non-trivial parameter dependence in the rigid WI operators: they become deformed in higher orders. Likewise follows the antighost eqn. as a consequence of local WI and ST. Had one fixed and a priori prescribed WI operators and the antighost eqn. as a postulate one could not have fixed all masses as physical ones – as poles of propagators.

As a overall consistency check one derives the Callan-Symanzik equation because it controls the motion of all parameters of the theory under renormalization. The outcome is as follows:

$$\mathscr{C}\Gamma = \text{soft} \cdot \Gamma \tag{80}$$

- exists IR-wise,
- contains β-functions for mass ratios,
- shows that θ_w and θ_G are independently renormalized.

This yields a consistent picture.
We collect the result:

- the algebra of WI operators (59), (60)
- rigid WI + ST + local WI
- on-shell normalization conditions

determine Γ uniquely to all orders. The rigid WI operators are deformed; as a new parameter enters the angle θ_G in the ghost sector. It goes along with non-diagonal transformations (67) of the antighost in higher orders.

The examples of this review demonstrate that normal products, combined with the action principle and used algebraically realize perturbation theory in its most powerful fashion.

Acknowledgements

It is a pleasure to devote this note to Prof. W. Zimmermann on the occassion of his seventieth birthday.

References

1. W. Zimmermann, Commun. Math. Phys. **11** (1968) 1
   W. Zimmermann, Commun. Math. Phys. **15** (1969) 208
2. J.H. Lowenstein, W. Zimmermann, Commun. Math. Phys. **44** (1975) 73
3. J.H. Lowenstein, Commun. Math. Phys. **47** (1976) 53

4. C. Becchi, A. Rouet, R. Stora, Commun. Math. Phys. **42** (1975) 127
C. Becchi, A. Rouet, R. Stora, Ann. Phys. (N.Y.) **98** (1976) 287
5. O. Piguet, Commun. Math. Phys. **37** (1974) 19
6. Y.M.P Lam, Phys. Rev. **D 6** (1972) 2145, 2161
J.H. Lowenstein, Commun. Math. Phys. **24** (1971) 1
J.H. Lowenstein, T.E. Clark, Nucl Phys. **B 113** (1976) 109
7. J. Wess, B. Zumino, Phys. Lett. **37 B** (1971) 95
8. J. Wess, B. Zumino, Phys.Lett. **49 B** (1974) 52
J. Wess, B. Zumino, Nucl. Phys. **B 78** (1974) 1
S. Ferrara, B. Zumino, Nucl. Phys. **B 79** (1974) 413
S. Ferrara, O. Piguet, Nucl. Phys. **B 93** (1975) 261
9. T.E. Clark, O. Piguet, K. Sibold, Ann. Phys. (NY) **109** (1977) 418
10. O. Piguet, M. Schweda, K. Sibold, Nucl. Phys. **B 174** (1980) 183
11. T.E. Clark, O. Piguet, K. Sibold, Nucl. Phys. **B 119** (1977) 292
12. W.A. Bardeen, O. Piguet, K. Sibold, Phys. Lett. **72 B** (1977) 231
13. G. Bandelloni, C. Becchi, A. Blasi, R. Collina, Commun. Math. Phys. **67** (1979) 147
14. O. Piguet, M. Schweda, K. Sibold, Nucl. Phys. **B 168** (1980) 337
15. T.E. Clark, O. Piguet, K. Sibold, Nucl. Phys. **B 174** (1980) 491
T.E. Clark, O. Piguet, K. Sibold, Nucl. Phys. **B 169** (1980) 77
16. O. Piguet, K. Sibold, Nucl. Phys. **B 247** (1984) 484
17. O. Piguet, K. Sibold, unpublished
18. O. Piguet, K. Sibold, Nucl. Phys. **B 248** (1984) 336
O. Piguet, K. Sibold, Nucl. Phys. **B 249** (1984) 396
19. O. Piguet, K. Sibold, *Renormalized Supersymmetry*, Birkhäuser Boston 1986
20. N. Maggiore, O. Piguet, S. Wolf, Nucl. Phys. **B 458** (1996) 403, **B 476** (1996) 329
21. J. Erdmenger, C. Rupp, K. Sibold, Nucl. Phys. **B 530** (1998) 501
22. E. Kraus, Ann. Phys. (NY) **262** (1998) 155
23. G. Bandelloni, C. Becchi, A. Blasi, R. Collina, Ann. Inst. Henri Poincaré **28** (1978) 225, 255

Semi Classical Aspects of Gauge Theories

Raymond Stora

Laboratoire de Physique Théorique LAPTH, URA 1436,
Chemin de Bellevue, B.P. 110, F-74941 Annecy-le-Vieux Cedex, France

It is a pleasure to greet Wolfhart on his seventieth birthday. Not because 70 is a magic number -many of us are still surprised by the beauties of numbers-, but because we owe him a lot as far as our understanding of something is concerned. I am certainly one of those who benefited most. I believe we first met (in 1964) in New York when he and Kurt Symanzik were at New York University, as I was returning from a Symposium in Boulder about the Lorentz and the Poincaré group, a subject I was introduced to by Hans Joos in Princeton, in 1961-62. The contribution I presented there, in collaboration with P. Moussa, and about which I gave a seminar in New York was so close to E.P Wigner's famous 1939-article on the representations of the Poincaré

group that it took no less than V. Bargmann to explain E.P. Wigner what I was talking about! We met again in the Paris area -Wolfhart was visiting IHES, and I was back in Saclay- on which occasion we found we had complementary knowledges about the renormalization of massive QED. Wolfhart then proposed that we should write a small paper together, also with C. de Calan with whom I had discussed these matters in CERN for very practical reasons. Indeed Wolfhart did write this short paper (Lettere Nuovo Cimento **1** (1969) 877) signed by the three of us ... I must say that cosigning a paper with one of the authors of L.S.Z. was an honour for young people as we were at the time. In that same year, I had long discussions with Wolfhart about his efforts to understand how to renormalize gauge theories. Somehow, the times were not ripe, and the Faddeev Popov guess quite hard to recover using standard field theory methods not appealing to ill defined functional integrals, a standing problem, as it seems! Later, in 1973-74, when Carlo Becchi, Alain Rouet and I tried to understand the renormalization of gauge theories performed by G.'t Hooft and M. Veltman thanks to the dimensional regularization -renormalization scheme in terms of standard field theory techniques, we were lucky to have at our disposal the artillery prepared by Wolfhart, John Lowenstein, Peter Lam and already successfully applied to massive QED by John Lowenstein and Bert Schroer, and to symmetry breaking by John Lowenstein, Wolfhart and two early thirds of the future Marseille team. The rest of that story has been described in Carlo Becchi's talk. In these times, it has been an immense pleasure -it always is- to visit MPI, and later to meet again in Schloss Ringberg.

So much for good memories and down to present reality even though it may not be as pleasing as the old days.

In the remaining of this talk I want to mention two topics concerning gauge theories which I have been looking at reasonably closely.

P. Breitenlohner and D. Maison (Eds.): Proceedings 1998, LNP 558, pp. 192–196, 2000.

The first one, developed in Tobias Hurth's contribution concerns an approach to the renormalization of gauge theories which is not completed yet, but there is sufficient experimental evidence that it eventually will. It is grounded on the consequences of locality as an essential ingredient in the renormalization program as described by N.N. Bogoliubov and D.V. Shirkov and streamlined by H. Epstein, V. Glaser. It aims at defining renormalized gauge theories as solutions of an operator deformation problem of abelian gauge fixed gauge theories, the deformation being parametrized by local interactions. This is a quantum version of the so called iterated Noether method (c.f., for instance S. Deser, Gen. Rel. and Grav. I, **1** (1970) 9). It is conceptually more economical than the usual construction which provides a formal power series in $\hbar$, starting from a classical gauge fixed gauge theory. Here, the deformation parameter is a set of gauge coupling constants associated with a gauge Lie algebra. The Lie algebra structure stems from locality and an innocent looking Ward identity

$$[\mathcal{Q}, \mathcal{L}] = \partial_\mu \, \mathcal{K}_\mu$$

where $\mathcal{Q}$ is the Kugo Ojima asymptotic -abelian- operator which implements the asymptotic Slavnov symmetry, $\mathcal{L}$ the local interaction Lagrangian, and $\mathcal{K}_\mu$ some local "current". Suitably generalized to time ordered products of the $\mathcal{L}$'s, this has so far reproduced much of the gauge theories, in all examples which have been looked at. This program has all chances to reproduce gauge theories as we know them perturbatively. It is actively pursued in Zürich in G. Scharf's group, and outside, since former students are now experts.

The second topic concerns the celebrated Faddeev Popov gauge fixing procedure which proceeds through the factorization of the volume of the gauge group from the assumed gauge invariant formal functional integral and leads to a sound perturbative expansion. Even if one keeps in mind that formal manipulations on functional integrals need to be subjected to checks of some mathematical rigour, even at the formal level, some of the assumptions made in the Faddeev Popov argument are known to be a priori violated. As soon as one puts the theory inside a thermodynamic box (i.e.one does it on a compact riemannian space time), one knows (Singer 1978) that there does not exist an everywhere defined gauge function: this is a geometrical version of the celebrated Gribov problem. Therefore the construction is at best suitable for perturbation theory (since gauge choices always exist locally in field space). Besides, factoring out the gloriously infinite volume of the gauge group creates a certain psychological discomfort. Consequently, the Slavnov symmetry which characterizes the classical action used as a starting point to define the perturbative expansion might well be limited to the perturbative regime. We are going to see that the situation is slightly better. Putting together an argument in J. Zinn Justin's book which produces the Faddeev Popov gauge fixed action without factoring the volume of the gauge group, and the notion of vertical (i.e. along the gauge group) differentiation described for instance in C. Becchi's Zürich notes, one can recast the Faddeev Popov construction into the following form.

Let $\mathcal{A}$ be the space of gauge fields, $\mathcal{G}$ the gauge group. Let us assume for the sake of definiteness that $\mathcal{A}, \mathcal{G}$ and $\mathcal{A}/\mathcal{G}$ are finite dimensional smooth, but that

$\mathcal{G}$ is non compact. Let μ_{inv} be a $\mathcal{G}$ invariant differential form of maximal degree (dim $\mathcal{A} = |\mathcal{A}|$) on $\mathcal{A}$. Let $\mathcal{O}_{inv}$ be a $\mathcal{G}$ invariant function on $\mathcal{A}$. The question is whether one can define

$$< \mathcal{O}_{inv} > \stackrel{?}{=} \frac{\int_{\mathcal{A}} \mu_{inv} \mathcal{O}_{inv}}{\int_{\mathcal{A}} \mu_{inv}} \tag{1}$$

Let e_α be a basis of Lie $\mathcal{G}$ (of dimension dim $\mathcal{G} = |\mathcal{G}|$) and X_α the corresponding vertical vector fields on $\mathcal{A}$ and let $\bigwedge\limits_\alpha e_\alpha$ be a $\mathcal{G}$-invariant volume on Lie $\mathcal{G}$.

Define

$$\mu_{\mathbf{RS}} = \bigwedge_\alpha i(X_\alpha)\mu_{inv} \tag{2}$$

R. S. stands for Ruelle Sullivan (D. Ruelle and D. Sullivan, Topology **14** (1975) 319).

Obviously

$$i(X_v)\mu_{\mathbf{RS}} = 0 \tag{3}$$

for any vertical vector field (a linear combination of the X_α's): $\mu_{\mathbf{RS}}$ is horizontal. If $\mathcal{G}$ is unimodular (the left invariant Haar measure equals the right invariant Haar measure), then $\mu_{\mathbf{RS}}$ is closed:

$$\delta_{\mathcal{A}}\ \mu_{\mathbf{RS}} = 0 \tag{4}$$

where $\delta_{\mathcal{A}}$ is the exterior differential on $\mathcal{A}$. It is therefore invariant:

$$\ell(X)\mu_{\mathbf{RS}} = [i(X), \delta_{\mathcal{A}}]_+\ \mu_{\mathbf{RS}} = 0 \tag{5}$$

where X represents Lie $\mathcal{G}$ for the right action of $\mathcal{G}$ on $\mathcal{A}$.

$\mu_{\mathbf{RS}}$ is thus basic and defines a differential form $\tilde{\mu}_{\mathbf{RS}}$ of maximal degree (dim $\mathcal{A}/\mathcal{G} = |\mathcal{A}/\mathcal{G}|$) on $\mathcal{A}/\mathcal{G}$.

Then a tentative definition for $< \mathcal{O}_{inv} >$ is

$$< \mathcal{O}_{inv} >= -\frac{\int_{\mathcal{A}/\mathcal{G}} \tilde{\mu}_{\mathbf{RS}} \tilde{\mathcal{O}}_{inv}}{\int \tilde{\mu}_{\mathbf{RS}}} \tag{6}$$

where $\tilde{\mathcal{O}}_{inv}$ is the function on $\mathcal{A}/\mathcal{G}$ defined by $\mathcal{O}_{inv}$. This definition is independent of the definition of the volume form on Lie $\mathcal{G}$, up to scaling.

Now, cover $\mathcal{A}/\mathcal{G}$ by open sets U_i and use a partition of unity θ_i subordinated to it ($\Sigma\theta_i = 1$). Choose local sections σ_i over U_i, with local defining equations $g_i = 0$ and use the Faddeev Popov identity

$$\int_{fiber} \delta(g_i) \wedge \delta_{\mathcal{A}}\ g_i = 1 \tag{7}$$

where the first δ denotes the Dirac function and "fiber"is the copy $\mathcal{G}_{\dot{a}}$ of $\mathcal{G}$ over $\dot{a} \in U_i$.

So insert into both integrals in Eq. 6

$$\sum_i \theta_i(\dot{a}) \int_{\mathcal{G}(\dot{a})} \delta(g_i(a)) \wedge \delta g_i(a) = 1. \tag{8}$$

Now, introduce a $\mathcal{G}$ connection $\tilde{\omega}$ on $\mathcal{A}$

$$\delta_{\mathcal{A}}\ g_i(a) = \frac{\delta g_i}{\delta a}\ \delta a \equiv \int \frac{\delta g_i}{\delta a} \left(\tilde{\psi} - D_a \tilde{\omega}\right) \tag{9}$$

where

$$\tilde{\psi} = \delta a + D_a \tilde{\omega} \tag{10}$$

is the horizontal part of δa :

$$i(X_v)\tilde{\psi} = 0 \tag{11}$$

Since μ_{RS} is basic of maximal degree the term involving $\tilde{\psi}$ in Eq. 9 will drop out upon insertion into Eq. 6 and only the Faddeev Popov contribution to Eq. 9 involving

$$m_i = \frac{\delta g_i}{\delta a}\ D_a \tag{12}$$

will remain. Introducing a Faddeev Popov field ω to reconstruct μ_{inv} from μ_{RS} yields

$$< \mathcal{O}_{inv} >= \frac{\int_{\mathcal{A}} \mathcal{D}\omega \mu_{inv}\ \chi(a,\omega)\mathcal{O}_{inv}(a)}{\int_{\mathcal{A}} \mathcal{D}\omega \mu_{inv} \chi(a,\omega)} \tag{13}$$

where

$$\chi(a,\tilde{\omega}) = \sum_i \theta_i(\dot{a})\delta(g_i) \wedge m_i\ \tilde{\omega} \tag{14}$$

is a gauge fixing form and

$$\chi(a,\omega) = \sum_i \theta_i(\dot{a})\delta(g_i) \wedge m_i\ \omega \tag{15}$$

is the representative of $\chi(a,\tilde{\omega})$ in the quotient $\Omega^*(\mathcal{A})/\mathcal{I}_h^+$ where $\mathcal{I}_h^+$ is the differential ideal (for $\delta_{\mathcal{A}}$, acting on $\Omega^*\mathcal{A}$, the differential forms on $\mathcal{A}$) generated by horizontal forms of strictly positive degrees. From the structure equations

$$\begin{aligned}
\delta_{\mathcal{A}} a\ &= \tilde{\psi} - D_a\ \tilde{\omega} \\
\delta_{\mathcal{A}}\ \tilde{\psi}\ &= D_a\ \tilde{\Omega} + [\tilde{\omega}, \tilde{\psi}] \\
\delta_{\mathcal{A}}\ \tilde{\omega}\ &= \tilde{\Omega} - \frac{1}{2}[\tilde{\omega},\ \tilde{\omega}] \\
\delta_{\mathcal{A}}\ \tilde{\Omega} &= -[\tilde{\omega}, \tilde{\Omega}]
\end{aligned} \tag{16}$$

the differential s induced by $\delta_{\mathcal{A}}$ on $\Omega^*(\mathcal{A})/\mathcal{I}_h^+$ is given by

$$\begin{aligned}
sa &= -D_a\omega \\
s\omega &= -\frac{1}{2}[\omega,\omega]
\end{aligned} \tag{17}$$

namely, the geometrical part of the Slavnov symmetry, which, as suspected, is a robust geometrical ingredient.

The non geometrical part

$$\begin{aligned} s\bar{\omega} &= -ib \\ sb &= 0 \end{aligned} \tag{18}$$

on the other hand is connected with the Fourier integral representation of $\delta(g_i) \cdot \wedge m_i\ \omega$ and disappears from the global treatment.

Note the defining property of $\chi(a, \tilde{\omega})$

$$\int_{fiber} \chi(a, \tilde{\omega}) = 1 \tag{19}$$

$\chi(a, \tilde{\omega})|_{fiber}$ has compact support.

Assuming that $\mathcal{G}$ is connected, it follows that two gauge forms differ by a coboundary :

$$\chi_1(a, \omega) - \chi_2(a, \omega) = s\ \chi_{12}(a, \omega) \tag{20}$$

As expected, the choice of a connection $\tilde{\omega}$ has disappeared from the final answer (by the argument that the difference of two connections is horizontal and μ_{RS} is horizontal of maximal degree).

Finally, the above construction shows that the space of gauge fixing forms is non empty and convex.

These remarks conclude our rewriting of the Faddeev Popov argument, which although not computable as it stands is rid of the Gribov problem.

Even though gauge fixing can be rightly considered as one of the nightmares of gauge theories, its necessity is required by the present formulation which uses locality in field space for which so far no equivalent has been found which could be formulated directly on orbit space.

At this point, it is in order to apologize for the remoteness of the level of rigour adopted here from that of BPHZ.

W. ZIMMERMANN
1958, Novembre
Il Nuovo Cimento
Serie X, Vol. 10, pag. 597-614

On the Bound State Problem in Quantum Field Theory (*).

W. ZIMMERMANN (+)

Institute for Advanced Study - Princeton, N. J.
Physics Department, University of California - Berkeley, Cal.

(ricevuto il 1° Agosto 1958)

Summary. — A causal and invariant scalar field involving a stable bound state is investigated. A formula for the S-matrix is derived and it is shown that the bound state can be described by a local and invariant field operator. For simplicity only the case of spin zero particles and bound states is considered; however, the extension to other cases is possible.

Introduction.

For a relativistic quantum mechanical system with an energy momentum operator P_μ the one particle states are defined as eigen states of $-P_\mu^2$ with a discrete non-vanishing rest mass (1,2). In general, there may be several kinds of particles, each characterized by a certain value of rest mass, spin, charge, etc. Usually a division is made between elementary and composite particles. But it seems to be hard to define this distinction in a convincing manner. In the conventional formulation of quantum field theory each of the elementary

(*) This research was supported in part by the United States Air Force under contract no. AF 49(638)-327 monitored by the AF-Office of Scientific Research of the Air Research and Development Command.

(+) On leave of absence from the Max-Planck Institut für Physik, Göttingen, Germany.

(1) For a discussion of the particle aspect in quantum field theory compare R. HAAG: *Dan. Mat. Fys. Medd.*, **29**, 12 (1955), especially chap. 1, no. 1.

(2) We exclude the case of particles with zero rest mass.

P. Breitenlohner and D. Maison (Eds.): Proceedings 1998, LNP 558, pp. 199–216, 2000.

particles is described by a basic field operator whereas the composite particles (like the deuteron in the ground state, etc.) appear as the stable bound states of the system. But this definition of a composite particle depends, of course, on the formalism. The same particle which is regarded as an elementary particle in one formalism may be a composite particle in another [3,4].

In this paper we want to investigate whether the principle of microscopic causality sheds any new light on this question. We consider a model of a causal and invariant scalar field $A(x)$ describing just two kinds of particles, an elementary particle of mass m and a composite particle of mass M, both of spin zero. It will be shown that it is possible to define a field operator $B(x)$ for the composite particle (explicitly expressed in terms of the original field $A(x)$) which is in some respect analogous to the field operator of an elementary particle. The invariant operator $B(x)$ satisfies the requirement of microscopic causality, furthermore the S-matrix can be expressed by the vacuum τ-functions [5] of the field operator $A(x)$ and $B(x)$ exactly in the same way as in the case of elementary particles only. Therefore the principle of microscopic causality offers no possibility of distinguishing between elementary and composite particles. With respect to the S-matrix the elementary particles as well as the composite particles of the model are described by invariant field operators which satisfy the requirement of microscopic causality [6,7].

The formalism developed in this paper is closely related to other recent investigations. Starting from similar requirements one can derive the same results by applying the method of strong operator convergence in the form developed by HAAG [8,9]. An equivalent formalism was also obtained from the recursion formulae for retarded functions recently proposed by NISHIJIMA as an axiomatic formulation of quantum field theory [10,11].

(3) Compare E. FERMI and C. N. YANG: *Phys. Rev.*, **76**, 1739 (1949). In this paper the hypothesis that π-mesons may be composite particles is discussed.

(4) In this connection an interesting example is Gürsey's model of a theory of elementary particles (F. GÜRSEY: to be published). There, following a suggestion of Heisenberg, all particles appear as composite and no elementary particles correspond to the basic fields.

(5) The vacuum τ-functions are defined as the vacuum expectation values of multiple, time ordered, operator products.

(6) This possibility was first mentioned by N. BOGOLJUBOV: unpublished lecture notes.

(7) The S-matrix of the model considered is causal according to the definition given in H. LEHMANN, K. SYMANZIK and W. ZIMMERMANN: *Nuovo Cimento*, **6**, 319 (1957).

(8) R. HAAG: *Proc. of the Lille Conference on Mathematical problems of the quantum theory of fields* (1957), in print.

(9) R. HAAG: preprint, to be published.

(10) K. NISHIJIMA: *Progr. Theor. Phys.*, **17**, 765 (1957).

(11) K. NISHIJIMA: preprint, to be published.

1. – General conditions.

We consider the model of a neutral scalar field described by a hermitian operator $A(x)$ assumed to be invariant under the inhomogeneous Lorentz group. We assume that the principle of microscopic causality

$$[A(x), A(y)] = 0 \qquad \text{for } (x-y)^2 > 0 \tag{1}$$

holds and that no negative eigenvalues appear in the energy and rest mass spectrum. The operators $A(x)$ will be supposed to form an irreducible operator ring. For simplicity we assume that there are just two non-vanishing discrete eigenvalues m^2 and M^2 of $-P_\mu^2$ and that

$$(\Omega, A(x)\Phi) \not\equiv 0 \qquad \text{if} \quad -P_\mu^2\Phi = m^2\Phi$$

but

$$\left.\begin{aligned}(\Omega, A(x)\Psi) &\equiv 0\\ (\Omega, A(x)A(y)\Psi) &\not\equiv 0\end{aligned}\right\} \quad \text{if} \quad -P_\mu^2\Psi = M^2\Psi .$$

In addition we suppose that the states Φ and Ψ have spin zero.

This situation may occur for example in the case of A^4-coupling if there is an elementary particle of mass m and a stable two particle bound state of the mass M.

In order to describe the bound states we introduce the operator ([12])

$$B(x, \xi) = TA(x+\xi)A(x-\xi) \tag{2}$$

and define ([13]) incoming fields, according to WIGHTMAN ([14]), by

$$\begin{cases} A_{in}(x) = A(x) + \int \Delta_{Ret}(m, x-x')j(x')\,dx' ,\\[2mm] B_{in}(x,\xi) = B(x,\xi) + \int \Delta_{Ret}(M, x-x')j(x',\xi)\,dx' ,\end{cases} \tag{3}$$

([12]) ξ may be taken as spacelike or timelike 4-vector, but it is supposed that $\xi^2 \neq 0$.

([13]) The integral expressions in eq. (3) are to be understood in the sense of weak operator convergence which means that

$$(\Phi, A_{in}(x)\Psi) = (\Phi, A(x)\Psi) + \int \Delta_{Ret}(m, x-x')(\Phi, j(x')\Psi)\,dx' ,$$

between any normalizable state vectors Φ and Ψ.

([14]) The following definition of the asymptotic fields $A_{in}(x)$, $A_{out}(x)$ was first suggested by WIGHTMAN: private communication.

with the current operators

$$\text{(3')}\qquad \begin{cases} j(x) = (\Box_x - m^2)A(x)\,, \\ j(x,\xi) = (\Box_x - M^2)B(x,\xi)\,, \end{cases} \qquad \Box_x = \sum \frac{\partial^2}{\partial x_\mu^2}\,.$$

The outgoing fields $A_{\text{out}}(x)$, $B_{\text{out}}(x,\xi)$ are correspondingly defined with the help of the advanced functions $\Delta_{\text{Adv}}(m,x)$ and $\Delta_{\text{Adv}}(M,x)$. The general conditions under which the so defined operators $A_{\substack{\text{out}\\\text{in}}}(x)$ exist have been given and investigated in details by GREENBERG and can easily be extended to the case of the operators $B_{\substack{\text{out}\\\text{in}}}(x,\xi)$ ([15,16]).

Finally we demand that the incoming field operators $A_{\text{in}}(x)$, $B_{\text{in}}(x,\xi)$ together form an irreducible operator ring and correspondingly the outgoing fields $A_{\text{out}}(x)$, $B_{\text{out}}(x,\xi)$.

So far we have listed the general requirements which we need in the work which follows. We conclude this section by deriving some simple properties of the operators $A_{\substack{\text{out}\\\text{in}}}(x)$, $B_{\substack{\text{out}\\\text{in}}}(x,\xi)$. As a consequence of definition (3) they are solutions of the Klein-Gordon equations for the masses m and M, respectively:

$$\text{(4)}\qquad \begin{cases} (\Box_x - m^2)A_{\substack{\text{out}\\\text{in}}}(x) = 0\,, \\ (\Box_x - M^2)B_{\substack{\text{out}\\\text{in}}}(x,\xi) = 0\,. \end{cases}$$

We have the invariance properties

$$\text{(5)}\qquad \frac{\partial A_{\substack{\text{out}\\\text{in}}}(x)}{\partial x_\mu} = -i[P_\mu, A_{\substack{\text{out}\\\text{in}}}(x)]\,, \qquad \frac{\partial B_{\substack{\text{out}\\\text{in}}}(x,\xi)}{\partial x_\mu} = -i[P_\mu, B_{\substack{\text{out}\\\text{in}}}(x,\xi)]\,,$$

$$A_{\substack{\text{out}\\\text{in}}}(Lx) = U(L)A_{\substack{\text{out}\\\text{in}}}(x)U(L)^{-1}, \qquad B_{\substack{\text{out}\\\text{in}}}(Lx, L\xi) = U(L)B_{\substack{\text{out}\\\text{in}}}(x,\xi)U(L)^{-1}\,,$$

for an arbitrary Lorentz transformation L. ($U(L)$ denotes the unitary operator transforming $A(x)$ into $A(Lx)$.) These invariance properties are easy to prove if (3) is written in momentum space.

As a consequence of (4), (5) the vacuum expectation values of the incoming fields vanish:

$$\text{(6)}\qquad (\Omega, A_{\substack{\text{out}\\\text{in}}}(x)\Omega) = 0\,, \qquad (\Omega, B_{\substack{\text{out}\\\text{in}}}(x,\xi)\Omega) = 0\,.$$

For example,

$$M^2(\Omega, B_{\text{in}}(x,\xi)\Omega) = \sum \frac{\partial^2}{\partial x_\mu^2}(\Omega, B_{\text{in}}(x,\xi)\Omega) = 0\,,$$

because $P_\mu\Omega = 0$.

([15]) O. W. GREENBERG and A. S. WIGHTMAN: preprint, to be published.

([16]) In Sect. 2 we will use somewhat more restrictive conditions than Greenberg, for details see reference ([20]).

Furthermore $A_{\substack{out\\in}}(x)$ and $B_{\substack{out\\in}}(x, \xi)$ satisfy the asymptotic conditions

$$(7)\qquad \begin{cases} \lim\limits_{t\to\pm\infty} (\Phi, A_f(t)\Psi) = (\Phi, A_{\substack{out\\in}, f}\Psi), \\ \lim\limits_{t\to\pm\infty} (\Phi, B_F(t, \xi)\Psi) = (\Phi, B_{\substack{out\\in}, F}(\xi)\Psi), \end{cases}$$

with

$$(7')\qquad \begin{aligned} B_F(t, \xi) &= -i\int\limits_{x_0=t} d_3x \left\{B(x, \xi)\frac{\partial}{\partial x_0}F^*(x) - F^*(x)\frac{\partial}{\partial x_0}B(x, \xi)\right\}, \\ B_{\substack{out\\in}, F}(\xi) &= -i\int d_3x \left\{B_{\substack{out\\in}}(x, \xi)\frac{\partial}{\partial x_0}F^*(x) - F^*(x)\frac{\partial}{\partial x_0}B_{\substack{out\\in}}(x, \xi)\right\}, \end{aligned}$$

(correspondingly the definition of A_f and $A_{\substack{out\\in}, f}$) for any normalizable solution $f(x)$, $F(x)$ of the Klein-Gordon equation

$$(\Box - m^2)\, f(x) = 0\,,$$

$$(\Box - M^2)\, F(x) = 0\,.$$

Eq. (7) can be proved by forming the integral expression (7′) for both sides of (3) and taking the limit $t \to \pm\infty$.

2. - Commutation relations for the asymptotic fields.

2·1. *Elementary particles.* – Our first aim is to derive commutation relations for the incoming (or outgoing) fields defined in Sect. 1. We begin with the case of elementary particles and prove the relation ([17])

$$(8)\qquad [A_{in}(x), A_{in}(y)] = [A_{out}(x), A_{out}(y)] = i\,\Delta(m, x-y)$$

if $A(x)$ is normalized in the following manner. Let Φ_k be a one-particle state with energy momentum eigenvalue k_μ

$$\begin{cases} P_\mu \Phi_k = k_\mu \Phi_k\,, \\ (\Phi_k, \Phi_{k'}) = 2k_0^m\,\delta_3(\boldsymbol{k} - \boldsymbol{k}')\,, \end{cases} \qquad -k_\mu^2 = m^2, \quad k_0^m = |\sqrt{\boldsymbol{k}^2 + m^2}|.$$

([17]) This was already shown by GREENBERG, reference ([16]) under the assumption that the equal time commutator $[A(x)\dot{A}(y)]_{x_0=y_0}$ is a c-number. Here we use a different method which can be extended to the case of bound states.

Then it follows from translation invariance that

$$(\Omega, A(x)\Phi_k) = c\,\frac{\exp[ikx]}{(2\pi)^{\frac{3}{2}}},$$

with the constant

$$c = (2\pi)^{\frac{3}{2}}(\Omega, A(0)\Phi_k)\,.$$

Now $A(x)$ shall be normalized by the condition $c = 1$.

It is easy to determine the matrix elements of $A_{\text{in}}(x)$ between the vacuum state and an arbitrary state vector. From the definition (3) it follows that

$$(\Omega, A_{\text{in}}(x)\Phi_k) = (\Omega, A(x)\Phi_k) = \frac{\exp[ikx]}{(2\pi)^{\frac{3}{2}}},$$

for the one-particle state Φ_k. On the other hand if Φ is an eigenstate of $-P^2_\mu$ with a rest mass $\varkappa^2 \neq m^2$ we have

$$(\Omega, A_{\text{in}}(x)\Phi) = 0$$

because

$$(m^2 - \varkappa^2)(\Omega, A_{\text{in}}(x)\Phi) = 0$$

follows from (4) and (5).

With this result we can calculate the vacuum expectation value of the commutator (8). We see that

$$(\Omega, A_{\text{in}}(x)A_{\text{in}}(y)\Omega) = \int\frac{d_3k}{2k_0^m}(\Omega, A_{\text{in}}(x)\Phi_k)(\Phi_k, A_{\text{in}}(y)\Omega) = i\,\Delta^+(m, x-y)\,.$$

Hence,

(9) $$(\Omega, [A_{\substack{\text{out}\\\text{in}}}(x)A_{\substack{\text{out}\\\text{in}}}(y)]\,\Omega) = i\,\Delta(m, x-y)\,.$$

Now we turn to the operator form (8) and using the conditions of Sect. 1 we shall show that

(i) the commutators of the incoming and the outgoing fields coincide:

(10) $$[A_{\text{out}}(x), A_{\text{out}}(y)] = [A_{\text{in}}(x), A_{\text{in}}(y)]$$

and that

(ii) the commutators of the incoming field is a c-number:

(11) $$[A_{\text{in}}(x), A_{\text{in}}(y)] = (\Omega, [A_{\text{in}}(x), A_{\text{in}}(y)]\Omega)\,.$$

Relation (8) is, of course, a consequence of the statements (i), (ii) and Eq. (9).

In order to prove (i) and (ii) it is convenient to expand the field operators with respect to a complete orthonormal system $\{f_\alpha(x)\}$ of the positive frequency solutions of $(\Box - m^2)\, f(x) = 0$:

$$A(x) = \sum f_\alpha(x) A^+_\alpha(x_0) + \sum f^*_\alpha(x) A^-_\mu(x_0)\,,$$

$$A_{\substack{\text{out}\\ \text{in}}}(x) = \sum f_\alpha(x) A^+_{\substack{\text{out}\\ \text{in}}\,\alpha} + \sum f^*_\alpha(x) A^-_{\substack{\text{out}\\ \text{in}}\,\alpha}\,.$$

For the proof of statement (i) we start with the identity [18]

$$(12)\quad \int \mathrm{d}x \int \mathrm{d}y\, f^*_\alpha(x)\, f_\beta(y) K^m_x K^m_y\, TA(x)A(y) = \int \mathrm{d}y \int \mathrm{d}x\, f^*_\alpha(x) f_\beta(y) K^m_x K^m_y\, TA(x)A(y)\,,$$

$$K^m_x = \sum \frac{\partial^2}{\partial x^2_\mu} - m^2\,.$$

This relation is not self evident because there are simple examples of pathological fields which do not satisfy eq. (12). But using the fact that as a consequence of causality and spectrum conditions the vacuum expectation values

$$\big(\Omega,\, A(x_1) \cdot\cdot A(x_n)\, T(A(x), A(y))\, A(y_1) \cdot\cdot A(y_m)\, \Omega\big)$$

are boundary values of Wightman's analytical functions [19] a more detailed investigation [20] justifies the interchanging of the x- and y-integration in

$$\int \mathrm{d}x \int \mathrm{d}y\, f^*_\alpha(x) f_\beta(y) K^m_x K^m_y (\Omega,\, A(x_1) \ldots A(x_n) T(A(x), A(y)) A(y_1) \ldots A(y_m)\Omega)\,.$$

Then relation (12) holds for every matrix element of $TA(x)A(y)$ because, according to the irreducibility of the operators $A(x)$, any state vector can be written as a linear superposition of vectors $A(x_n) \ldots A(x_1)\Omega$.

Rearranging the integral on the left hand side with the help of Green's theorem we get

$$-i \int \mathrm{d}y\, f_\beta(y) K^m_y\, TA(x)A(y) = -i \int \mathrm{d}y_0 \frac{\partial}{\partial y_0} \int \mathrm{d}_3 y\, TA(x)A(y) \overset{\leftrightarrow}{\frac{\partial}{\partial y_0}} f_\beta(y) =$$

$$= A(x) A^-_{\text{in}\,\beta} - A^-_{\text{out}\,\beta} A(x)\,,$$

$$\left(f(x) \overset{\leftrightarrow}{\frac{\partial}{\partial x_0}} g(x) = f(x) \frac{\partial g(x)}{\partial x_0} - g(x) \frac{\partial f(x)}{\partial x_0} \right).$$

---

(18) In the sense of weak operator convergence (compare footnote (13)), correspondingly for all following operator expressions containing time integration.

(19) A. S. WIGHTMAN: *Phys. Rev.*, 101, 860 (1956).

(20) W. ZIMMERMANN: *Order of integrations in reduction formulae*, unpublished manuscript.

Carrying out the integration over x in the same way, we obtain

$$\int \mathrm{d}x \int \mathrm{d}y f_\alpha^*(x) f_\beta(y) K_x^m K_y^m TA(x)A(y) = A_{\mathrm{out}\beta}^- A_{\mathrm{out}\alpha}^+ - A_{\mathrm{out}\beta}^- A_{\mathrm{in}\alpha}^+ - A_{\mathrm{out}\alpha}^+ A_{\mathrm{in}\beta}^- + A_{\mathrm{in}\alpha}^+ A_{\mathrm{in}\beta}^- .$$

For the integral on the right hand side of (12) we have, correspondingly

$$\int \mathrm{d}y \int \mathrm{d}x f_\alpha^*(x) f_\beta(y) K_x^m K_y^m TA(x)A(y) = A_{\mathrm{out}\alpha}^+ A_{\mathrm{out}\beta}^- - A_{\mathrm{out}\alpha}^+ A_{\mathrm{in}\beta}^- - A_{\mathrm{out}\beta}^- A_{\mathrm{in}\alpha}^+ + A_{\mathrm{in}\beta}^- A_{\mathrm{in}\alpha}^+ .$$

Inserting these expressions into (12), we get

$$[A_{\mathrm{out}\alpha}^+, A_{\mathrm{out}\beta}^-] - [A_{\mathrm{in}\alpha}^+, A_{\mathrm{in}\beta}^-] = 0 .$$

With the corresponding relations between $A_{\substack{\mathrm{out}\\ \mathrm{in}}\alpha}^+$ and $A_{\substack{\mathrm{out}\\ \mathrm{in}}\beta}^+$ or $A_{\substack{\mathrm{out}\\ \mathrm{in}}\alpha}^-$ and $A_{\substack{\mathrm{out}\\ \mathrm{in}}\beta}^-$ statement (i) follows:

$$[A_{\mathrm{in}}(x), A_{\mathrm{in}}(y)] = [A_{\mathrm{out}}(x), A_{\mathrm{out}}(y)] .$$

The second statement (ii) may be derived from the identity

$$(13) \quad \int \mathrm{d}x \int \mathrm{d}y f_\alpha^*(x) f_\beta(y) K_x^m K_y^m TA(x)\, A(y)\, A(z) =$$
$$= \int \mathrm{d}y \int \mathrm{d}x f_\alpha^*(x) f_\beta(y) K_x^m K_y^m TA(x)\, A(y)\, A(z) ,$$

which can again be rearranged with the help of Green's theorem. The final result is

$$A(z)[A_{\mathrm{in}\alpha}^+, A_{\mathrm{in}\beta}^-] = [A_{\mathrm{out}\alpha}^+, A_{\mathrm{out}\beta}^-] A(z) .$$

Using statement (i), we get

$$[[A_{\mathrm{in}\alpha}^+, A_{\mathrm{in}\beta}^-]\, A(z)] = 0 ,$$

and this shows that $[A_{\mathrm{in}\alpha}^+, A_{\mathrm{in}\beta}^-]$ is a c-number because we have assumed that the $A(x)$ form an irreducible operator ring. With the corresponding results for $[A_{\mathrm{in}\alpha}^+, A_{\mathrm{in}\beta}^+]$, $[A_{\mathrm{in}\alpha}^-, A_{\mathrm{in}\beta}^-]$, we have proved the statement (ii).

2·2. *Bound states.* – In the next step we want to derive commutation relations for the field operators $B_{\substack{\mathrm{out}\\ \mathrm{in}}}(x, \xi)$. We begin with the proof of the following statements:

(i) The commutators of the incoming and outgoing fields coincide:

(14) $$[B_{\rm out}(x,\xi)\,B_{\rm out}(y,\eta)] = [B_{\rm in}(x,\xi)\,B_{\rm in}(y,\eta]\,.$$

(ii) The commutator of the incoming field is a c-number:

(15) $$[B_{\rm in}(x,\xi)\,B_{\rm in}(y,\eta)] = (\Omega, [B_{\rm in}(x,\xi)\,B_{\rm in}(x,\eta)]\,\Omega)\,.$$

We expand the operators B and $B_{\substack{\rm out\\ \rm in}}$ with respect to a complete orthonormal system $\{F_\alpha(x)\}$ of the positive frequency solutions of $(\Box - M^2)F(x) = 0$:

$$B(x,\xi) \quad = \sum F_\alpha(x)\,B^+_\alpha(x_0,\xi) + \sum F^*_\alpha(x)\,B^-_\alpha(x_0,\xi)\,,$$

$$B_{\substack{\rm out\\ \rm in}}(x,\xi) = \sum F_\alpha(x)\,B^+_{\substack{\rm out\\ \rm in}\,\alpha}(\xi) + \sum F^*_\alpha(x)\,B^-_{\substack{\rm out\\ \rm in}\,\alpha}(\xi)\,.$$

The coefficients are determined by ([21])

$$B^\pm_\alpha(x_0,\xi) = \mp\, i\!\int\! d_3x \left\{B(x,\xi)\,\frac{\partial}{\partial x_0}\,F^\mp_\alpha(x) - F^\mp_\alpha(x)\,\frac{\partial}{\partial x_0}\,B(x,\xi)\right\},$$

$$F^+_\alpha(x) = F_\alpha(x),\; F^-_\alpha(x) = F^*_\alpha(x),$$

(and similarly for $B^\pm_{\substack{\rm out\\ \rm in}\,\alpha}(\xi)$).

As a consequence of causality and spectrum conditions we have the identity

(16) $$\int\! dx \!\int\! dy\, F^*_\alpha(x) F_\beta(y) K^M_x K^M_y T A(x+\xi) A(x-\xi) A(y+\eta) A(y-\eta) =$$

$$= \int\! dy \!\int\! dx\, F^*_\alpha(x) F_\beta(y) K^M_x K^M_y\, T A(x+\xi) A(x-\xi) A(y+\eta) A(y-\eta)\,.$$

After integration over x_0 and y_0, this relation yields

$$[B^+_{{\rm out}\,\alpha}(\xi),\, B^-_{{\rm out}\,\beta}(\eta)] - [B^+_{{\rm in}\,\alpha}(\xi),\, B^-_{{\rm in}\,\beta}(\eta)] = 0\,,$$

and similarly for $[B^\pm_{{\rm out}\,\alpha}(\xi),\, B^\pm_{{\rm out}\,\beta}(\eta)]$. So we have proved that

$$[B_{\rm out}(x,\xi),\, B_{\rm out}(y,\eta)] = [B_{\rm in}(x,\xi),\, B_{\rm in}(y,\eta)]\,.$$

([21]) If ξ is spacelike we have

$$B^-_\alpha(x_0,\xi) = B^+_\alpha(x_0,\xi)^*\,.$$

This relation is no longer true if ξ is a timelike vector.

Statement (ii) may be derived from

(17) $$\int dx \int dy F_\alpha^*(x) F_\beta(y) K_x^M K_y^M TA(x+\xi)A(x-\xi)A(y+\eta)A(y-\eta)A(z) =$$
$$= \int dy \int dx F_\alpha^*(x) F_\beta(y) K_x^M K_y^M TA(x+\xi)A(x-\xi)A(y+\eta)A(y-\eta)A(z)\,,$$

which yields

$$[B_{\text{out}\,\alpha}^+(\xi)\, B_{\text{out}\,\beta}^-(\eta)]A(z) = A(z)[B_{\text{in}\,\alpha}^+(\xi)\, B_{\text{in}\,\beta}^-(\eta)]$$

or

$$[[B_{\text{in}\,\alpha}^+(\xi)\, B_{\text{in}\,\beta}^-(\eta)]A(z)] = 0\,.$$

Therefore the commutator $[B_{\text{in}}(x, \xi)\, B_{\text{in}}(y, \eta)]$ is a c-number (statement ii).

Now we can determine the commutator (15) by calculating the vacuum expectation value. For this purpose we determine first the matrix elements of $B_{\text{in}}(x, \xi)$ between the vacuum state and an arbitrary state vector.

If Φ_k is an eigenstate of P_μ belonging to the eigenvalue k_μ and the rest mass $-k_\mu^2 = M^2$ we have

$$(\Omega, TA(x+\xi)A(x-\xi)\Phi_k) = \frac{\exp[ikx]}{(2\pi)^{\frac{3}{2}}} F_k(\xi)\,,$$

with

(18) $$F_k(\xi) = (2\pi)^{\frac{3}{2}} (\Omega, TA(\xi)A(-\xi)\Phi_k)$$

and

$$(\Box_x + M^2)(\Omega, TA(x+\xi)A(x-\xi)\Phi_k) = 0\,.$$

From this and the definition (3) of $B_{\text{in}}(x, \xi)$ it follows that

(19) $$(\Omega, B_{\text{in}}(x, \xi)\Phi_k) = (\Omega, B(x, \xi)\Phi_k) = \frac{\exp[ikx]}{(2\pi)^{\frac{3}{2}}} F_k(\xi)\,.$$

On the other hand if Φ is an eigenstate belonging to the eigenvalue k_μ with the rest mass $-k_\mu^2 \neq M^2$ we get from (5):

$$(k_\mu^2 + M^2)(\Omega, B_{\text{in}}(x, \xi)\Phi) = -(\Box_x - M^2)(\Omega, B_{\text{in}}(x, \xi)\Phi) = 0$$

hence

$$(\Omega, B_{\text{in}}(x, \xi)\Phi) = 0\,.$$

Therefore we have

$$(\Omega, B_{\text{in}}(x, \xi)B_{\text{in}}(y, \eta)\Omega) = \int \frac{d_3k}{2k_0^M} (\Omega, B_{\text{in}}(x, \xi)\Phi_k)(\Phi_k, B_{\text{in}}(y, \eta)\Omega) =$$
$$= \frac{1}{(2\pi)^3} \int \frac{d_3k}{2k_0^M} F_k(\xi)\bar{F}_k(\eta) \exp[ik(x-y)]\,,$$

where we have used the invariance under space time reflection for

$$(\Phi_k, TA(\eta)A(-\eta)\Omega) = (\Omega, TA(\eta)A(-\eta)\Phi_k) = F_k(\eta) . \tag{20}$$

Introducing the Fourier transform $B_{in}(k, \xi)$, we have

$$B_{in}(x, \xi) = \frac{1}{(2\pi)^{\frac{3}{2}}} \int d_4 k \exp [ikx] \delta(k^2 + M^2) B_{in}(k, \xi) ,$$

and defining creation and annihilation operators

$$B_{in}^+(\boldsymbol{k}, \xi) = B_{in}(k, \xi) , \qquad \text{for } k_0 = +|\sqrt{\boldsymbol{k}^2 + M^2}| ,$$

$$B_{in}^-(\boldsymbol{k}, \xi) = B_{in}(-k, \xi) , \qquad \text{for } k_0 = -|\sqrt{\boldsymbol{k}^2 + M^2}| ,$$

we get from (19) and (15) the final commutation relations

$$[B_{in}^+(\boldsymbol{k}, \xi) B_{in}^-(\boldsymbol{p}, \eta)] = F_k(\xi) F_p(\eta) 2k_0'' \delta_3(\boldsymbol{k} - \boldsymbol{p}) , \tag{21}$$

$$[B_{in}^\pm(\boldsymbol{k}, \xi) B_{in}^\pm(\boldsymbol{p}, \eta)] = 0 . \tag{22}$$

Finally we only mention that the same methods used in this Section to prove the commutation relations (8) and (21) may be applied in order to derive

$$[A_{in}(x), B_{in}(y, \eta)] = 0 . \tag{23}$$

3. – Derivation of an S-matrix formula.

The operators $B_{\substack{out\\in}}(x, \xi)$ depend on the relative co-ordinate ξ of the bound state. For the S-matrix we are only interested in the center of mass motion of the bound states and want to carry out the limit $\xi \to 0$. Therefore we define operators $B_{\substack{out\\in}}(x)$ by

$$\begin{cases} B_{\substack{out\\in}}(x) = \lim\limits_{\xi \to 0} \dfrac{B_{\substack{out\\in}}(x, \xi)}{F_0(\xi)} , \\ F_0(\xi) = (2\pi)^{\frac{3}{2}} (\Omega, TA(\xi)A(-\xi)\Phi_0) , \end{cases} \tag{24}$$

where Φ_0 is the bound state at rest.

In order to prove the existence of the limits $\xi \to 0$ in (24) we divide (21) by $F_k(\xi)$ and get

$$\left[\frac{B_{in}^+(\boldsymbol{k}, \xi)}{F_k(\xi)}, B_{in}^-(\boldsymbol{p}, \eta)\right] = F_k(\eta) \delta_3(\boldsymbol{k} - \boldsymbol{p}) .$$

Differentiation with respect to ξ_μ yields

$$\left[\frac{\partial}{\partial\xi_\mu}\frac{B^+_{\text{in}}(k,\xi)}{F_k(\xi)},\ B^-_{\text{in}}(p,\eta)\right]=0\,,$$

and correspondingly from (22), (23)

$$\left[\frac{\partial}{\partial\xi_\mu}\frac{B^+_{\text{in}}(k,\xi)}{F_k(\xi)},\ B^+_{\text{in}}(p,\eta)\right]=0\,,$$

$$\left[\frac{\partial}{\partial\xi_\mu}\frac{B^+_{\text{in}}(k,\xi)}{F_k(\xi)},\ A^\pm_{\text{in}}(p)\right]=0\,.$$

Thus $(\partial/\partial\xi_\mu)(B^+_{\text{in}}(k,\xi)/(F_k(\xi))$ commutes with all operators $A_{\text{in}}(x)$ and $B_{\text{in}}(y,\eta)$. Since we have assumed that the operators $A_{\text{in}}(x)$, $B_{\text{in}}(y,\eta)$ together form an irreducible operator ring $(\partial/\partial\xi_\mu)(B^+_{\text{in}}(k,\xi)/F_k(\xi))$ is a c-number:

$$\frac{\partial}{\partial\xi_\mu}\frac{B^+_{\text{in}}(k,\xi)}{F_k(\xi)}=\frac{\partial}{\partial\xi_\mu}\frac{(\Omega,B^+_{\text{in}}(k,\xi)\Omega)}{F_k(\xi)}=0\,,$$

because of (6). Therefore the expression

$$\frac{B^+_{\text{in}}(k,\xi)}{F_k(\xi)}=B^+_{\text{in}}(k)\,,$$

is independent of ξ and

(25) $$B_{\text{in}}(k)=\lim_{\xi\to 0}\frac{B_{\text{in}}(k,\xi)}{F_0(\xi)}\,,$$

does exist because

$$\frac{F_k(\xi)}{F_0(\xi)}=\frac{F(k\xi,\xi^2)}{F(0,\xi^2)}\,,$$

is in the limit $\xi\to 0$ independent of k:

$$\lim_{\xi\to 0}\frac{F_k(\xi)}{F_0(\xi)}=1\,.$$

From (21), (22), (23) we get:

$$[B^+_{\text{in}}(\boldsymbol{k}),\ B^-_{\text{in}}(\boldsymbol{p})]=2k_0''\delta_3(\boldsymbol{k}-\boldsymbol{p})\,,$$

$$[B^\pm_{\text{in}}(\boldsymbol{k}),\ B^\pm_{\text{in}}(\boldsymbol{p})]=0\,,$$

$$[B^\pm_{\text{in}}(\boldsymbol{k}),\ A_{\text{in}}(y)]=0\,.$$

Hence

$$[B_{\substack{\text{out}\\ \text{in}}}(x),\ B_{\substack{\text{out}\\ \text{in}}}(y)] = i\Delta(M,\ x-y)\,, \tag{26}$$

$$[B_{\substack{\text{out}\\ \text{in}}}(x),\ A_{\substack{\text{out}\\ \text{in}}}(y)] = 0\,. \tag{27}$$

Further we have

$$(\Box - M^2)\, B_{\substack{\text{out}\\ \text{in}}}(x) = 0$$

and

$$(\Omega,\ B_{\substack{\text{out}\\ \text{in}}}(x)\Phi_k) = \frac{\exp[ikx]}{(2\pi)^{\frac{3}{2}}}\,, \qquad k^2 + M^2 = 0,$$

if Φ_k is the bound state with four momentum k_μ. $B_{\substack{\text{out}\\ \text{in}}}(x)$ is given in terms of $A(x)$ by

$$B_{\substack{\text{out}\\ \text{in}}}(x) = \lim_{\xi\to 0} \frac{1}{F_0(\xi)} \left\{ TA(x+\xi)A(x-\xi) + \int dx'\, \Delta_{\substack{\text{Adv}\\ \text{Ret}}}(M,\ x-x')J(x',\ \xi) \right\}, \tag{28}$$

$$J(x,\ \xi) = (\Box_x - M^2)\, TA(x+\xi)A(x-\xi)\,.$$

According to (25) $B^{+}_{\text{in}}(k,\ \xi)$ is a multiple of $B_{\text{in}}(k)$. From this fact we conclude that the operators $A_{\text{in}}(x)$, $B_{\text{in}}(y)$ together already form an irreducible operator ring. Therefore the whole Hilbert space $\mathcal{H}$ can be built up by the creation operators

$$A^{-}_{\substack{\text{out}\\ \text{in}}}(k)\,, \qquad B^{-}_{\substack{\text{out}\\ \text{in}}}(k)$$

of incoming or outgoing elementary particles or bound states of momentum k. The state vectors

$$\Phi^{k_1\ldots k_n}_{\text{in}} = A^{-}_{\text{out}}(\boldsymbol{k}_1)\ldots A^{-}_{\text{in}}(\boldsymbol{k}_l)\, B^{-}_{\text{in}}(\boldsymbol{k}_{l+1})\ldots B^{-}_{\text{in}}(\boldsymbol{k}_n)\,\Omega \tag{29}$$

as well as

$$\Phi^{k_1\ldots k_n}_{\text{out}} = A^{-}_{\text{out}}(\boldsymbol{k}_1)\ldots A^{-}_{\text{out}}(\boldsymbol{k}_l)\, B^{-}_{\text{out}}(\boldsymbol{k}_{l+1})\ldots B^{-}_{\text{out}}(\boldsymbol{k}_n)\,\Omega, \tag{30}$$
$$(k_i^2 = -m^2 \quad \text{for} \quad i < l\,, \quad k_i^2 = -M^2 \quad \text{for } i > l)$$

form a complete orthonormal system of $\mathcal{H}$. The S-matrix is defined as the operator transforming the incoming into the outgoing states:

$$(\Phi^{p_1\ldots p_k}_{\text{in}},\ S\Phi^{q_1\ldots q_l}_{\text{in}}) = (\Phi^{p_1\ldots p_k}_{\text{out}},\ \Phi^{q_1\ldots q_l}_{\text{in}})\,. \tag{31}$$

Since both systems (29) and (30) form a complete basis of the Hilbert space the S-matrix is unitary.

It is possible to express the S-matrix by the vacuum expectation values of T-products only. To show this we derive from (7) the following reduction formula ([22])

$$(32)\qquad [ST(x_1 \dots x_k),\, A^{\pm}_{\text{in}\,\lambda}] = \pm i\int dz f^{\mp}_{\lambda}(z) K^m_z ST(x_1 \dots x_k z)\,,$$

$$(33)\qquad [ST(x_1 \dots x_k),\, B^{\mp}_{\text{in}\,\alpha}(\zeta)] = \pm i\int dz F^{\mp}_{\alpha}(z) K^M_z ST(x_1 \dots x_k,\, z-\zeta,\, z+\zeta)\,,$$

with

$$T(x_1 \dots x_n) = TA(x_1) \dots A(x_n), \qquad f^{+}_{\lambda}(z) = f_{\lambda}(z), \quad f^{-}_{\lambda}(z) = f^{*}_{\lambda}(z).$$

If we now insert plane waves instead of the wave packets $f_\lambda(x)$ we get

$$(34)\qquad [ST(x_1 \dots x_k),\, A^{*}_{\text{in}}(k)] = -\frac{i\varepsilon(k)}{(2\pi)^{\frac{3}{2}}}\int dz \exp[ikz] K^m_z ST(x_1 \dots x_k z)\,,$$

$$(35)\qquad [ST(x_1 \dots x_k),\, B^{*}_{\text{in}}(k)] = -\frac{i\varepsilon(k)}{(2\pi)^{\frac{3}{2}} F_k(\zeta)}\int dz \exp[ikz] K^M_z ST(x_1 \dots x_k,\, z-\zeta,\, z+\zeta)\,.$$

Since the state vectors (29) form a complete basis S can be expanded with respect to the incoming fields:

$$(36)\qquad S = \sum_{n,m=0}^{\infty} \frac{(-i)^{m+n}}{m!\,n!}\int dk_1 \dots dk_m\, dl_1 \dots dl_n\, c(k_1 \dots k_m;\, l_1 \dots l_n)\cdot \prod_{i=1}^{m}\delta(k_i^2+m^2)\prod_{j=1}^{n}\delta(k_j^2+M^2) :A_{\text{in}}(k_1)\dots A_{\text{in}}(k_m) B_{\text{in}}(l_1)\dots B_{\text{in}}(l_n):\,,$$

where the coefficients are

$$\begin{cases} c(k_1 \dots k_m;\, l_1 \dots l_n) = \\ = \varepsilon(k_1)\dots\varepsilon(k_m)\varepsilon(l_1)\dots\varepsilon(l_n)(\Omega,\, [\dots[S,\, A^*_{\text{in}}(k_1)]\dots A^*_{\text{in}}(k_m)]B^*_{\text{in}}(l_1)]\dots B^*_{\text{in}}(l_n)]\Omega) = \\ = \dfrac{1}{(2\pi)^{\frac{3}{2}(m+n)} F_{l_1}(\zeta_1)\dots F_{l_n}(\zeta_n)}\displaystyle\int dy_1 \dots dy_m\, dz_1 \dots dz_n \exp\left[i\left(\sum k_i y_i + \sum l_j z_j\right)\right]\cdot \\ \cdot K^m_{y_1}\dots K^m_{y_m} K^M_{z_1}\dots K^M_{z_n}(\Omega,\, T(y_1 \dots y_m,\, z_1+\zeta_1, \dots, z_n+\zeta_n,\, z_1-\zeta_1, \dots, z_n-\zeta_n)\Omega)\,. \end{cases}$$

([22]) All reduction formulae in H. LEHMANN, K. SYMANZIK and W. ZIMMERMANN: *Nuovo Cimento*, 1, 425 (1955), Sect. 2 are contained in the single formula (32) which was given by K. SYMANZIK: unpublished.

The last line of (37) follows by iterating (34), (35) and taking the vacuum expectation value. Transforming (37) into co-ordinate space we get ([23])

$$(38)\quad \begin{cases} S = \sum\limits_{m,n=0}^{\infty} \dfrac{(-i)^{m+n}}{m!\,n!} \lim\limits_{\zeta\to 0} \dfrac{1}{F_0(\zeta)^n} \displaystyle\int dy_1 \dots dy_m dz_1 \dots dz_n K^m_{y_1} \dots K^m_{y_m} K^M_{z_1} \dots K^M_{z_n} \cdot \\ \cdot (\Omega, T(y_1 \dots y_m, z_1+\zeta, \dots, z_n+\zeta, z_n-\zeta, \dots z_n-\zeta)\Omega) : \\ \qquad : A_{\text{in}}(y_1) \dots A_{\text{in}}(y_m) B_{\text{in}}(z_1) \dots B_{\text{in}}(z_n) : . \end{cases}$$

The function $F_k(\zeta)$ which enters in the expansions (37) and (38) can easily be expressed by vacuum τ-functions (*i.e.*, vacuum expectation values of T-products) if we take the vacuum expectation value of reduction formula (35) for $k=2$, $x_1=\xi$, $x_2=-\xi$:

$$(39)\quad F_k(\xi)F_k(\eta) = -i\int dy \exp[iky] K^M_y (\Omega, TA(\xi)A(-\xi)A(y+\eta)A(y-\eta)\Omega) .$$

Putting $\xi=\eta$ we get

$$(40)\quad F_k(\xi)^2 = -i\int dx \exp[ikx] K^M_x (\Omega, TA(\xi)A(-\xi)A(x+\xi)A(x-\xi)\Omega) .$$

Inserting (40) in (37) or (38) the S-matrix is completely given by the vacuum τ-functions.

4. - Local field operators for bound states.

In this section we want to take the limit $\xi \to 0$ of the operator

$$B(x;\xi) = TA(x+\xi)A(x-\xi)$$

itself. We assume the existence of

$$(41)\quad B(x) = \lim_{\xi\to 0} \frac{TA(x+\xi)A(x-\xi) - (\Omega, TA(\xi)A(-\xi)\Omega)}{(2\pi)^2(\Omega, TA(\xi)A(-\xi)\Phi_0)} ,$$

([23]) One gets corresponding expansions of the T-product replacing S by $ST(x_1 \dots x_k)$ and

$$T(y_1 \dots y_m, z_1+\zeta_1, \dots, z_n+\zeta_n, z_1-\zeta_1, \dots, z_n-\zeta_n) ,$$

by

$$T(x_1 \dots x_k, y_1 \dots y_m, z_1+\zeta_1, \dots, z_n+\zeta_n, z_1-\zeta_1, \dots, z_n-\zeta_n) ,$$

in the expansions (37) and (38).

which in a formal sense can be written as ([24])

$$B(x) = a^{-\frac{1}{2}} A(x)^2 - a^{-\frac{1}{2}} b \tag{42}$$

with the two (probably divergent) renormalization constants a and b

$$\begin{cases} a = -i \int \exp[-iMx_0](\Omega, TA(0)^2 A(x)^2 \Omega)\, dx_0 , \\ b = (\Omega, A(0)^2 \Omega) . \end{cases} \qquad \text{(according to (40))} \tag{43}$$

The operator $B(x)$ is, of course, invariant under the inhomogeneous Lorentz group and commutes with $B(y)$ and $A(y)$ if $(x-y)^2 > 0$. Furthermore we get from (7) or (28) the asymptotic conditions

$$\lim_{t \to \pm\infty} (\Phi, B_F(t) \Psi) = (\Phi, B_{\substack{\text{out}\\ \text{in}}, F} \Psi) \tag{44}$$

with

$$B_F(t) = -i \int\limits_{x_0 = t} d_3 x\, B(x) \overset{\leftrightarrow}{\frac{\partial}{\partial x_0}} F^*(x) ,$$

for any normalizable solution of $(\square_x - M^2) F(x) = 0$.

Therefore the field operators $A(x)$ and $B(x)$ together satisfy the three principles of invariance, causality and asymptotic conditions exactly in the same formulation which was given in an earlier paper for elementary particles only ([25]). From these principles we get in the usual way an expansion of the S-matrix:

$$\begin{cases} S = \sum\limits_{m,n=0}^{\infty} \dfrac{(-i)^{m+n}}{m!\, n!} \int dy_1 \dots dy_m dz_1 \dots dz_n K_{y_1}^m \dots K_{y_m}^m K_{z_1}^M \dots K_{z_n}^M \cdot \\ \cdot (\Omega, TA(y_1) \dots A(y_m) B(z_1) \dots B(z_n) \Omega) : A_{\text{in}}(y_1) \dots A_{\text{in}}(y_m) B_{\text{in}}(z_1) \dots B_{\text{in}}(z_n) : , \end{cases} \tag{45}$$

which may also be obtained directly from (38) taking the limit $\zeta \to 0$.

We remark that there remains at least one formal difference between the case of two independent elementary particles and the case of one elementary and one composite particle. If $B(x)$ belongs to a bound state composed of elementary particles described by a field $A(x)$, it is possible to represent $B(x)$

([24]) For the case of A^4-coupling at least in perturbation theory the matrix elements of $B(x)$ can be made finite to any order of the coupling constant if a and b are determined by (43).

([25]) H. Lehmann, K. Symanzik and W. Zimmermann: *Nuovo Cimento*, 1, 425 (1955). Although there only the case of one scalar field was considered the generalization to several fields is obvious.

—roughly speaking—as a polynomial in $A(x)$ (in our model Eq. (42)). We may impose this relationship as an additional condition beyond the principles of invariance, causality and the asymptotic condition in order to exclude the case of two elementary particles.

5. - Examples.

In this Section we shall give simple examples for the expansions of the S-matrix which we have derived in the last sections. Instead of the model of one scalar field we consider the more interesting case of a charged scalar « nucleon » field $\psi(x)$ interacting with a neutral scalar « meson » field $A(x)$. We assume that there are just three non-vanishing discrete eigenvalues m^2, $\varkappa^2$ and M^2 of $-P_\mu^2$: m^2 belongs to the one nucleon states, $\varkappa^2$ to the one meson states and M^2 to a stable bound state Ψ of charge two and spin zero (« deuteron » state). As a consequence of charge conservation

$$(\Omega,\ \psi(x)\Psi) = 0$$

vanishes identically for a deuteron state Ψ. Under the corresponding assumptions, as formulated in Sect. 1 for the case of one scalar field, we can apply the methods developed in the last sections. As an example for the S-matrix expansions (36), (38) and (45) we consider the special scattering process with one deuteron and one meson of momentum q_μ resp k_μ in the initial state and two outgoing nucleons with momentum p_μ^1 and p_μ^2 in the final state:

$$(\Phi_{\text{in}}^{p_1,p_2},\ S\Phi_{\text{in}}^{q,k}) = \frac{1}{(2\pi)^6 F_q(\zeta)}\int \mathrm{d}x_1\,\mathrm{d}x_2\,\mathrm{d}y\,\mathrm{d}z \exp\left[i(qz+ky-p_1x_1-p_2x_2)\right]\cdot \cdot K_{x_1}^m K_{x_2}^m K_y^\varkappa K_z^M (\Omega,\ T\psi(x_1)\psi(x_2)A(y)\overline{\psi}(z+\zeta)\overline{\psi}(z-\zeta)\Omega) \tag{36'}$$

$$= \lim_{\zeta\to 0}\frac{1}{(2\pi)^6 F_0(\zeta)}\int \mathrm{d}x_1\,\mathrm{d}x_2\,\mathrm{d}y\,\mathrm{d}z \exp\left[i(qz+ky-p_1x_1-p_2x_2)\right]\cdot \cdot K_{x_1}^m K_{x_2}^m K_z^\varkappa K_y^M (\Omega,\ T\psi(x_1)\psi(x_2)A(y)\overline{\psi}(z+\zeta)\overline{\psi}(z-\zeta)\Omega)\,, \tag{38'}$$

$$= \int \mathrm{d}x_1\,\mathrm{d}x_2\,\mathrm{d}y\,\mathrm{d}z \exp\left[i(qz+ky-p_1x_1-p_2x_2)\right]\cdot \cdot K_{x_1}^m K_{x_2}^m K_y^\varkappa K_z^M (\Omega,\ T\psi(x_1)\psi(x_2)A(y)\overline{B}(z)\Omega)\,, \tag{45'}$$

$$p_1^2 = p_2^2 = -m^2\,,\qquad q^2 = -M^2\,,\qquad k^2 = -\varkappa^2\,.$$

The function $F_q(\zeta)$ is given by

$$F_q(\zeta) = (2\pi)^{\frac{3}{2}}(\Omega,\ T\psi(\zeta)\psi(-\zeta)\Phi_q) = (2\pi)^{\frac{3}{2}}(\Phi_q,\ T\psi(\zeta)\psi(-\zeta)\Omega)$$

and satisfies

$$F_q(\zeta)^2 = -i(2\pi)^4 \int dz \exp[iqz] K_z^M (\Omega, T\psi(\zeta)\psi(-\zeta)\overline{\psi}(z+\zeta)\overline{\psi}(z-\zeta)\Omega) .$$

The deuteron field operator $B(z)$ is defined by

$$B(z) = \lim_{\zeta \to 0} \frac{T\psi(z+\zeta)\psi(z-\zeta)}{F_0(\zeta)} ,$$

where $F_0(\zeta)$ belongs to the deuteron state Ψ_0 at rest. In a formal sense $B(z)$ may be written as

$$B(z) = c^{-\frac{1}{2}} \psi(z)^2$$

with the renormalization constant

$$c = (2\pi)^3(\Omega, \psi(0)^2\Psi_0)^2 = -i\int \exp[-iMx_0](\Omega, T\psi(0)^2\psi(x)^2\Omega)\, dx .$$

Finally we remark that expressions (36′), (38′) and (45′) hold also if the model contains additional stable bound states. In this case the incoming nucleon, meson and deuteron fields are, of course, not irreducible in the whole Hilbert space, and the S-matrix expansions contain in addition the incoming fields belonging to the higher bound states. But if the state vectors $\Phi_{\text{in}}^{k_1 \dots k_s}$ and $\Phi_{\text{out}}^{p_1 \dots p_s}$ contain only nucleon, meson and deuteron states, expressions for the S-matrix elements $(\Phi_{\text{out}}^{p_1 \dots p_s}, \Phi_{\text{in}}^{k_1 \dots k_s})$ can be derived from the asymptotic properties of $A_{\text{in}}^{\text{out}}$, $\Psi_{\text{in}}^{\text{out}}$ and $B_{\text{in}}^{\text{out}}$ in a similar way as in Sect. 3.

* * *

I would like to thank Professor OPPENHEIMER for the kind hospitality of the Institute for Advanced Study, and I am grateful to the International Co-operation Administration, Washington, for a grant. I am indebted to many physicists in Göttingen, Princeton and Berkeley for helpful discussions.

RIASSUNTO (*)

Si esamina un campo scalare causale e invariante coinvolgente uno stato stabile legato. Si deriva una formula per la matrice S e si dimostra che lo stato legato si può descrivere per mezzo di un operatore di campo locale e invariante. Per semplicità si considera solo il caso di particelle di spin nullo e di stati legati; l'estensione ad altri casi è tuttavia possibile.

(*) *Traduzione a cura della Redazione.*

Commun. math. Phys. 15, 208—234 (1969)

Convergence of Bogoliubov's Method of Renormalization in Momentum Space

W. ZIMMERMANN*

Institut des Hautes Etudes Scientifiques, Bures sur Yvette

and

Istituto di Fisica Teorica, Mostra d'Oltremare, Napoli

Received June 25, 1969

Abstract. Bogoliubov's method of renormalization is formulated in momentum space. The convergence of the renormalized Feynman integrand is proved by an application of the power counting theorem.

1. Introduction

A general theory of renormalization has been developed by Bogoliubov for arbitrary local and invariant interactions. It was shown by Hepp that the renormalized Feynman integrals constructed according to Bogoliubov's rules converge to well defined distributions when the regularization is removed [1–4].

In a recent paper [5] a different formulation of Bogoliubov's method was used which works in momentum space and does not refer to a regularization. The starting point of this approach is the integrand I_Γ of the unrenormalized Feynman integral

$$J_\Gamma(p_1 \cdots p_r) = \lim_{\varepsilon \to +0} \int dk_1 \cdots dk_m \, I_\Gamma(k_1 \cdots k_m, p_1 \cdots p_r) \tag{1.1}$$

in momentum space. The integrand R_Γ of the finite part of (1.1)

$$F_\Gamma(p_1 \cdots p_r) = \lim_{\varepsilon \to +0} \int dk_1 \cdots dk_m \, R_\Gamma(k_1 \cdots k_m, p_1 \cdots p_r) \tag{1.2}$$

is defined as a rational function of the internal and external momenta by substracting appropriate counter terms from I_Γ. The method is thus an extension of the original work of Dyson and Salam [6–8] [1]. For handling the overlapping divergencies Bogoliubov's combinatorial technique is used which applies to renormalizable as well as non-renormalizable theories.

[1] For some references of other methods of renormalization see [9—12].

* On leave of absence from Courant Institute of Mathematical Sciences, New York University, New York.

P. Breitenlohner and D. Maison (Eds.): Proceedings 1998, LNP 558, pp. 217–243, 2000.

In this paper the convergence of the renormalized integral (1.2) is proved by a simple application of the power counting theorem [13–15]. The main problem will be to verify the hypothesis of the power counting theorem, i.e. to check that the dimension of (1.2) and any sub-integral along an arbitrary hyperplane in k-space has negative dimension. The power counting theorem for Minkowski metric then implies that the integrals are absolutely convergent for $\varepsilon > 0$ and yield co-variant distributions in the limit $\varepsilon \to 0$ [15].

In Section 2 the definition of the finite part (1.2) is discussed. An explicit formula for R_Γ is derived in Section 3. Section 4 contains the proof that the renormalized Feynman integral meets the requirements of the power counting theorem.

2. The Finite Part of an Arbitrary Feyman Integral

We consider a Feynman diagram Γ with N vertices $V_1, \ldots, V_N$. The lines connecting the vertices V_a, V_b will be denoted by $L(V_a, V_b, \sigma)$ or $L_{ab\sigma}$ ($\sigma = 1, \ldots, \nu(ab)$). V_a, V_b are called endpoints of $L_{ab\sigma}$. Lines connecting a vertex with itself are excluded, ($a = b$). Each vertex is supposed to be endpoint of at least one line. No restriction is placed on the number of lines joining at a vertex.

$\mathscr{L}(\Gamma)$ denotes the set of lines, $\mathscr{V}(\Gamma)$ the set of vertices.

$$\sum_{ab\sigma}{}^{\Gamma}, \quad \prod_{ab\sigma}{}^{\Gamma}$$

denote sum or product resp. over all lines $L_{ab\sigma}$ of the diagram Γ.

$$\sum_{a}{}^{\Gamma}, \quad \prod_{a}{}^{\Gamma}$$

denote sum or product resp. over all vertices V_a of the diagram Γ.

To each line $L_{ab\sigma}$ we assign an internal momentum

$$l_{ab\sigma} = -l_{ba\sigma}. \tag{2.1}$$

To each vertex V_a a momentum q_a is assigned. In general the q_a will be linear combinations of external momenta $p_1, \ldots, p_r$

$$q_a = q_a(p_1 \cdots p_r). \tag{2.2}$$

The internal and external momenta are subject to the relation of momentum conservation at each vertex

$$\sum_{b\sigma}{}^{a}_{\Gamma} l_{ab\sigma} = q_a \qquad a = 1, \ldots, N. \tag{2.3}$$

Here $\sum_{b\sigma}{}^{a}_{\Gamma}$ denotes the sum over all lines $L_{ab\sigma}$ of the diagram Γ having V_a as one of its endpoints. As consequence of (2.1), (2.3) the external momenta must satisfy momentum conservation

$$\sum_{a}^{\Gamma_\nu} q_a = 0 \qquad \nu = 1, \ldots, c \tag{2.4}$$

with $\Gamma_1, \ldots, \Gamma_c$ denoting the connected components of Γ.

To each line $L_{ab\sigma}$ we assign a propagator

$$\Delta_F^{ab\sigma} = P_{ab\sigma}(l_{ab\sigma})\,(l_{ab\sigma}^2 - \mu_{ab\sigma}^2 + i\varepsilon(\boldsymbol{l}_{ab\sigma}^2 + \mu_{ab\sigma}^2))^{-1}, \; \varepsilon > 0, \; \mu_{ab\sigma}^2 > 0 \tag{2.5}$$

where $P_{ab\sigma}$ denotes a polynomial in the components of $l_{ab\sigma}$. To each vertex V_a we assign a polynomial

$$P_a = P_a(l_{ab_1 1}, \ldots, l_{ab_\tau \nu(ab_\tau)}) \tag{2.6}$$

in the components of the vectors $l_{ab1}, \ldots$ Here $V_{b_1}, \ldots, V_{b_\tau}$ denote the vertices which are connected to V_a by an internal line. With these insertion rules the corresponding unrenormalized integral becomes

$$J(p_1, \ldots, p_r) = \lim_{\varepsilon \to +0} \int dk_1 \cdots dk_m \prod_{ab\sigma}^{\Gamma} \Delta_F^{ab\sigma} \prod_{a}^{\Gamma} P_a . \tag{2.7}$$

The internal momenta in $\Delta_F^{ab\sigma}$, P_a(Eq. (2.5–2.6)) are of the form

$$l_{ab\sigma} = k_{ab\sigma} + q_{ab\sigma} . \tag{2.8}$$

The $q_{ab\sigma}$ are linear combinations

$$q_{ab\sigma} = q_{ab\sigma}(q_1, \ldots, q_N) \tag{2.9}$$

of $q_1, \ldots, q_N$ and form a particular solution of

$$\sum_{b\sigma}^{a}{}^{\Gamma} q_{ab\sigma} = q_a, \qquad q_{ab\sigma} + q_{ba\sigma} = 0 . \tag{2.10}$$

The $q_{ab\sigma}$ are called basic internal momenta. The $k_{ab\sigma}$ are linear combinations

$$k_{ab\sigma} = k_{ab\sigma}(k_1, \ldots, k_m) \tag{2.11}$$

of the integration variables $k_1, \ldots, k_m$ and represent the general solution of the homogeneous equations

$$\sum_{b\sigma}^{a}{}^{\Gamma} k_{ab\sigma} = 0, \qquad k_{ab\sigma} + k_{ba\sigma} = 0 . \tag{2.12}$$

m of the forms $k_{ab\sigma}$ are chosen as independent four vectors $k_1, \ldots, k_m$. We introduce the following abbreviations

$$k = (k_1, \ldots, k_m), \quad q = (q_1, \ldots, q_N), \quad p = (p_1, \ldots, p_r) . \tag{2.13}$$

K denotes the set

$$K = \{k_{ab\sigma}\}_{L_{ab\sigma} \in \mathscr{L}(\Gamma)}, \tag{2.14}$$

of four vectors $k_{ab\sigma}$ satisfying (2.12). Equations (2.2), (2.11) are written as

$$q = q(p)\,, \quad K = K(k)\,. \tag{2.15}$$

In this notation the Feynman integral becomes

$$J(p) = \lim_{\varepsilon \to +0} \int dk_1 \cdots dk_m\, I_\Gamma(K(k), q(p)) \tag{2.16}$$

where

$$I_\Gamma(K, q) = \prod_{ab\sigma}{}_\Gamma \Delta_F^{ab\sigma} \prod_{a}{}_\Gamma P_a\,. \tag{2.17}$$

$\Delta_F^{ab\sigma}$, P_a are given by (2.5–2.6) with the substitutions

$$l_{ab\sigma} = l_{ab\sigma}(K\,q) = k_{ab\sigma} + q_{ab\sigma}(q)\,. \tag{2.18}$$

In this section the finite part of (2.16) will be defined in the form[2]

$$F_\Gamma(p) = \lim_{\varepsilon \to +0} \int dk_1 \cdots dk_m\, R_\Gamma(K(k), q(p)) \tag{2.19}$$

where the modified integrand $R_\Gamma(kq)$ is obtained from the original integrand $I_\Gamma(kq)$ by a suitable number of subtractions. Unfortunately the definition of R_Γ will depend on the choice of the basic internal momenta $q_{ab\sigma}(q)$. Though the final integral (2.19) is the same for a large class of basic momenta, we will – for the sake of definiteness – make a unique choice of $q_{ab\sigma}$ in the definition of the finite part. To this end we define the canonical momenta as that solution of (2.10) for which the quadratic form

$$\sum_{ab\sigma}{}_\Gamma q_{ab\sigma}^2 \tag{2.20}$$

is stationary under the constraints (2.10). With the Lagrange multipliers $u_1, \ldots, u_N$ of the constraints (2.10) the $q_{ab\sigma}$ become uniquely determined by

$$\begin{gathered} q_{ab\sigma} = u_a - u_b\,, \\ \sum_{b\sigma}{}_\Gamma^a q_{ab\sigma} = q_a\,, \quad \sum_{a}{}_{\Gamma_\nu} q_a = 0\,. \end{gathered} \tag{2.21}$$

These are $N-c$ independent equations for the $N-c$ independent differences $u_a - u_b$ if c is the number of connected components of Γ.

We introduce some combinatorial concepts which will be needed later on.

Let $\mathscr{L}(\Lambda)$ denotes the set of lines and $V(\Lambda)$ denote the set of vertices of a diagram Λ. To any set $\mathscr{L} \subseteq \mathscr{L}(\Gamma)$ we define a subdiagram Λ of Γ by the lines $L \in \mathscr{L}$ and the vertices which are endpoints of a line in $\mathscr{L}$.

[2] The integral does not depend on the choice of $k_1, \ldots, k_m$. For, any two sets $k_1, \ldots, k_m$ and $k'_1, \ldots, k'_m$ of m linearly independent internal momenta $k_{ab\sigma}$ are related by an orthogonal transformation.

We say that the diagram Δ is spanned by the set $\mathscr{L}$ of lines. The following definitions concern subdiagrams of a given diagram Γ. We define $\Delta = \Delta_1 \cap \Delta_2$ as the diagram spanned by

$$\mathscr{L}(\Delta) = \mathscr{L}(\Delta_1) \cap \mathscr{L}(\Delta_2).$$

Δ_1 is called a subdiagram of Δ_2, i.e. $\Delta_1 \subseteqq \Delta_2$ if $\mathscr{L}(\Delta_1) \subseteqq \mathscr{L}(\Delta_2)$. If $\Delta_1 \subseteqq \Delta_2$, the diagram $\Delta_2 \backslash \Delta_1$ is defined as the diagram spanned by $\mathscr{L}(\Delta_2) \backslash \mathscr{L}(\Delta_1)$.[3]

Let δ be a subdiagram of Δ with connected components $\delta_1, \ldots, \delta_c$. We form the reduced diagram

$$\bar{\Delta} = \Delta/\delta = \Delta/\delta_1 \cdots \delta_c \tag{2.22}$$

by contracting each line of δ to a point. More precisely the reduced diagram Δ is defined by

$$\mathscr{L}(\bar{\Delta}) = \mathscr{L}(\Delta) \backslash \mathscr{L}(\delta), \quad \mathscr{V}(\bar{\Delta}) = \mathscr{V}(\Delta) \backslash \mathscr{V}(\delta) \cup \{\bar{V}_1, \ldots, \bar{V}_c\}. \tag{2.23}$$

Here

$$\bar{V}_1 = \mathscr{V}(\delta_1), \ldots, \bar{V}_c = \mathscr{V}(\delta_c) \tag{2.24}$$

serve as new vertices of $\bar{\Delta}$ replacing the vertices of the reduced diagrams $\delta_1, \ldots, \delta_c$. Two vertices $V, V' \in \mathscr{V}(\Delta) \backslash \mathscr{V}(\delta)$ are connected in $\bar{\Delta}$ by the same lines as in Δ. $V \in \mathscr{V}(\Delta) \backslash \mathscr{V}(\delta)$ and $\bar{V}_a$ are in $\bar{\Delta}$ connected by all lines of $\mathscr{L}(\Delta) \backslash \mathscr{L}(\delta)$ which in Δ connect V with any vertex of $\bar{V}_a$. $\bar{V}_a$ and $\bar{V}_b$ are connected by all lines of $\mathscr{L}(\Delta) \backslash \mathscr{L}(\delta)$ which in Δ connect a vertex of δ_a with a vertex of δ_b.

In the work that follows we will consider subdiagrams as well as reduced diagrams of our original Feynman diagram Γ. Let $\gamma \subseteqq \Gamma$ be a subdiagram of Γ. $m(\gamma)$ denotes the number of independent internal momenta, $N(\gamma)$ the number of vertices of γ. The unrenormalized integrand I_γ is defined by the same insertion rules as for I_Γ. Explicitly

$$\begin{gathered} I_\gamma(K^\gamma q^\gamma) = \prod_{\gamma}{}_{aba} \Delta_F^{aba} \prod_{\gamma}{}_{a} P_a, \\ K^\gamma = \{k^\gamma_{aba}\}_{l_{aba} \in \mathscr{L}(\gamma)}, \qquad q^\gamma = \{q^\gamma_a\}_{V_a \in \mathscr{V}(\gamma)}. \end{gathered} \tag{2.25}$$

Δ_F^{aba}, P_a are given by (2.5–2.6) with the substitutions

$$l_{aba} = l^\gamma_{aba}(K^\gamma, q^\gamma) = k^\gamma_{aba} + q^\gamma_{aba}(q^\gamma). \tag{2.26}$$

Here q^γ_{aba} denote the canonical momenta defined in reference to the subdiagram γ. I.e. q^γ_{aba} is that solution of

$$\sum_{\gamma}{}^{a}_{aba} q^\gamma_{aba} = q_a, \quad q^\gamma_{aba} + q^\gamma_{baa} = 0, \quad V_a \in \mathscr{V}(\gamma) \tag{2.27}$$

[3] $A \backslash B$ denotes the difference of the two sets A, B.

for which

$$\sum_{\substack{\gamma \\ ab\sigma}} (q^{\gamma}_{ab\sigma})^2 \tag{2.28}$$

is stationary. The four vectors $k^{\gamma}_{ab\sigma}$ satisfy

$$\sum_{\substack{\gamma \\ ab\sigma}}^{a} k^{\gamma}_{ab\sigma} = 0\,, \quad k^{\gamma}_{ab\sigma} + k^{\gamma}_{ba\sigma} = 0\,, \quad V_a \in \mathscr{V}(\gamma)\,. \tag{2.29}$$

We next define $k^{\gamma}_{ab\sigma}$, q^{γ}_a as linear combinations of $k_{ab\sigma}$, q_a by requiring

$$l^{\gamma}_{ab\sigma}(K^{\gamma} q^{\gamma}) \equiv l_{ab\sigma}(K q) \tag{2.30}$$

for all $L_{ab\sigma} \in \mathscr{L}(\gamma)$. According to

$$q^{\gamma}_a(K q) = \sum_{\substack{\gamma \\ ab\sigma}}^{a} l_{ab\sigma}(K q)\,, \quad k^{\gamma}_{ab\sigma}(K q) = l_{ab\sigma}(K q) - q^{\gamma}_{ab\sigma}(K q) \tag{2.31}$$

the $k^{\gamma}_{ab\sigma}$, q^{γ}_a are uniquely determined by this requirement. Equation (2.31) implies

$$q^{\gamma}_a(K q) = -\sum_{\substack{\Gamma \setminus \gamma \\ ab\sigma}}^{a} l_{ab\sigma}(K q) = -\sum_{\substack{\Gamma \setminus \gamma \\ ab\sigma}}^{a} (k_{ab\sigma} + q_{ab\sigma}(q))\,. \tag{2.32}$$

It can further be shown that the $k^{\gamma}_{ab\sigma}$ depend only on the $k_{ab\sigma}$. For the Eqs. (2.21) and (2.31) imply

$$\begin{gathered} q_{ab\sigma}(q) = u_a - u_b \quad \text{for} \quad V_a, V_b \in \mathscr{V}(\gamma)\,, \\ \sum_{\substack{\gamma \\ ab\sigma}}^{a} q_{ab\sigma}(q) = q^{\gamma}_a(0, q)\,. \end{gathered} \tag{2.33}$$

These equations, however, determine the functions $q^{\gamma}_{ab\sigma}(0, q)$.

Accordingly

$$q^{\gamma}_{ab\sigma}(0 q) = q_{ab\sigma}(q)$$

and

$$k^{\gamma}_{ab\sigma}(0 q) = l_{ab\sigma}(0 q) - q^{\gamma}_{ab\sigma}(0 q) = 0$$

which proves the assertion. Hence we have the result that K^{γ}, q^{γ} are linear combinations of the form

$$K^{\gamma} = K^{\gamma}(K)\,, \quad q^{\gamma} = q^{\gamma}(K q)\,. \tag{2.34}$$

With the substitutions (2.26), (2.34) the unrenormalized integral I_{γ} becomes a function of K and q.

Let μ be a subdiagram of $\gamma \subset \Gamma$. In an analogous way we introduce the function

$$K^{\mu} = K^{\mu}_{\gamma}(K^{\gamma})\,, \quad q^{\mu} = q^{\mu}_{\gamma}(K^{\gamma} q^{\gamma}) \tag{2.35}$$

by the requirement

$$l^{\mu}_{ab\sigma}(K^{\mu} q^{\mu}) \equiv l^{\gamma}_{ab\sigma}(K^{\gamma}, q^{\gamma})\,. \tag{2.36}$$

For later use we note the relation corresponding to (2.32)

$$q^{\mu}_a(K^{\gamma} q^{\gamma}) = -\sum_{\substack{\gamma \setminus \mu \\ ab\sigma}}^{a} (k^{\gamma}_{ab\sigma} + q^{\gamma}_{ab\sigma}(q^{\gamma}))\,. \tag{2.37}$$

If γ is proper we define the dimension $d(\gamma)$ by

$$d(\gamma)=\sum_{aba}^{\gamma} d(L_{aba})+\sum_{a}^{\gamma} d(V_a)+4m(\gamma)\,. \tag{2.38}$$

Here $d(L_{aba})$ is the degree of the propagator corresponding to L_{aba} with respect to the components of l_{aba}, $d(V_a)$ is the degree of the polynomial assigned to V_a. Apparently $d(\gamma)$ is the dimension of the unrenormalized Feynman integral of γ. Proper subdiagrams of dimension $d(\gamma)\geqq 0$ are called renormalization parts.

We next study reduced diagrams of Γ. Let $\gamma_1,\dots\gamma_c$ be subdiagrams of Γ, proper and mutually disjoint. We consider the reduced diagram

$$\bar\Gamma=\Gamma/\gamma_1\dots\gamma_c\,. \tag{2.39}$$

We define the unrenormalized integrand $I_{\bar\Gamma}$ by the same insertion rules as for I_Γ except that the factor 1 is assigned to the reduced vertices $V_1,\dots,V_c$. Explicitly

$$I_{\bar\Gamma}(q\,K)=\prod_{aba}^{\bar\Gamma}\Delta_F^{aba}\prod_{a}^{\bar\Gamma}P_a \tag{2.40}$$

with (2.5–2.6) and (2.18). Here $\prod_{aba}^{\bar\Gamma}$, $\prod_{a}^{\bar\Gamma}$ denotes the product over all lines L_{aba} or vertices V_a resp. of Γ which do not belong to $\gamma_1,\dots,\gamma_c$. Apparently (2.40) agrees with the definition (2.17) if applied to the reduced diagram $\bar\Gamma$.

After these preparations we now give the definition of the finite part of a Feynman integral. The integrand $R_\Gamma(K,q)$ of the finite part (2.19) is defined by

$$R_\Gamma(K,q)=I_\Gamma(K,q)+\sum_{\gamma_1\dots\gamma_c} I_{\Gamma/\gamma_1\dots\gamma_c}(K,q)\prod_{\tau=1}^{c}O_{\gamma_\tau}(K^{\gamma_\tau},q^{\gamma_\tau}) \tag{2.41}$$

with

$$K^{\gamma_\tau}=K^{\gamma_\tau}(K)\,,\quad q^{\gamma_\tau}=q^{\gamma_\tau}(K,q)\,. \tag{2.42}$$

The sum extends over all sets $s=(\gamma_1,\dots,\gamma_c)$ of renormalization parts of Γ which are mutually disjoint

$$\gamma_\tau\cap\gamma_\sigma=\theta\quad\text{for}\quad\tau\neq\sigma\,.$$

This includes the case that s consists of Γ itself provided Γ is a renormalization part. The functions O_γ are recursively defined for every renormalization part γ of Γ by

$$O_\gamma(K^\gamma q^\gamma)=-t_{q^\gamma}^{d(\gamma)}\Big\{I_\gamma(K^\gamma q^\gamma)+\sum_{\gamma_1\dots\gamma_c}{}'^{\gamma} I_{\Gamma/\gamma_1\dots\gamma_c}(K^\gamma q^\gamma)\prod_{\tau=1}^{c}O_{\gamma_\tau}(K^{\gamma_\tau},q^{\gamma_\tau})\Big\} \tag{2.43}$$

with

$$K^{\gamma_\tau}=K_\gamma^{\gamma_\tau}(K^\gamma)\,,\quad q^{\gamma_\tau}=q_\gamma^{\gamma_\tau}(K^\gamma,q^\gamma)\,. \tag{2.44}$$

$\sum'^{\gamma}$ denotes the sum over all sets $s=(\gamma_1, \ldots, \gamma_c)$ of renormalization parts $\gamma_a \neq \gamma$ of γ which are mutually disjoint. The $'$ indicates that $s=\{\gamma\}$ is excluded in the sum. $t_q^{\nu_\gamma}$ applied to a function $F(q^\gamma)$ of $q_1^\gamma, \ldots, q_M^\gamma$ denotes the Taylor series in the components of the vectors q_i^γ up to the order ν. This completes the definition of the finite part.

We further introduce a function $\bar{R}_\gamma$ by

$$\bar{R}_\gamma(K^\gamma q^\gamma) = I_\gamma(K^\gamma q^\gamma) + \sum_{\gamma_1 \ldots \gamma_c}'^{\gamma} I_{\Gamma/\gamma_1 \ldots \gamma_c}(K q) \prod_{\tau=1}^{c} O_{\gamma_\tau}(K^{\gamma_\tau}, q^{\gamma_\tau}). \quad (2.45)$$

Apparently one has

$$O_\gamma = -t_{q^\gamma}^{d(\gamma)} \bar{R}_\gamma \quad (2.46)$$

R_Γ is related to $\bar{R}_\Gamma$ by

$$R_\Gamma = \bar{R}_\Gamma \quad (2.47)$$

if Γ is no renormalization part and

$$R_\Gamma = \bar{R}_\Gamma + O_\Gamma = (1 - t_q^d)\, \bar{R}_\Gamma \quad (2.48)$$

if Γ is a renormalization part.

The definition of the finite part can be generalized in various respect. First of all we remark that it is sometimes necessary to consider other sets of basic internal momenta besides of the canonical momenta. We give the appropriate definitions for a sufficiently general class of basic internal momenta.

Let $q_{ab\sigma}^\gamma$ be basic internal momenta given for every subdiagram γ of Γ. We consider the set S of all $q_{ab\sigma}^\gamma$. Again we can define q_a^γ, $k_{ab\sigma}^\gamma$ as linear combinations of q, k requiring that

$$l_{ab\sigma}^\gamma(K^\gamma, q^\gamma) \equiv l_{ab\sigma}(K, q) \quad (2.49)$$

with

$$\begin{aligned} l_{ab\sigma}^\gamma &= k_{ab\sigma}^\gamma + q_{ab\sigma}^\gamma(q^\gamma), \\ l_{ab\sigma} &= k_{ab\sigma} + q_{ab\sigma}(q). \end{aligned} \quad (2.50)$$

The set S is called admissable if the momenta K^γ depend only on K or K^μ resp.

$$K^\gamma = K^\gamma(K), \quad K^\gamma = K_\mu^\gamma(K^\mu) \quad \text{for any} \quad \mu \supset \gamma. \quad (2.51)$$

The canonical internal momenta are an example of an admissable set of basic internal momenta[4].

Let $\{q_{ab\sigma}^\gamma\}$ be a set of basic internal momenta for Γ. Then a set of basic internal momenta $\{\tilde{q}_{ab\sigma}^\lambda\}$ for the reduced diagram $\Gamma/\delta_1 \ldots \delta_c$

[4] It can be shown that the finite part (2.19) is the same for any admissable set of basic internal momenta.

may be introduced in the following way. For

$$L_{ab\sigma} \in \mathscr{L}(\lambda), \ \lambda \subseteqq \Gamma/\delta_1 \ldots \delta_c, \ \lambda = \gamma/\delta_1 \ldots \delta_c, \gamma \subseteqq \Gamma. \tag{2.52}$$

Set

$$\tilde{q}^{\lambda}_{ab\sigma} = q^{\gamma}_{ab\sigma}. \tag{2.53}$$

$\{\tilde{q}^{\lambda}_{ab\sigma}\}$ is called the set of basic internal momenta induced by $\{q^{\gamma}_{ab\sigma}\}$ in $\Gamma/\delta_1 \ldots \delta_c$. If $\{q^{\gamma}_{ab\sigma}\}$ is admissable $\{\tilde{q}^{\lambda}_{ab\sigma}\}$ is also admissable. Apparently the function $I_{\bar{\Gamma}}(K, q)$ as defined by (2.40) is constructed by using the basic internal momenta of $\bar{\Gamma}$ which are induced by the canonical momenta of Γ.

Another generalization concerns the number of substractions. Sometimes it is convenient to take more substractions than would actually be necessary for convergence. For self-energy diagrams, for instance, one will always take at least two substractions even if the diagram should be convergent. To include this possibility we introduce a function $d(\gamma)$ which assigns to every proper subdiagram γ of Γ an integer larger or equal to the dimension of γ

$$d(\gamma) \geqq \sum_{\substack{\gamma \\ ab\sigma}} d(L_{ab\sigma}) + \sum_{\substack{\gamma \\ a}} d(V_a) + 4m(\gamma). \tag{2.54}$$

$d(\gamma)$ is called the degree of γ*. Proper diagrams of non-negative degree are called renormalization parts relative to $d(\gamma)$.

Relative to $d(\gamma)$ and an admissable set of basic internal momenta the finite part and the functions R_γ, O_γ, $\bar{R}_\gamma$ are then defined by the same equations (2.41–2.44).

3. Explicit Form of the Finite Part

The integrand R_Γ of the finite part was defined recursively by Eqs. (2.41–2.44). In this section we will derive explicit formulae for the function R_Γ. We begin with some combinatorial definitions concerning subdiagrams of a given diagram Γ. The diagrams γ_1, γ_2 are said to overlap

$$\gamma_1 \circ \gamma_2$$

if none of the following three relations holds

$$\gamma_1 \subseteqq \gamma_2, \ \gamma_2 \subseteqq \gamma_1, \ \gamma_1 \cap \gamma_2 = \emptyset.$$

Otherwise γ_1, γ_2 are called non-overlapping

$$\gamma_1 \oslash \gamma_2.$$

Let Γ be any diagram. A Γ-forest U is a set of diagrams satisfying the following conditions

* In addition we require $d(\gamma) \geqq d(\bar{\gamma}) + \sum d(\gamma_a)$ for any reduced diagram $\bar{\gamma} = \gamma/\gamma_1 \ldots \gamma_c$ with $d(\bar{\gamma})$ defined by (2.38).

(i) the elements of U are proper subdiagrams of Γ,
(ii) any two elements γ', γ'' are non-overlapping

$$\gamma' \oslash \gamma'' ,$$

(iii) U may also be the empty set.

If in addition each element of U is a renormalization part we call U a restricted Γ-forest.

Any subset U' of a Γ-forest U is again a Γ-forest. All possible Γ-forests are partially ordered by $\subset$. A Γ-forest U is called maximal if there is no other Γ-forest U' such that $U \subset U'$. Let $U_1, \ldots, U_c$ be the maximal Γ-forests. Then all possible Γ-forests are given by the subsets of any U_a.

We will next be concerned with the structure of a given Γ-forest U. An element γ of U is called maximal (minimal) if there is no other $\gamma' \in U$ such that $\gamma \subset \gamma'$ or $\gamma' \subset \gamma$ resp. Let γ', γ'' be two maximal elements of U. Since $\gamma' \subset \gamma''$ and $\gamma'' \subset \gamma'$ are excluded we must have

$$\gamma' \cap \gamma'' = \theta$$

for maximal elements of U.

Let γ be any diagram of U. Denote by $U(\gamma)$ the set of all $\gamma' \in U$ satisfying $\gamma' \subset \gamma$. $U(\gamma)$ is a Γ-forest as well as a γ-forest.

Let $\gamma_1, \ldots, \gamma_c$ be the maximal elements of $U(\gamma)$. Then we define

$$\bar{\gamma}(U) = \gamma / \gamma_1 \ldots \gamma_c . \tag{3.1}$$

A Γ-forest U containing Γ itself is called full, a Γ-forest U not containing Γ is called normal. If Γ is no renormalization part all restricted Γ-forests are normal. If Γ is a renormalization part then there is a one-to-one correspondence between full restricted Γ-forests T and normal restricted Γ-forests U given by

$$T = U \cup \{\Gamma\} . \tag{3.2}$$

Note that the empty set $U = \theta$ corresponds to $T = \{\Gamma\}$.

Let U be a normal Γ-forest and γ be an element of U. By $P(\gamma)$ we denote the set of all $\gamma' \in U$ with

$$\gamma' \supseteqq \gamma .$$

Since

$$\gamma' \cap \gamma'' \neq \theta$$

for any two elements of $P(\gamma)$ the set $P(\gamma)$ is totally ordered by $\subset$.

We now define the position $n(\gamma)$ of γ in U by the number of elements contained in $P(\gamma)$. Any two renormalization parts with the same position in U are disjoint.

Let T be a full Γ-forest. Then we assign the position 0 to Γ in T. For any other element $\gamma \in T$ we define the position in T by the position which γ has in

$$U = T \setminus \{\Gamma\}.$$

Let $\mathscr{L}(U)$ denote the set of all lines which belong to at least one diagram of U. All elements of U containing a given line $L_{ab\sigma} \in \mathscr{L}(U)$ are totally ordered by $\subset$. Hence there is a uniquely determined element $\gamma_{ab\sigma}$ with

$$L_{ab\sigma} \in \mathscr{L}(\gamma_{ab\sigma})$$

such that

$$L_{ab\sigma} \notin \mathscr{L}(\gamma) \quad \text{if} \quad \gamma \subset \gamma_{ab\sigma}, \ \gamma \in U .$$

Therefore $\mathscr{L}(U)$ is partitioned into mutually disjoint sets $\mathscr{L}(\bar{\gamma}(U))$

$$\begin{aligned} \mathscr{L}(U) &= \bigcup_{\gamma \in U} \mathscr{L}(\bar{\gamma}(U)), \\ \bar{\gamma}(U) \cap \bar{\gamma}'(U) &= \emptyset \quad \text{for} \quad \gamma \neq \gamma' . \end{aligned} \tag{3.3}$$

The discussion of the recursive Eq. (2.43) can considerably be simplified by introducing substitution operators of the following kind. S_μ denotes the substitution operator

$$S_\mu : K^\gamma \to K^\gamma(K^\mu), \ q^\gamma \to q^\gamma(K^\mu, q^\mu) \quad \text{for} \quad \gamma \subset \mu . \tag{3.4}$$

S_Γ denotes the substitution operator

$$S_\Gamma : K^\gamma \to K^\gamma(K), \ q^\gamma \to q^\gamma(K, q) . \tag{3.5}$$

More precisely S_μ is defined as follows. Let f be a function of the variables

$$K, K^\gamma, q, q^\gamma$$

where γ runs over all renormalization parts of Γ. Then $S_\mu f$ denotes the function which is obtained from f by substituting

$$K_\mu^\gamma(K^\mu), \ q_\mu^\gamma(K^\mu, q^\mu)$$

for all variable K^γ, q^γ with $\gamma \subset \mu$. With the notation

$$K^\Gamma = K , \ q^\Gamma = q$$

this definition holds for S_Γ too. In addition we use the abbreviation

$$t^\mu = t_{q^\mu}^{d(\mu)} . \tag{3.6}$$

With this notation the defining Eqs. (2.41) and (2.43) for R_Γ and O_γ become

$$R_\Gamma(K,q)=I_\Gamma(Kq)+S_\Gamma \sum_{\gamma_1\ldots\gamma_c} I_{\Gamma/\gamma_1\ldots\gamma_c}(Kq)\prod_{\tau=1}^{c} O_{\gamma_\tau}(K^{\gamma_\tau}q^{\gamma_\tau}), \tag{3.7}$$

$$O_\gamma(K^\gamma q^\gamma)=-t^\gamma I_\gamma(K^\gamma q^\gamma)-t^\gamma S_\gamma \sum_{\gamma_1\ldots\gamma_c}{}'_\gamma\, I_{\gamma/\gamma_1\ldots\gamma_c}\prod_{\tau=1}^{c} O_{\gamma_\tau}(K^{\gamma_\tau}q^{\gamma_\tau}). \tag{3.8}$$

The following lemma states an explicit formula for the O-functions.

Lemma 3.1. *The O-function of a renormalization part γ of Γ may explicitly be written as*

$$O_\gamma(K^\gamma q^\gamma)=-t^\gamma S_\gamma \sum_{U\in\mathcal{N}_\gamma}\prod_{\lambda\in U}(-t^\lambda S^\lambda)\, I_\gamma(U) \tag{3.9}$$

or

$$O_\gamma(K^\gamma q^\gamma)=\sum_{T\in\mathcal{F}_\gamma}\prod_{\lambda\in T}(-t^\lambda S_\lambda\, I_\gamma(T)\,. \tag{3.10}$$

The sum in (3.9) *extends over the set $\mathcal{N}_\gamma$ of all normal restricted γ-forests U including the empty set. The sum* (3.18) *extends over the set $\mathcal{F}_\gamma$ of all full restricted γ-forests. $I_\gamma(U)$ is essentially the function I_γ but with a special choice of the variables. We define*

$$I_\gamma(U)=\prod_{ab\sigma}{}_\gamma\, \Delta_F^{ab\sigma}\prod_a P_a \tag{3.11}$$

with (2.5–2.6) *and the substitutions*

$$l_{ab\sigma}=l^\gamma_{ab\sigma}(K^\gamma q^\gamma)\quad \textit{if}\quad L_{ab\sigma}\in\mathcal{L}(\bar\gamma) \tag{3.12}$$

and

$$l_{ab\sigma}=l_{ab\sigma}(K,q)\quad \textit{if}\quad L_{ab\sigma}\notin\mathcal{L}(U)\,. \tag{3.13}$$

In the product

$$\prod_{\gamma\in U}(-t^\gamma S_\gamma)\,,$$

the factors $-t_\gamma S^\gamma$ are ordered from left to right according to increasing position. For elements of equal position in U the order is irrelevant since

$$t^\gamma S_\gamma t^{\gamma'} S_{\gamma'}=t^{\gamma'} S_{\gamma'} t^\gamma S_\gamma \quad \textit{for}\quad \gamma\cap\gamma'=\emptyset\,.$$

Proof. We use the notation

$$D(U)=\prod_{\gamma\in U}(-t^\gamma S_\gamma)\,. \tag{3.14}$$

First we note that the function

$$\tilde O_\gamma(Kq)=-t^\gamma S_\gamma \sum_{U\in\mathcal{N}} D(U)\, I_\gamma(U) \tag{3.15}$$

may also be written as

$$\tilde{O}_\gamma(Kq) = \sum_{T \in \mathscr{F}} D(T)\, I_\gamma(T) \tag{3.16}$$

using the relation (3.2) between full and normal forests. We prove $\tilde{O}_\gamma = O_\gamma$ by showing that $\tilde{O}_\gamma$ solves the recursive Eq. (2.39). First we rewrite the right hand side of (3.15)

$$\sum_{U \in \mathscr{N}} D(U)\, I_\gamma(U) = -t^\gamma I_\gamma(K^\gamma q^\gamma) - t^\gamma S_\gamma \sum_{\gamma_1 \dots \gamma_c}{}'_\gamma \sum_{U \in K(\gamma_1 \dots \gamma_c)} D(U)\, I_\gamma(U)\,. \tag{3.17}$$

Here $K(\gamma_1 \dots \gamma_c)$ is the class of all normal restricted γ-forests having the maximal elements $\gamma_1, \dots, \gamma_c$. The first term of the r.h.s. of (3.17) corresponds to $U = \theta$. Now any $U \in K(\gamma_1 \dots \gamma_c)$ has the form

$$U = T_1(\gamma_1) \cup \dots \cup T_c(\gamma_c) \tag{3.18}$$

where $T_\tau(\gamma_\tau)$ is the set of all $\gamma \in U$ with $\gamma \subseteqq \gamma_\tau$. $T_\tau(\gamma_\tau)$ is a full restricted γ_τ-forest. On the other hand any set $T_1, \dots, T_c$ of full restricted γ_τ-forests defines a $U \in K(\gamma_1 \dots \gamma_c)$ by

$$U = T_1 \cup \dots \cup T_c\,.$$

Hence

$$\begin{aligned} \sum_{U \in K(\gamma_1 \dots \gamma_c)} D(U)\, I_\gamma(U) &= \prod_{\tau=1}^{c} \sum_{T_\tau \in \mathscr{F}_\tau} D(T_\tau)\, I_\gamma(U) \\ &= I_{\gamma/\gamma_1 \dots \gamma_c}(K^\gamma q^\gamma) \prod_{\tau=1}^{c} \sum_{T_\tau \in \mathscr{F}_\tau} D(T_\tau)\, I_{\gamma_\tau}(T_{\tau_\tau}) \end{aligned} \tag{3.19}$$

where $\mathscr{F}_\tau$ denotes the set of all full γ_τ-forests. Using (3.16–3.17) we get

$$\tilde{O}_\gamma(K^\gamma, q^\gamma) = -t^\gamma I_\gamma(K^\gamma, q^\gamma) - t^\gamma S_\gamma \sum_{\gamma_1 \dots \gamma_c}{}'^\gamma I_{\gamma/\gamma_1 \dots \gamma_c}(K^\gamma, q^\gamma) \prod_{\tau=1}^{c} \tilde{O}_{\gamma_\tau}(K^{\gamma_\tau}, q^{\gamma_\tau})$$

which proves $\tilde{O}_\gamma = O_\gamma$.

Theorem 3.1. *The function R_Γ is given explicitly by*

$$R_\Gamma(Kq) = S_\Gamma \sum_{U \in \mathscr{U}_\Gamma} \prod_{\gamma \in U} (-t^\gamma S_\gamma)\, I_\Gamma(U) \tag{3.21}$$

with the sum extending over the set $\mathscr{U}_\Gamma$ of all restricted Γ-forests.

Proof. From (3.7) and the explicit formula (3.10) for O_γ we obtain

$$\begin{aligned} R_\Gamma(Kq) &= I_\Gamma(Kq) + S_\Gamma \sum_{\gamma_1 \dots \gamma_c} I_{\Gamma/\gamma_1 \dots \gamma_c}(Kq) \prod_{\tau=1}^{c} \sum_{T_{\gamma_\tau} \in \mathscr{F}_\tau} D(T_{\gamma_\tau})\, I_{\gamma_\tau}(T_{\gamma_\tau}) \\ &= I_\Gamma(Kq) + S_\Gamma \sum_{\gamma_1 \dots \gamma_c} \sum_{U \in K(\gamma_1 \dots \gamma_c)} D(U)\, I_\Gamma(U)\,. \end{aligned}$$

Formula (3.21) can considerably be simplified by using the identity

$$\prod_{\gamma \in U_0} (1 - t^\gamma) S_\gamma I_\Gamma(U_0) = \sum_{U \subsetneq U_0} \prod_{\gamma \in U} (-t^\gamma S_\gamma) I_\gamma(U) \tag{3.22}$$

which holds for any Γ-forest U_0. Let now Γ be a diagram with no overlapping divergencies, i.e.

$$\gamma_1 \circledS \gamma_2$$

for any two renormalization parts of Γ. Then the set U of all renormalization parts is a Γ-forest. The subsets of U_0 form all possible restricted Γ-forests. Using (3.21) and (3.22) we obtain the following theorem.

Theorem 3.2. *Let Γ be a Feynman diagram with no overlapping renormalization parts. Then the integral of the finite part is given by*

$$R_\Gamma(Kq) = S_\Gamma \prod_{\gamma \in U_0} (1 - t^\gamma) S_\gamma I_\Gamma(U_0) \tag{3.23}$$

where the product extends over all renormalization parts of Γ.

Formula (3.23) represents Dyson's prescription for removing non-overlapping divergencies [16]. Using the power counting theorem it is not difficult to prove that the corresponding integral (2.19) is absolutely convergent. A generalization of (3.23) to the case of overlapping divergencies can be given. The formula obtained, however, is not useful for proving convergence. We therefore quote the result only.

Theorem 3.3. *Let $U_1, \ldots, U_c$ be the maximal restricted Γ-forests of a diagram Γ. Form the intersections*

$$U_{i_1 \ldots i_\nu} = U_{i_1} \cap \cdots \cap U_{i_\nu} \tag{3.24}$$

for all subsets

$$(i_1 \ldots i_\nu) \subsetneq (1, \ldots, c)$$

(some of the intersection (3.24) *may be empty). The integrand of the finite part is then given by*

$$R_\Gamma(Kq) = S_\Gamma \sum_{i_1 \ldots i_\nu} (-1)^{\nu+1} \prod_{\gamma \in U_{i_1} \ldots U_{i_\nu}} (1 - t^\gamma) S_\gamma I_\Gamma(U_{i_1 \ldots i_\nu}). \tag{3.25}$$

For the convergence proof of the following section it is convenient to use Eq. (3.21) in a more general form given by the following theorem.

Theorem 3.4. *The function R_Γ is given explicitly by*

$$R_\Gamma(Kq) = S_\Gamma \sum_{U \in \mathscr{U}} \prod_{\gamma \in U} (-t^\gamma S_\gamma) I_\Gamma(U) \tag{3.26}$$

with the sum extending over the set U of all Γ-forests. Here the convention is used that

$$t^{\gamma}=0 \quad of \quad d(\gamma)<0. \tag{3.27}$$

The proof is trivial since on account of (3.27) all non-restricted Γ-forests give zero contribution to (3.26).

4. Convergence Proof

In this section it will be shown that the finite part

$$\int R_{\Gamma}(K(k),q)\,dk_1 \dots dk_m \tag{4.1}$$

satisfies the requirements of the power counting theorem. In a previous paper the power counting theorem was proved for integrals of the form

$$\int dk \frac{P(k,q)}{\prod_{j=1}^{m}(l_{j0}^2-l_j^2-\mu_j^2+i\varepsilon(l_j^2+\mu_j^2))} \tag{4.2}$$

where $P(k,q)$ is a polynomial in k and q. Clearly (4.1) is of the form (4.2) since $R_{\Gamma}(K,q)$ may be written as

$$\begin{aligned}
&R_{\Gamma}=\frac{A}{B_1\,B_2},\\
&B_1=\prod_{aba}{}_{\Gamma}(l_{aba}^2-\mu_{aba}^2+i\varepsilon(l_{aba}^2+\mu_{aba}^2)),\\
&B_2=\prod_{\gamma}\prod_{aba}{}_{\gamma}\{k_{aba}^{\gamma 2}-\mu_{aba}^2+i\varepsilon(k_{aba}^{\gamma 2}+\mu_{aba}^2)\}^{c(\gamma aba)},\\
&l_{aba}=k_{aba}+q_{aba}(q),\quad k_{aba}^{\gamma}=k_{aba}^{\gamma}(K)
\end{aligned} \tag{4.3}$$

where A is a polynomial in K and q. The product $\prod_{\gamma}$ extends over all renormalization parts γ of Γ.

The hypothesis of the power counting theorem is contained in the following theorem

Theorem 4.1. *The finite part of a Feynman integral*

$$\int_H dk\, R_{\Gamma}(K(k),q) \tag{4.4}$$

has negative dimension for R_{4m} and any hyperplane H described by a set of linear equations

$$\begin{aligned}
&k=k(t)=a+bt,\\
&a=(a_i),\ t=(t_j),\ b=(b_{ij}),\\
&i=1,\dots,m;\ j=1,\dots,h.
\end{aligned} \tag{4.5}$$

With this result the power counting theorem (Theorem 2 of Ref. [6]) *implies that* (4.1) *is absolutely convergent for* $\varepsilon > 0$ *and approaches a well defined distribution in the limit* $\varepsilon \to +0$.

For the proof of the theorem we begin with a couple of definitions. A Γ-forest U is called complete on H if $\Gamma \in U$ and if for any $\gamma \in U$
either (i) all lines $L_{ab\sigma} \in \mathscr{L}(\bar{\gamma})$ are variable on H relative to γ
or (ii) all lines $L_{ab\sigma} \in \mathscr{L}(\bar{\gamma})$ are constant on H relative to γ.
A line $L_{ab\sigma} \in \mathscr{L}(\Gamma)$ is called constant on H relative to γ if

$$k^{\gamma}_{ab\sigma} = \text{const on } H\,, \quad \text{i.e. } k^{\gamma}_{ab\sigma}(T) = \text{const.}$$

Let U be an arbitrary forest of Γ. We are going to define a completion $\bar{U}$ of U which will be shown to be the unique minimal complete forest containing U.

We begin defining $\bar{U}$ for a full U, i.e. $\Gamma \in U$. Let $W(U)$ be the set of all $\gamma \in U$ with the property that at least one line of $\bar{\gamma}(U)$ is constant relative to γ. For any $\gamma \in W(U)$ let $s(\gamma U)$ be the subdiagram of γ which is spanned by the set of constant lines of $\bar{\gamma}(U)$ relative to γ [5]. Let $\delta_1, \ldots, \delta_c$ be the connected components of $\gamma \backslash s(\gamma U)$ [3]. We first show that each δ_a is proper.

Lemma 4.1. *Each connected component of* $\gamma \backslash s(\gamma U)$ *is proper.*

Proof. Assume that $L_{ab\sigma} \in \mathscr{L}(\gamma)$ is an improper line of $\gamma \backslash s(\gamma U)$. Then momentum conservation at each vertex implies

$$k^{\gamma}_{ab\sigma} = \sum c_i k^{\gamma}_{a_i b_i \sigma_i}\,, \qquad L_{a_i b_i \sigma_i} \in \mathscr{L}(s(\gamma U))\,.$$

By definition of $s(\gamma U)$ the momenta $k^{\gamma}_{a_i b_i \sigma_i}$ are constant on H, hence also $k^{\gamma}_{ab\sigma}$ is constant on H. If $L_{ab\sigma} \in \mathscr{L}(\bar{\gamma})$ we have a contradiction because $s(\gamma U)$ is the set of all constant lines of $\bar{\gamma}(U)$. Therefore $L \notin \mathscr{L}(\bar{\gamma})$, i.e. $L \in \mathscr{L}(\varphi)$ with $\varphi \in U$, $\varphi \subset \gamma$. Since φ is connected we have $\varphi \subsetneqq \delta_a$. If L were an improper line of δ_a it would also be an improper line of φ which is impossible. This completes the proof that $\gamma \backslash s(\gamma U)$ does not contain improper lines i.e. each δ_a must be proper.

We define $\mathscr{A}(U)$ as the set of all diagrams $\tau \notin U$ which are connected components δ_a of $\gamma \backslash s(\gamma U)$ with $\gamma \in W(U)$. The completion of a full forest U is then defined by

$$\bar{U} = U \cup \mathscr{A}(U)\,. \tag{4.6}$$

Our first aim is to show that $\bar{U}$ is a forest. We begin with

Lemma 4.2. *If* τ *is a connected component of* $\gamma \backslash s(\gamma U)$ *and* $\gamma' \in U$, $\gamma' \subset \gamma$ *we have*

$$\gamma' \cap \tau = \emptyset \quad \text{or} \quad \gamma' \subsetneqq \tau\,.$$

[5] Note that $s(\gamma(U))$ is defined as a subdiagram of γ but not of $\bar{\gamma}(U)$. That means that no vertices of are identified in $s(\gamma(U))$.

Proof. Since all $\gamma' \subset \gamma$, $\gamma' \in U$ are connected and $\gamma' \subsetneqq \gamma \backslash s(\gamma U)$ it follows $\gamma' \subsetneqq \delta_a$ where δ_a is a connected component of $\gamma \backslash s(\gamma U)$. Hence $\gamma' \subsetneqq \delta_a \neq \tau$ or $\gamma' \subsetneqq \tau$. In the first case $\gamma' \cap \tau = \emptyset$, in the second case $\gamma' \subsetneqq \tau$.

Lemma 4.3. *U is a forest.*

Proof. (i) We first prove that any elements $\tau \in \mathscr{A}(U)$ and $\gamma' \in U$ do not overlap. Let τ be a connected component of $\gamma \backslash s(\gamma U)$. We have

$$\gamma' \supset \gamma, \ \gamma' \cap \gamma = \emptyset \quad \text{or} \quad \gamma' \subset \gamma .$$

$\gamma' \supset \gamma$ implies $\gamma' \supset \tau$. $\gamma' \cap \gamma = \emptyset$ implies $\gamma' \cap \tau = \emptyset$. $\gamma' \subset \gamma$ implies $\gamma' \subset \tau$ or $\gamma' \cap \tau = \emptyset$ (Lemma 4.2). Hence $\gamma' \oslash \tau$ for any $\tau \in \mathscr{A}(U)$ and $\gamma' \in U$.

(ii) We next show that any two different elements $\tau_1, \tau_2 \in \mathscr{A}$ do not overlap.

(a) Let τ_1, τ_2 both be connected components of $\gamma \backslash s(\gamma U)$. Then $\tau_1 \cap \tau_2 = \emptyset$.

(b) Let τ_1 be a connected component of $\gamma_1 \backslash s(\gamma_1 U)$, τ_2 be a connected component of $\gamma_2 \backslash s(\gamma_2 U)$. If $\gamma_1 \subset \gamma_2$ Lemma 4.2 implies $\gamma_1 \subset \gamma_2$ or $\tau_2 \cap \tau_2 = \emptyset$. Hence $\tau_1 \subset \tau_2$ or $\tau_1 \cap \tau_2 = \emptyset$.

If on the other hand $\gamma_1 \cap \gamma_2 = \emptyset$ then also $\tau_1 \cap \tau_2 = \emptyset$. This completes the proof that U is a forest.

Lemma 4.4. *Let $\tau \in \mathscr{A}(U)$ be a connected component of $\gamma \backslash s(\gamma U)$. Let $U(\tau)$ be the set of all $\sigma \in U$ with $\sigma \subset \tau$ and $\gamma_1, \ldots, \gamma_c$ be the maximal element of $U(\tau)$. Then all lines of*

$$\bar{\tau}(\bar{U}) = \tau/\gamma_1 \ldots \gamma_c$$

are variable relative to τ.

Proof. Let $L_{ab\sigma} \in \mathscr{L}(\bar{\tau}(\bar{U}))$. By definition

$$\begin{aligned} k^{\tau}_{ab\sigma} + q^{\tau}_{ab\sigma}(q^{\tau}) &= k^{\gamma}_{ab\sigma} + q^{\gamma}_{ab\sigma}(q^{\gamma}), \\ q^{\tau} = q^{\tau}(K^{\gamma} q^{\gamma}), \ & k^{\tau}_{ab\sigma} = k^{\tau}_{ab\sigma}(K^{\gamma}). \end{aligned} \tag{4.7}$$

Setting $q^{\gamma} = 0$ we obtain

$$k^{\tau}_{ab\sigma} + q^{\tau}_{ab\sigma}(q^{\tau}) = k^{\gamma}_{ab\sigma}, \ q^{\tau} = q^{\tau}(K^{\gamma}, 0), \ k^{\tau}_{ab\sigma} = k^{\tau}_{ab\sigma}(K^{\gamma}). \tag{4.8}$$

Since τ is a connected component of $\gamma \backslash s(\gamma U)$ we have

$$q^{\tau}_{a}(K^{\gamma}, 0) = \sum c_i k^{\gamma}_{a_i b_i \sigma_i}, \quad L_{a_i b_i \sigma_i} \in \mathscr{L}(s(\gamma U)). \tag{4.9}$$

Hence all $q^{\tau}_{a}(K^{\gamma}, 0)$ are constant on H. If $k^{\tau}_{ab\sigma}$ is constant on H Eq. (4.8) implies that $k^{\gamma}_{ab\sigma}$ is also constant on H in contradiction to $L_{ab\sigma} \in \mathscr{L}(\bar{\tau}) \subsetneqq \mathscr{L}(\bar{\gamma} \backslash s(\gamma U))$.

Lemma 4.5. *U is complete.*

Proof. (i) $\Gamma \in U$.

(ii) Let $\gamma \in U$, $\gamma \notin W(U)$. All lines of $\bar{\gamma}(\bar{U})$ are variable relative to γ since $\bar{\gamma}(\bar{U}) = \bar{\gamma}(U)$.

(iii) Let $\gamma \in W(U)$. Then $\mathscr{L}(\bar{\gamma}(\bar{U})) = \mathscr{L}(s(\gamma U))$ i.e. all lines of $\bar{\gamma}(U)$ are constant relative to γ.

(iv) Let $\tau \in \mathscr{A}(U)$. Then $\tau(\bar{U}) = \tau/\gamma_1 \ldots \gamma_c$ where $\gamma_1 \ldots \gamma_c$ are the maximal elements of $U(\tau)$. Lemma 4.4 implies that all lines of $\bar{\tau}(\bar{U})$ are variable relative to τ.

Lemma 4.6. *Let U be a full forest. Any set V with*

$$U \subsetneqq V \subsetneqq \bar{U} \tag{4.10}$$

is a forest with completion $\bar{U}$.

Proof. Clearly V is a forest. We will show that $W(V) = W(U)$.

(i) $\gamma \in W(U)$ implies $\gamma \in W(V)$ since $s(\gamma U) = s(\gamma V)$.

(ii) $\gamma \in U$, $\gamma \in W(V)$ implies $\gamma \in W(U)$ since

$$\mathscr{L}(\bar{\gamma}(V)) \subsetneqq \mathscr{L}(\bar{\gamma}(U)).$$

(iii) If $\tau \in V \backslash U \subsetneqq \mathscr{A}(U)$ all lines of $\bar{\tau}(V) = \bar{\tau}(U)$ are variable.

Combining (i)–(iii) we obtain $W(V) = W(U)$. Hence

$$\mathscr{A}(V) = \mathscr{A}(U) \backslash (V \backslash U)$$

and $\bar{V} = \bar{U}$.

Next we will define the base $\underline{U}$ of a forest U which will turn out to be the minimal forest among all forests with the completion $\bar{U}$. The set $\mathscr{B}(U)$ is defined as the set of all diagrams $\tau \in U$ satisfying

(i) $\tau \notin W(U)$.

(ii) τ is a connected component of $\gamma \backslash s(\gamma U)$ with $\gamma \in W(U)$. Let U be full. The base $\underline{U}$ of U is defined by

$$\underline{U} = U \backslash \mathscr{B}(U) \backslash \{\Gamma\} \tag{4.11}$$

$\underline{U}$ is a forest. Furthermore define a full forest $\underline{U}'$ by

$$\underline{U}' = U \backslash \mathscr{B}(U). \tag{4.12}$$

Lemma 4.7. *If U is a complete forest on H the sets $W(U)$ and $\mathscr{B}(U)$ are given by the following conditions:*

$W(U)$ is the set of all $\gamma \in U$ for which all lines of $\bar{\gamma}(U)$ are constant relative to γ.

If $\gamma \notin W(U)$ all lines of $\bar{\gamma}(U)$ are variable on H relative to γ.

$\mathscr{B}(U)$ is the set of all diagrams $\tau \in U$ satisfying

(i) *$\tau \notin W(U)$.*

(ii) *τ is a maximal element of $U(\gamma)$ with $\gamma \in W(U)$.*

Proof. The first and the second statement follow immediately from the definition of $W(U)$.

If U is complete we have

$$\mathscr{L}(s(\gamma U)) = \mathscr{L}(\bar{\gamma}(U)) \quad \text{for} \quad \gamma \in W(U) \tag{4.13}$$

or

$$\mathscr{L}(\gamma \backslash s(\gamma U)) = \mathscr{L}(\gamma_1 \cup \cdots \cup \gamma_c), \tag{4.14}$$

where $\gamma_1 \ldots \gamma_c$ are the maximal elements of $U(\gamma)$. Hence the connected components of $\gamma \backslash s(\gamma U)$ are identical with the maximal elements γ_a of $U(\gamma)$. This proves the last statement.

Lemma 4.8. *For any full forest U holds*

$$W(\underline{U}') = W(U). \tag{4.15}$$

Proof. (i) If $\tau \in U \backslash \underline{U}' = \mathscr{B}(U)$ the diagram $\bar{\tau}(U)$ does not contain any constant lines relative to γ. Hence

$$\tau \in U \backslash \underline{U}' \quad \text{implies} \quad \tau \notin W(U). \tag{4.16}$$

(ii) Let $\gamma \in \underline{U}'$, $\gamma \in W(U)$. Then $\bar{\gamma}(U)$ contains a line $L_{ab\sigma}$ which is constant relative to γ. Since $\mathscr{L}(\bar{\gamma}(U)) \subsetneqq \mathscr{L}(\bar{\gamma}(\underline{U}'))$ the line $L_{ab\sigma}$ belongs to $\bar{\gamma}(\underline{U}')$. Hence

$$\gamma \in \underline{U}', \ \gamma \in W(U) \quad \text{implies} \quad \gamma \in W(\underline{U}'). \tag{4.17}$$

(iii) Let $\gamma \in \underline{U}'$, $\gamma \notin W(U)$. Then $\bar{\gamma}(U)$ does not contain a line which is constant relative to γ. By definition $\bar{\gamma}(U) = \gamma/\gamma_1 \ldots \gamma_c$ where $\gamma_1 \ldots \gamma_c$ are the maximal elements of $U(\gamma)$. Since $\gamma \notin W(U)$ each γ_a belongs to $\underline{U}'$, therefore $\bar{\gamma}(\underline{U}') = \bar{\gamma}(U)$. Hence

$$\gamma \in \underline{U}', \ \gamma \notin W(U) \quad \text{implies} \quad \gamma \notin W(\underline{U}'). \tag{4.18}$$

Combining (4.16–4.18) we obtain the statement of the lemma.

Lemma 4.9. *For each $\gamma \in W(U') = W(U)$ the two sets $s(\gamma U)$, $s(\gamma \underline{U}')$ are equal*

$$s(\gamma U) = s(\gamma \underline{U}') \tag{4.19}$$

Proof. Let $\gamma \in W(U)$. Any line of $\bar{\gamma}(U)$ also belongs to $\bar{\gamma}(\underline{U}')$. Hence

$$s(\gamma U) \subsetneqq s(\gamma \underline{U}').$$

Suppose that $L_{ab\sigma} \in \mathscr{L}(s(\gamma \underline{U}')) \backslash s(\gamma U)$. $L_{ab\sigma}$ must belong to a $\tau \in \mathscr{B}(U)$ which is a connected component of $\gamma \backslash s(\gamma U)$ with $\gamma \in W(U)$. It cannot belong to any $\sigma \in U$ with $\sigma \subset \tau$ since $\mathscr{L}(s(\gamma \underline{U}')) \subsetneqq \mathscr{L}(\bar{\gamma}(\underline{U}'))$. Hence $L_{ab\sigma} \in \bar{\tau}(U)$. Using Eqs. (4.7–4.9) of Lemma 4.4 we find that

$$k^{\gamma}_{ab\sigma} = \text{const on } H \text{ implies } k^{\tau}_{ab\sigma} = \text{const on } H$$

in contradiction to the requirement on the elements of $\mathscr{B}$.

We now extend the definition of completion and base to normal forests. If U is a normal forest we define the completion $\bar{U}$ and the base $\underline{U}$ by

$$\bar{U} = \bar{V}, \; \underline{U} = \underline{V} \quad \text{where} \quad V = U \cup \{\Gamma\}.$$

Lemma 4.10. *Let U be a given forest with completion $\bar{U}$ and base $\underline{U}$. $\bar{U}$ is the completion of $\underline{U}$ and $\underline{U}$ is the base of $\bar{U}$.*

Proof. It is sufficient to consider a full forest U. Lemma 4.8 and 4.9 imply

$$\mathscr{A}(\underline{U}') = \mathscr{B}(U) \cup \mathscr{A}(U).$$

With the notation $B = \underline{U}'$, $C = \bar{U}$

$$\bar{B} = B \cup \mathscr{B}(U) \cup \mathscr{A}(U) = U \cup \mathscr{A}(U) = C.$$

Lemma 4.4 implies

$$\mathscr{B}(C) = \mathscr{A}(U) \cup \mathscr{B}(U).$$

Hence $\underline{C} = C - \mathscr{B}(C) = B$.

Theorem 4.2. *Let C be a given complete forest with base B. The set of all forests U with the completion C is given by the condition*

$$B \subseteqq U \subseteqq C. \tag{4.21}$$

Proof. Let U have the completion C. Then $U \subseteqq C$ and $\underline{U} = B$ (Lemma 4.10). Hence $B \subseteqq U \subseteqq C$.

Since C is the completion of B (Lemma 4.10) any U satisfying (4.21) has the completion C.

Theorem 4.3. *The finite part of the integrand of a Feynman integral is given by*

$$R_\Gamma(K\,q) = S_\Gamma \sum_{U \in \mathscr{C}} X_U, \tag{4.22}$$

$$X_U = \prod_{\gamma \in U} (t(\gamma)\, S_\gamma)\, I_\gamma(U), \tag{4.23}$$

$$f(\gamma) = 1 - t^\gamma \quad \textit{if} \quad \gamma \in \mathscr{B}(U), f(\gamma) = -t^\gamma \quad \textit{if} \quad \gamma \notin \mathscr{B}(U). \tag{4.24}$$

The sum in (4.22) *extends over the set $\mathscr{C}$ of all complete forests of Γ. The set $\mathscr{B}(U)$ is given by Lemma* 4.7.

Proof. Two forests are called equivalent if they have the same completion. According to the Theorem 4.2 the corresponding equivalence classes are given by the condition $\underline{C} \subseteqq U \subseteqq C$ where C is a complete forest with base $\underline{C}$. This partition of the set of all forests into equivalence classes leads to the following formula (see Eq. (3.21))

$$R_\Gamma = \sum_{C \in \mathscr{C}} X_C, \quad X_C = S_\Gamma \sum_{\underline{C} \subseteq U \subsetneq C} \prod_{\gamma \in U} (-t^\gamma S_\gamma) I_\Gamma(U). \tag{4.25}$$

An alternative formula for X_C is

$$X_C = S_\Gamma \prod_{\gamma \in C} (f(\gamma) S_\gamma) I_\Gamma(C). \tag{4.26}$$

An equivalent definition of $f(\gamma)$ is

$$f(\gamma) = 1 - t^\gamma \quad \text{for} \quad \gamma \in C - \underline{C} = \mathscr{B}(C),$$
$$f(\gamma) = -t^\gamma \quad \text{for} \quad \gamma \in \underline{C}.$$

In order to show the equivalence of the two formula for X_C we work out the products of factors $(1 - t^\gamma)$ in (4.26)

$$X_C = S_\Gamma \sum_{Q \subseteq \underline{C} - C} \prod_{\gamma \in Q} (g_Q(\gamma) S_\gamma) I_\Gamma(C), \tag{4.27}$$
$$g_Q(\gamma) = -t^\gamma \quad \text{for} \quad \gamma \in Q, \quad g_Q(\gamma) = 1 \quad \text{for} \quad \gamma \notin Q.$$

The sum extends over all subsets Q of $C - \underline{C}$. Introducing $V = \underline{C} + Q$ as new variable of summation we obtain (4.26) which can be rewritten in the form (4.23).

The next aim is to give upper bounds for the degree of the function X_U with respect to the parameters T of the hyperplane H. This will eventually lead to the desired result that the dimension of the renormalized Feynman integral is always negative.

We first state a recursion formula determining X_U which follows easily from the definition (4.23). The subscript U will be omitted in the work that follows. For the sets $W(U)$, $\mathscr{B}(U)$ given by Lemma 4.7 we will use the notation

$$\mathscr{B} = \mathscr{B}(U), \quad W = W(U).$$

Lemma 4.11. *For a given complete forest U the function X_U is determined by*

$$X_U = X = (1 - t^\Gamma) Y_\Gamma \tag{4.28}$$

where for any $\gamma \in U$

$$Y_\gamma = I_{\gamma/\gamma_1 \ldots \gamma_c} S_\gamma f_{\gamma_1} Y_{\gamma_1} \ldots f_{\gamma_c} Y_{\gamma_c}. \tag{4.29}$$

$\gamma_1, \ldots, \gamma_c$ denote the maximal elements of $U(\gamma)$. If γ is minimal we set $Y_\gamma = I_\gamma$. f_γ is defined by (4.24).

The function Y_γ has the general form

$$Y_\gamma(K^\gamma q^\gamma) = \frac{A}{B_1 B_2} \tag{4.30}$$

where A is a polynomial in K^γ, q^γ and

$$B_1 = \prod_{ab\sigma}{}_\gamma \left(l^2_{ab\sigma} - \mu^2_{ab\sigma} + i\varepsilon(l^2_{ab\sigma} + \mu^2_{ab\sigma})\right),$$

$$B_2 = \prod_{\varphi \in U(\gamma)} \prod_{ab\sigma}{}_\varphi \left(k^{\varphi 2}_{ab\sigma} - \mu^2_{ab\sigma} + i\varepsilon(k^{\varphi 2}_{ab\sigma} + \mu^2_{ab\sigma})\right)^{c(\gamma ab\sigma)}, \tag{4.31}$$

$$l_{ab\sigma} = k^\gamma_{ab\sigma} + q^\gamma_{ab\sigma},$$

$$k^\varphi_{ab\sigma} = k^\varphi_{ab\sigma}(k^\gamma).$$

We next want to determine the degree of the function Y_γ. To this end the following lemma will be useful.

Lemma 4.12. *Let F be a function of the form*

$$Y = \frac{A}{C}, \quad C = \prod_\alpha \left(l_\alpha^2 - \mu_\alpha^2 + i\varepsilon(l_\alpha^2 + \mu_\alpha^2)\right) \tag{4.32}$$

where A is a polynomial in $t_1, \ldots, t_\varrho$ and

$$l_\alpha = a_\alpha + \sum c_{\alpha\beta} t_\beta + \sum d_{\alpha\beta} q_\beta$$

with all

$$\sum c_{\alpha\beta} t_\beta \not\equiv 0 .$$

Then the relation

$$\mathrm{degr}_{t,q} F \leqq l \tag{4.33}$$

implies

$$\mathrm{degr}_t (1 - t^d_q) F \leqq l - d - 1 \tag{4.34}$$

$\mathrm{degr}_x F$ denotes the degree of a rational function F with respect to the variables $x = (x_1 \ldots x_n)$.

Proof. We decompose the polynomial

$$A = \sum_{\gamma=0}^{g} A_\gamma \tag{4.35}$$

such that A_γ is homogeneous in q of degree γ. (4.29) implies

$$\mathrm{degre}_{t,q} A \leqq e + \delta \tag{4.36}$$

if

$$\mathrm{degr}_{t,q} C = \delta . \tag{4.37}$$

From (4.35) we get

$$\mathrm{degr}_t A_\gamma \leqq e + \delta - \gamma . \tag{4.38}$$

Since A_γ is homogeneous in q

$$(1 - t^d_q) \frac{A_\gamma}{C} = A_\gamma (1 - t_p^{d-\gamma}) \frac{1}{C} . \tag{4.39}$$

Here

$$\operatorname{degr}(1 - t_q^{d-\gamma}) \frac{1}{C} \leqq -\delta - d + \gamma - 1 . \tag{4.40}$$

Hence

$$\begin{aligned} \operatorname{degr}_t(1 - t_q^d) \frac{A_\gamma}{C} &\leqq \operatorname{degr}_t A_\gamma + \operatorname{degr}_t(1 - t_q^{d-\gamma}) \frac{1}{C} \\ &\leqq e - d - 1 \end{aligned} \tag{4.41}$$

using (4.34) and (4.36).

It is convenient to introduce the following integer

$$M(\gamma) = 4 \sum_{\mu}^{\gamma} m(\bar{\mu}) \tag{4.42}$$

where the sum extends over all μ satisfying the condition

$$\mu \in U , \ \mu \subseteqq \gamma \quad \text{and} \quad \mu \notin W .$$

$m(\bar{\mu})$ is the number of independent internal momenta of $\bar{\mu}$. Apparently $M(\Gamma)$ is the number of independent integration variables on H. For $\bar{\mu}$ contains only variable lines if $\mu \notin W$ while $\bar{\mu}$ contains only constant lines if $\mu \in W$.

Lemma 4.13. *The following inequalities hold*

$$\operatorname{degr}_{t,q}^{\gamma} \ Y_\gamma(K^\gamma(K), q^\gamma) \leqq d(\gamma) - M(\gamma) \quad \textit{for} \quad \gamma \in U, \ \gamma \notin W, \tag{4.43}$$

$$\operatorname{degr}_t \ Y_\gamma(K^\gamma(K), q^\gamma) < -M(\gamma) \quad \textit{for} \quad \gamma \in W. \tag{4.44}$$

Proof. As hypothesis of induction we assume the inequalities to be valid for all maximal elements γ_a of $U(\gamma)$. It will then be shown that the inequalities also hold for γ itself.

(i) Case $\gamma \in W$. Then the recursion formula holds with

$$f_{\gamma_a} = -t^{\gamma_a} \quad \text{for} \quad \gamma_a \in W .$$

$$f_{\gamma_a} = 1 - t^{\gamma_a} \quad \text{for} \quad \gamma_a \notin W .$$

We will find the following relations

$$\operatorname{degr}_t I_{\gamma/\gamma_1 \ldots \gamma_c} = 0 , \tag{4.45}$$

$$\operatorname{degr}_t S_\gamma t^{\gamma_a} Y_{\gamma_a} < -M(\gamma_a) \quad \text{if} \quad \gamma_a \in W , \tag{4.46}$$

$$\operatorname{degr}_t S_\gamma (1 - t^{\gamma_a}) Y_{\gamma_a} < -M(\gamma_a) \quad \text{if} \quad \gamma_a \notin W \tag{4.47}$$

(4.45–4.47) imply the inequality (4.44)

$$\operatorname{degr}_t Y_\gamma \leqq -\sum M(\gamma_a) = -M(\gamma) .$$

We next prove the relations (4.45–4.47):

(α) Relation (4.45) follows since all lines of $\bar{\gamma}(U)$ are constant relative to γ (Lemma 4.7).

(β) Proof of relation (4.46). According to the hypothesis of induction we have

$$\operatorname{degr}_t Y_{\gamma_\alpha}(K^{\gamma_\alpha}(K), q^{\gamma_\alpha}) < -M(\gamma_\alpha).$$

This implies

$$\operatorname{degr}_t t_{\gamma_\alpha} Y_{\gamma_\alpha}(K^{\gamma_\alpha}(K), q^{\gamma_\alpha}) < -M(\gamma_\alpha).$$

For it is

$$t_{\gamma_\alpha} Y_{\gamma_\alpha} = t_\varrho^{d(\gamma)} Y_{\gamma_\alpha}(K^{\gamma_\alpha}, \varrho q^{\gamma_\alpha})|_{\varrho=1},$$

$$\operatorname{degr}_t t_\varrho^{d(\gamma_\alpha)} Y_{\gamma_\alpha}(K^{\gamma_\alpha}(K), \varrho q^{\gamma_\alpha}) \leqq \operatorname{degr}_t Y_{\gamma_\alpha}(K^{\gamma_\alpha}(K), \varrho q^{\gamma_\alpha}).$$

By definition of the substitution operator S_γ

$$(S_\gamma t^{\gamma_\alpha} Y_{\gamma_\alpha})(K^\gamma q^\gamma) = (t^{\gamma_\alpha} Y_{\gamma_\alpha})(K^{\gamma_\alpha}(K^\gamma), q^{\gamma_\alpha}(K^\gamma, q^\gamma)).$$

In $q^{\gamma_\alpha}(K^\gamma, q^\gamma)$ only those $k^\gamma_{ab\sigma}$ occur with $L_{ab\sigma} \in \mathscr{L}(\bar{\gamma})$ which are constant on H relative to γ. Hence

$$\operatorname{degr}_t(S_\gamma t^{\gamma_\alpha} Y_{\gamma_\alpha})(K^\gamma(K) q^\gamma) = \operatorname{degr}_t(t^{\gamma_\alpha} Y_{\gamma_\alpha})(K^{\gamma_\alpha}(K), q^{\gamma_\alpha}).$$

(γ) Proof of (4.47). Let $\gamma_\alpha \notin W$. According to the hypothesis of induction.

$$\operatorname{degr}_{t, q^{\gamma_\alpha}} Y_{\gamma_\alpha}(K^{\gamma_\alpha}(K(T)), q^{\gamma_\alpha}) \leqq d(\gamma_\alpha) - M(\gamma_\alpha).$$

Applying Lemma 4.12

$$\operatorname{degr}_t(1 - t_{\gamma_\alpha}) Y_{\gamma_\alpha}(K^{\gamma_\alpha}(K), q^{\gamma_\alpha}) < -M(\gamma_\alpha).$$

Applying the substitution operator we obtain

$$(S_\gamma(1 - t_{\gamma_\alpha}) Y_{\gamma_\alpha})(K^\gamma(K), q^\gamma) = ((1 - t_{\gamma_\alpha} Y_{\gamma_\alpha})(K^{\gamma_\alpha}(K), q^{\gamma_\alpha})(K^\gamma(K), q^\gamma).$$

Again q^{γ_α} depends only on components $k^\gamma_{ab\sigma}$ of K^γ which are constant on H.

Hence

$$\operatorname{degr}_t(S_\gamma(1 - t_{\gamma_\alpha}) Y_{\gamma_\alpha})(K^\gamma(K), q^\gamma) = \operatorname{degr}_t(1 - t^{\gamma_\alpha}) Y_{\gamma_\alpha})(K^{\gamma_\alpha}(K), q^{\gamma_\alpha})$$

and

$$\operatorname{degr}_t(S_\gamma(1 - t_{\gamma_\alpha}) Y_{\gamma_\alpha})(K^\gamma(K), q^\gamma) < -M(\gamma_\alpha).$$

(ii) Case $\gamma \notin W$.

In that case the recursion formula reads

$$Y_\gamma = I_{\gamma/\gamma_1 \dots \gamma_c} S_\gamma(-t_{\gamma_1}) Y_{\gamma_1} \dots (-t_{\gamma_c}) Y_{\gamma_c}.$$

The relations

$$\operatorname{degr}_{tq^\gamma} I_{\gamma/\gamma_1 \dots \gamma_\alpha} = d(\bar{\gamma}) - 4m(\bar{\gamma}), \tag{4.48}$$

$$\operatorname{degr}_{tq^\gamma} S_\gamma t_{\gamma_\alpha} Y_{\gamma_\alpha} \leqq d(\gamma_\alpha) - M(\gamma_\alpha) \quad \text{if} \quad \gamma_\alpha \notin W, \tag{4.49}$$

$$\operatorname{degr}_{tq^\gamma} S_\gamma t_{\gamma_\alpha} Y_{\gamma_\alpha} < d(\gamma_\alpha) - M(\gamma_\alpha) \quad \text{if} \quad \gamma_\alpha \in W \tag{4.50}$$

imply

$$\begin{aligned} \operatorname{degr}_{t q^\gamma} Y_\gamma &\leqq d(\bar\gamma) - 4m(\bar\gamma) + \sum_\alpha (d(\gamma^\alpha) - M(\gamma^\alpha)) \\ &\leqq d(\gamma) - M(\gamma) \end{aligned} \tag{4.51}$$

since $d(\bar\gamma) + \sum d(\gamma_\alpha) \leqq d(\gamma)$.

(4.51) is the inequality (4.42) stated in the lemma.

We now prove the relations (4.48–4.50).

(α) The relation (4.48) follows from the definitions of I.

(β) The hypothesis of induction

$$\operatorname{degr}_{t q^{\gamma_\alpha}} Y_{\gamma^\alpha}(K^{\gamma_\alpha}(K), q^{\gamma_\alpha}) \leqq d(\gamma_\alpha) < M(\gamma_\alpha)$$

implies

$$\operatorname{degr}_{t q^{\gamma_\alpha}} t^{\gamma_\alpha} Y_{\gamma_\alpha}(K^{\gamma_\alpha}(K), q^{\gamma_\alpha}) \leqq d(\gamma_\alpha) - M(\gamma_\alpha)\,.$$

Application of the substitution operator S_γ to $t^{\gamma_\alpha} Y_{\gamma_\alpha}$ yields

$$(S_\gamma t^{\gamma_\alpha} Y_{\gamma_\alpha})(K^\gamma(K), q^\gamma) = (t^{\gamma_\alpha} Y_{\gamma_\alpha})(K^{\gamma_\alpha}(K), q^{\gamma_\alpha})(K^\gamma(K), q^\gamma)\,.$$

Since $t^{\gamma_\alpha} Y_{\gamma_\alpha}$ is a polynominal in q^{γ_α} the substitution

$$q^{\gamma_\alpha} \to q^{\gamma_\alpha}(K^\gamma(K), q^\gamma)$$

can only decrease the degree with respect to T, q^γ

$$\begin{aligned} &\operatorname{degr}_{t q^\gamma}(S_\gamma t^{\gamma_\alpha} Y_{\gamma_\alpha})(K^\gamma(K), q^\gamma) \\ &\qquad \leqq \operatorname{degr}_{t q^{\gamma_\alpha}}(t^{\gamma_\alpha} Y_{\gamma_\alpha})(K^{\gamma_\alpha}(K), q^{\gamma_\alpha}) \leqq d(\gamma_\alpha) - M(\gamma_\alpha)\,, \end{aligned}$$

(γ) Proof of relation (4.50). We assume $\gamma_\alpha \in W$. According to the hypothesis of induction

$$\operatorname{degr}_t Y_{\gamma_\alpha}(K^{\gamma_\alpha}(K), q^{\gamma_\alpha}) < -M(\gamma^\alpha)\,.$$

This implies

$$\begin{aligned} &\operatorname{degr}_{t q^{\gamma_\alpha}} t^{\gamma_\alpha} Y_{\gamma_\alpha}(K^{\gamma_\alpha}(K), q^{\gamma_\alpha}) < d(\gamma_\alpha) - M(\gamma_\alpha)\,, \\ &\operatorname{degr}_{t q^\gamma}(S_\gamma t_{\gamma_\alpha} Y_{\gamma_\alpha})(K^\gamma(K), q^\gamma) \\ &\qquad = \operatorname{degr}_{t q^\gamma}(t_{\gamma_\alpha} Y_{\gamma_\alpha})(K^{\gamma_\alpha}(K), q^{\gamma_\alpha})(K^\gamma(K), q^\gamma) \\ &\qquad \leqq \operatorname{degr}_{t q^{\gamma_\alpha}}(t_{\gamma_\alpha} Y_{\gamma_\alpha})(K^{\gamma_\alpha}(K), q^{\gamma_\alpha}) < d(\gamma_\alpha) - M(\gamma_\alpha)\,. \end{aligned}$$

This completes the proof of the lemma.

The results obtained in Lemma 4.13 will now be used in order to show that the dimension of the integral (4.1) is negative. We have

$$X = (1 - t_\Gamma)\, Y_\Gamma\,. \tag{4.52}$$

First let $\Gamma \in W$. Then

$$\operatorname{degr}_t Y_\Gamma(K q) < -M(\Gamma)$$

implies

$$\mathrm{degr}_t (1 - t_\Gamma)\, Y_\Gamma(Kq) < -M(\Gamma).$$

Let $\Gamma \notin W$. Then

$$\mathrm{degr}_{t,q}\, Y_\Gamma(Kq) \leqq d(\Gamma) - M(\Gamma)$$

implies

$$\mathrm{degr}_t (1 - t_\Gamma)\, Y_\Gamma(Kq) < -M(\Gamma).$$

In any case

$$\mathrm{degr}_t\, X < -M(\Gamma) \tag{4.53}$$

and therefore

$$\mathrm{degr}_t\, R_\Gamma(Kq) < -M(\Gamma). \tag{4.54}$$

Since M_Γ is the number of independent parameters of the hyperplane H it follows

$$\dim \int_H dk\, R_\Gamma < 0.$$

This completes the proof of the theorem.

Appendix

We shortly indicate the proof of the following generalized form of Eq. (4.54)

$$\mathrm{degr}_{t,M}\, R_\Gamma(Kq) < -M(\Gamma) \tag{A.1}$$

which is useful for checking the equivalence of Bogoliubov's original definition of the renormalized Feynman integral to the one used in this paper [17] M denotes a subset

$$M \subseteqq \{\mu_{ab\sigma}\}$$

of the mass parameters.

First we note that under the hypothesis of Lemma 4.12

$$\mathrm{degr}_{tM}(1 - t_q^d)\, F \leqq \mathrm{degr}_{tMq}\, F - d - 1 \tag{A.2}$$

can be derived which is a generalization of (4.33–4.34). The Eq. (4.43–4.44) can be generalized to

$$\mathrm{degr}_{tMq^\gamma}\, Y_\gamma(K^\gamma(K), q^\gamma) \leqq d(\gamma) - M(\gamma) \quad \text{for} \quad \gamma \in U,\ \gamma \notin W, \tag{A.3}$$

$$\mathrm{degr}_{tM}\, Y_\gamma(K^\gamma(K), q^\gamma) < -M(\gamma) \quad \text{for} \quad \gamma \in W. \tag{A.4}$$

These relations are derived from

$$\mathrm{degr}_{tM}\, I_{\gamma/\gamma_1 \ldots \gamma_c} \leqq 0, \tag{A.5}$$

$$\mathrm{degr}_{tM}\, S_\gamma t^{\gamma_\alpha} Y_{\gamma_\alpha} < -M(\gamma_\alpha) \quad \text{if} \quad \gamma_\alpha \notin W, \tag{A.6}$$

$$\mathrm{degr}_{tM}\, S_\gamma (1 - t^{\gamma_\alpha})\, Y_{\gamma_\alpha} < -M(\gamma_\alpha) \quad \text{if} \quad \gamma_\alpha \notin W \tag{A.7}$$

if $\gamma \in W$ and from

$$\operatorname{degr}_{t M q^\gamma} I_{\gamma/\gamma_1 \ldots \gamma_c} \leqq d(\bar{\gamma}) - 4m(\bar{\gamma}), \tag{A.8}$$

$$\operatorname{degr}_{t M q^\gamma} S_\gamma t^{\gamma_\alpha} Y_{\gamma_\alpha} \leqq d(\gamma_\alpha) - M(\gamma_\alpha) \quad \text{if} \quad \gamma_\alpha \notin W, \tag{A.9}$$

$$\operatorname{degr}_{t M q^\gamma} S_\gamma t^{\gamma_\alpha} Y_{\gamma_\alpha} < d(\gamma_\alpha) - M(\gamma_\alpha) \quad \text{if} \quad \gamma_\alpha \in W \tag{A.10}$$

if $\gamma \notin W$. The result (A.1) follows if the Eqs. (A.3–A.4) are applied to (4.52).

I am grateful to Dr. K. Hepp for many useful discussions and I wish to thank Dr. G. Dell'Antonio and Dr. L. Motchane for the kind hospitality extended to me at the I.H.E.S. and the Istituto di Fisica Teorica.

Bibliography

1. Bogoliubov, N. N., and D. V. Shirkov: Introduction to the theory of quantized fields. New York: Interscience 1959.
2. —, and O. S. Parasiuk: Acta Math. **97**, 227 (1957).
3. Hepp, K.: Commun. Math. Phys. **6**, 161 (1967).
4. Speer, E.: J. Math. Phys. **9**, 1404 (1968).
5. Zimmermann, W.: Commun. Math. Phys. **6**, 161 (1967).
6. Dyson, F. J.: Phys. Rev. **75**, 1736 (1949)
7. Salam, A.: Phys. Rev. **82**, 217 (1951).
8. — Phys. Rev. **84**, 426 (1951).
9. Wu, T. T.: Phys. Rev. **125**, 1436 (1962).
10. Steinmann, O.: Ann. Phys. **36**, 267 (1966).
11. Yennie, D. R.: Cargese Lectures (1967).
12. Caianiello, L. R., F. Guerra et M. Marinaro: Nuovo Cimento **60** A, 713 (1969).
13. Weinberg, S.: Phys. Rev. **118**, 838 (1960).
14. Hahn, Y., and W. Zimmermann: Commun. Math. Phys. **10**, 330 (1968).
15. Zimmermann, W.: Commun. Math. Phys. **11**, 1 (1969).
16. See for instance Bjorken, J., and S. Drell: Relativistic quantum fields. New York: McGraw-Hill 1962.
17. Hepp, K., and W. Zimmermann: Unpublished.

W. Zimmermann
Courant Institute
New York University
251, Mercer St.
New York, N. Y. 10012, USA

Reprinted from ANNALS OF PHYSICS
All Rights Reserved by Academic Press, New York and London

Vol. 77, No. 1-2, May 1, 1973
Printed in Belgium

Composite Operators in the Perturbation Theory of Renormalizable Interactions*†

WOLFHART ZIMMERMANN

Department of Physics, New York University, New York, 10003

Received May 23, 1972

In perturbation theory the Green's functions of composite operators are constructed by applying Bogoliubov's method of renormalization. Normalization conditions for the composite operators are derived, as well as identities which relate composite operators of different degree.

1. INTRODUCTION

The purpose of this paper is to define composite operators and discuss some of their properties in the perturbation theory of renormalizable interactions. As an example we consider the model of a scalar field with a mixed A^3— and A^4-coupling

$$\mathscr{L}\{A\} = \mathscr{L}_0\{A\} + \mathscr{L}_{\text{int}}\{A\} \tag{1.1}$$

$$\mathscr{L}_0\{A\} = (1/2)\, \partial_\mu A \partial^\mu A - (m^2/2)\, A^2 \tag{1.2}$$

$$\mathscr{L}_{\text{int}}\{A\} = -\frac{g}{3!} A^3 - \frac{\lambda}{4!} A^4 + \frac{a}{2} A^2 + b\mathscr{L}_0\{A\} \tag{1.3}$$

$$\lambda \neq 0 \tag{1.4}$$

$\mathscr{L}\{A\}$ will be used as an effective Lagrangian in the following sense. m, g and λ represent the renormalized mass and coupling constants. The parameters a and b are power series in g and λ with finite coefficients depending on m. The coefficients are recursively defined by the condition that the propagator has a pole at m^2 of residue i. In contradistinction to the Lagrangian of conventional renormalization theory $\mathscr{L}\{A\}$ does not contain infinite parameters.

* This work was supported in part by the National Science Foundation Grant No. GP-25609.

† A preliminary version of the material contained in this paper appeared in W. Zimmermann, Renormalization and Composite Field Operators, in "Lectures on Elementary Particles and Quantum Field Theory" (S. Deser, M. Grisaru, and H. Pendleton, Eds.), 1970 Brandeis Summer Institute in Theoretical Physics, Vol. 1, MIT Press, Cambridge, Mass. (1971).

P. Breitenlohner and D. Maison (Eds.): Proceedings 1998, LNP 558, pp. 244–277, 2000.

The time ordered Green's functions of the theory are defined by the Gell-Mann Low expansion [1]

$$\langle TA(x_1)\cdots A(x_n)\rangle = \mathrm{Fin}(\langle Te^{i\int \mathscr{L}^0_{\mathrm{int}}dz} A^0(x_1)\cdots A^0(x_n)\rangle_0/\langle Te^{i\int \mathscr{L}^0_{\mathrm{int}}dz}\rangle_0) \quad (1.5)$$
$$\mathscr{L}^0_{\mathrm{int}} = \mathscr{L}_{\mathrm{int}}\{A^0(z)\}$$

A^0 denotes a free scalar field of mass m, $\langle\rangle_0$ the free field vacuum expectation value. The symbol 'Fin' indicates that the finite part of the contribution from each Feynman diagram should be taken. Various methods for separating the finite part from a divergent Feynman integral are available [2–7][1]. In the original version of Bogoliubov's method the finite part is first defined for the regularized contribution from a given diagram [2]. As was shown by Hepp the finite part thus defined approaches a well defined distribution when the regularization is removed [3]. In this paper Bogoliubov's method is used in the version of [5]. Without regularization the finite part is separated by taking suitable subtractions for the integrand of a Feynman integral in momentum space, as in the original work of Dyson and Salam [8–9]. The appropriate subtraction terms are obtained by applying Bogoliubov's R-operation to the unrenormalized integrand. The resulting integrals can be shown to be convergent as a consequence of the power counting theorem [5].

Composite operators $B(x)$ will be defined through their Green's function

$$\langle TB(x)\, A(y_1)\cdots A(y_n)\rangle \quad (1.6)$$

The matrix elements of $B(x)$ between incoming and outgoing states can be expressed in terms of the Green's functions (1.6) by using the reduction formulas. It is not obvious, however, that the time ordered Green's functions of the operator thus defined are identical to the original functions (1.6). For this certain consistency conditions[2] on the corresponding retarded functions are required which also yield the locality of B. This point will not be discussed in the present paper.

Let

$$M = M\{A(x)\} \quad (1.7)$$

be a monomial of the field A and its derivatives. Our aim is to define an operator B which can be interpreted as the finite part of (1.7). Formally the time ordered Green's functions of M are given by the Gell-Mann Low expansion

$$\langle TM\{A(x)\}\, A(y_1)\cdots A(y_n)\rangle$$
$$= (\langle Te^{i\int \mathscr{L}^0_{\mathrm{int}}dz} M\{A^0(x)\}\, A^0(y_1)\cdots A^0(y_n)\rangle_0/\langle Te^{i\int \mathscr{L}^0_{\mathrm{int}}dz}\rangle_0) \quad (1.8)$$

This suggests defining the Green's functions

$$\langle TB(x)\, A(y_1)\cdots A(y_n)\rangle \quad (1.9)$$

[1] The Epstein-Glaser method [6] including some applications is reviewed in ref. [7].

[2] For a formulation of the consistency conditions see [10].

of the composite operator B by taking the finite part of (1.8). B may be called a normal product of (1.7), in generalization of Wick's definition of a normal product in the free field case [11].

The Feynman diagrams appearing in the expansion (1.8) contain a new vertex x which requires additional subtractions. Depending on the number of subtractions employed for removing the new divergencies we obtain different versions of the normal product. The minimal normal product of (1.7) is constructed by making only those subtractions which are required for removing all divergencies. In this paper a family

$$N_a[M\{A(x)\}] \tag{1.10}$$

of normal products will be introduced. The normal product (1.10) of degree a is constructed by making $a - r + 1$ over-all subtractions for any proper diagram which contains x and has r external lines. It will be shown that this operator is finite provided

$$a \geqslant d(M) \tag{1.11}$$

where $d(M)$ is the dimension of the monomial M. If the degree equals the dimension we use the notation

$$\begin{gathered} N[M\{A(x)\}] = N_a[M\{A(x)\}], \\ a = d(M), \end{gathered} \tag{1.12}$$

and say that the normal product is of minimal degree. If the A^3-coupling is absent, i.e., $g = 0$, the operators (1.12) represent the minimal normal products. For $g \neq 0$ the operators (1.12) are not minimal. In the general case the minimal operators do not seem to be very useful and will not be considered in this paper.

Alternative methods of defining composite operators have been developed by Epstein–Glaser [6, 7] and Wilson [12]. Wilson's method uses a renormalized version of the relations

$$\begin{aligned} \frac{\partial A(x)}{\partial \lambda} &= -i \int_{-\infty}^{x_0} dy_0 \int dy\, [\mathscr{L}_{\text{int}}(y), A(x)] \\ \frac{\partial \mathscr{L}_{\text{int}}(x)}{\partial \lambda} &= -i \int_{-\infty}^{x_0} dy_0 \int dy\, [\mathscr{L}_{\text{int}}(y)\, \mathscr{L}_{\text{int}}(x)] \end{aligned} \tag{1.13}$$

On the basis of these relations the short distance expansion and composite operators are constructed recursively in perturbation theory.

In the method of Epstein and Glaser composite operators are first constructed for a model where the coupling constant is replaced by an external source $\lambda(x)$.

The adiabatic limit $\lambda(x) \to \lambda$ is then shown to exist [6, 7]. One advantage is that composite operators are constructed directly as local operators—without reference to Green's functions. On the other hand the method described in the present paper seems to be more suitable for investigating anomalies [18, 19, 26].

In Section 2 some definitions of Green's functions are collected. After a review of the renormalization of Feynman integrals (Section 3) the Green's functions of composite operators are constructed and their normalization conditions derived (Section 4). The integrands of Feynman integrals contributing to composite operators satisfy certain algebraic identities which will be derived in Section 5. These identities imply that any normal product

$$N_b[M] \tag{1.14}$$

may be written as a linear combination of normal products of degree a, for any given integer $a > b$

$$N_b[M] = \sum c_{ba}(M, M')\, N_a[M'] \tag{1.15}$$

(Section 6). The sum contains only a finite number of terms, it extends over all monomials M' for which the dimension $d(M')$ is less or equal than the degree a,

$$d(M') \leqslant a. \tag{1.16}$$

As a consequence of (1.15) any composite operator (1.10) may be linearly expressed in terms of the operators (1.12).

We finally review some applications of the normal product algorithm [17–19, 24–29]. The field operator A satisfies a finite field equation which has the form of the Euler–Lagrange equation derived from the effective Lagrangian (1.1–3), but with normal products N_3 applied to the interaction terms. The Wilson expansion may be used to express the normal product in terms of A by the point-splitting method. In this way Valatin's form of the field equation is obtained [13–17][3].

Lowenstein [18] constructed a finite energy–momentum tensor [19–22][4]

$$T_{\mu\nu} = N_4[\theta_{\mu\nu}] + c(\partial_\mu\partial_\nu - g_{\mu\nu}\Box)\, N_2[A^2] \tag{1.17}$$

in perturbation theory which has all required properties, in particular the appropriate Ward identities are satisfied. $\theta_{\mu\nu}$ is the formal energy–momentum tensor

$$\theta_{\mu\nu} = (\partial\mathscr{L}/\partial\partial^\mu A)\, \partial_\nu A - g_{\mu\nu}\mathscr{L}$$

of the Lagrangian (1.1–3), c is an arbitrary constant. As was shown by Lowenstein

[3] A derivation of local field equations in the framework of the Epstein-Glaser method was given by Stora [7].

[4] The improved energy-momentum tensor introduced in [21] for a regularized theory with A^4-coupling does not have a finite limit when the regularization is removed. Symanzik [22] found a corrected form of this tensor containing one free parameter which in the limit approaches (1.17).

the trace of the tensor (1.17) is never soft in perturbation theory, no matter how c is chosen.

In a paper by Schroer it was shown that independent of perturbation theory the trace of the tensor (1.17) is soft for suitable c provided a Gell-Mann Low eigenvalue of the coupling constant exists. This would imply the asymptotic conformal invariance of the Green's functions [23].

Lowenstein [25] found an elegant proof of the Callan–Symanzik equation [23] and derived a differential equation of the renormalization group. Anomalous Ward identities and gauge invariance problems were studied for various cases [26]. Field equations were derived for a very general class of interactions [27, 29]. Lam proposed a generalization of the subtraction scheme which is particularly useful for studying nonlinear transformations [27]. Finally, the normal product algorithm has been extended by Lowenstein to Green's functions involving several composite operators [25].

2. Notation of Green's Functions

In this section we collect some definitions concerning Green's functions. For the Fourier transform $\tilde{f}$ of a function f of n four vectors $x_1, \dots, x_n$ we use the unsymmetric definition

$$\tilde{f}(p_1 \cdots p_n) = \int dx_1 \cdots dx_n \, e^{i\Sigma p_j x_j} f(x_1 \cdots x_n)$$

$$f(x_1 \cdots x_n) = \int \frac{dp_1}{(2\pi)^4} \cdots \int \frac{dp_n}{(2\pi)^4} \, e^{-i\Sigma p_j x_j} \tilde{f}(p_1 \cdots p_n) \tag{2.1}$$

Let $TO_1(x_1) \cdots O_n(x_n)$ be a suitably defined time ordered product of components $O_1, \dots, O_n$ of local field operators such as the scalar field A, derivatives thereof or composite field operators. The Fourier transform or partial Fourier transform of $TO_1(x_1) \cdots O_n(x_n)$ will be denoted by

$$T\tilde{O}_1(p_1) \cdots \tilde{O}_n(p_n) = \int dx_1 \cdots dx_n \, e^{i\Sigma p_j x_j} TO_1(x_1) \cdots O_n(x_n) \tag{2.2}$$

$$TO_1(x_1) \cdots O_r(x_r) \, \tilde{O}_{r+1}(p_{r+1}) \cdots \tilde{O}_n(p_n)$$
$$= \int dx_{r+1} \cdots dx_n \, e^{i\Sigma_{j=r+1}^{n}} TO_1(x_1) \cdots O_n(x_n) \tag{2.3}$$

For the Fourier transform $\tilde{\Delta}_F'(p)$ of the propagator

$$\Delta_F'(z) = \langle TA(x) \, A(y) \rangle, \qquad z = x - y \tag{2.4}$$

we have

$$\langle T\tilde{A}(p_1) \, \tilde{A}(p_2) \rangle = (2\pi)^4 \, \delta(p_1 + p_2) \, \tilde{\Delta}_F'(p_1) \tag{2.5}$$

and

$$\tilde{\Delta}_F'(p) = \langle TA(0)\,\tilde{A}(p)\rangle \tag{2.6}$$

$\langle\,\rangle$ denotes the vacuum expectation value.

The connected part of a time ordered vacuum expectation value is recursively defined by

$$\langle TO_1(x_1)\cdots O_n(x_n)\rangle = \sum \langle TO_{i_{11}}(x_{i_{11}})\cdots\rangle^{\text{conn}}\cdots\langle TO_{i_{A1}}(x_{i_{A1}})\cdots\rangle^{\text{conn}} \tag{2.7}$$

The sum extends over all partitions of $x_1,\dots, x_n$ into classes

$$(x_{i_{11}}, x_{i_{21}},\dots), (x_{i_{21}}, x_{i_{22}},\dots),\dots, (x_{i_{A1}}, x_{i_{A2}},\dots)$$

The Fourier transform of the connected part is of the form

$$\langle T\tilde{O}_1(p_1)\cdots\tilde{O}_n(p_n)\rangle^{\text{conn}} = (2\pi)^4\,\delta\left(\sum p_j\right)\langle T\tilde{O}_1(p_1)\cdots\tilde{O}_n(p_n)\rangle^{\text{trunc}} \tag{2.8}$$

$\langle\,\rangle^{\text{trunc}}$ is called the truncated part. A Fourier transformation with respect to p_1 yields

$$\langle T_1\tilde{O}(p_1)\cdots\tilde{O}_n(p_n)\rangle^{\text{trunc}} = \langle T\tilde{O}_1(0)\,\tilde{O}_2(p_2)\cdots\tilde{O}_n(p_n)\rangle^{\text{conn}} \tag{2.9}$$

The generalized Wick product

$$:\tilde{O}_1(x_1)\cdots\tilde{O}_n(x_n):$$

is defined recursively by

$$TO_1(x_1)\cdots O_n(x_n) = :O_1(x_1)\cdots O_n(x_n): + \sum\langle TO_{i_{01}}(x_{i_{01}})\cdots\rangle : O_{i_{11}}(x_{i_{11}})\cdots. \tag{2.10}$$

The sum extends over all partitions of $(x_1,\dots, x_n)$ into classes

$$(x_{i_{01}}, x_{i_{02}},\dots), (x_{i_{11}}, x_{i_{12}},\dots)$$

where the second class may be empty.

The mixed product

$$T:O_1(x_1)\cdots O_n(x_n): O_1'(y_1)\cdots O_m'(y_m)$$

is defined by

$$\begin{aligned} &TO_1(x_1)\cdots O_n(x_n)\,O_1'(y_1)\cdots O_m'(y_m)\\ &\quad = T:O_1(x_1)\cdots O_n(x_n):O_1'(y_1)\cdots O_m'(y_m)\\ &\quad + \sum\langle TO_{i_{01}}(x_{i_{01}})\cdots\rangle\, T:O_{i_{11}}(x_{i_{11}})\cdots :O_1'(y_1)\cdots O_m'(y). \end{aligned} \tag{2.11}$$

Next we introduce the factorization by

$$\begin{aligned} &\langle T\{O_1(x_1)\cdots O_n(x_n)\}\, O_1'(y_1)\cdots O_m'(y_m)\rangle^{\text{fact}} \\ &\quad = \sum \langle TO'_{i_{01}}(y_{i_{01}})\cdots\rangle\langle TO_1(x_1)\, O'_{i_{11}}(y_{i_{11}})\cdots\rangle^{\text{conn}} \times \cdots \\ &\quad\quad \times \langle TO_n(x_n)\, O'_{i_{n1}}(y_{i_{n1}})\cdots\rangle^{\text{conn}} \end{aligned} \tag{2.12}$$

The sum extends over all ordered partitions of $y_1 ,..., y_m$ into $n + 1$ classes

$$(y_{i_{01}}, y_{i_{02}},\ldots), (y_{i_{11}}, y_{i_{12}},\ldots),\ldots, (y_{i_{n1}}, y_{i_{n2}},\ldots)$$

$(y_{i_{j1}}\cdots)$ is assigned to $x_j (j = 1,..., n)$. $(y_{i_{01}}\cdots)$ may be empty.

We give some simple properties of the factorization. For $m = n$

$$\langle T\{A(x_1)\cdots A(x_n)\}\, A(y_1)\cdots A(y_n)\rangle^{\text{fact}} = \sum \Delta_F'(x_{i_1} - y_1)\cdots\Delta_F'(x_{i_n} - y_n) \tag{2.13}$$

with the sum taken over all permutations of the x_j.

$$\begin{aligned} \langle T\{O_1(x_1)\cdots O_n(x_n)\}\, O_1'(y_1)\cdots O_m'(y_m)\rangle^{\text{fact}} &= 0 \qquad \text{for} \quad n > m \\ \langle T\{O(x)\}\, O_1(y_1)\cdots O_m(y_m)\rangle^{\text{fact}} &= \langle TO(x)\, O_1(y_1)\cdots O_m(y_m)\rangle \end{aligned} \tag{2.14}$$

Let $G_m(x_1 ,..., x_n ; y_1 ,..., y_m)$ be a set of Green's functions which are symmetric in $y_1 ,..., y_m$. We may take, for instance,

$$G_m(x_1\cdots x_n ; y_1\cdots y_m) = \langle T :A(x_1)\cdots A(x_n): A(y_1)\cdots A(y_m)\rangle \tag{2.15}$$

or

$$G_m(x; y_1\cdots y_m) = \langle TB(x)\, A(y_1)\cdots A(y_m)\rangle \tag{2.16}$$

where $B(x)$ is a composite operator. We introduce the proper part G_m^{prop} of G_m by the recursion formula

$$\begin{aligned} G_m(x_1\cdots x_n ; y_1\cdots y_m) = \sum_{a=1}^{m} \frac{1}{a!}\int dz_1\cdots dz_a\, G_a(x_1\cdots x_n ; z_1\cdots z_a)^{\text{prop}} \\ \times \langle T\{A(z_1)\cdots A(z_a)\}\, A(y_1)\cdots A(y_m)\rangle^{\text{fact}} \end{aligned} \tag{2.17}$$

For a unique definition we must require G_m^{prop} to be symmetric in $y_1 ,..., y_m$. We give a few examples. For the set

$$G_m(x; y_1\cdots y_m) = \langle TA(x)\, A(y_1)\cdots A(y_m)\rangle$$

of Green's functions we have

$$\begin{aligned} \langle TA(x)\, A(y)\rangle^{\text{prop}} &= \delta(x - y) \\ \langle TA(x)\, A(y_1)\cdots A(y_m)\rangle^{\text{prop}} &= 0 \qquad \text{if} \quad m > 1. \end{aligned} \tag{2.18}$$

For (2.15) we find

$$\langle T{:}\tilde{A}(p_1) \cdots \tilde{A}(p_n){:}\tilde{A}(q)\rangle^{\text{prop}} = \frac{\langle T{:}\tilde{A}(p_1) \cdots \tilde{A}(p_n){:}\tilde{A}(q)\rangle}{\tilde{\Delta}_F'(p)} \tag{2.19}$$

For later reference we give the free field value of the proper parts.

$$\langle T{:}A_0(x_1) \cdots A_0(x_n){:}A_0(y_1) \cdots A_0(y_m)\rangle_0^{\text{prop}} = \delta_{mn} \sum_P \delta(x_1 - y_{i_1}) \cdots \delta(x_n - y_{i_n}) \tag{2.20}$$

with the sum extending over all permutations P of $y_1, \ldots, y_n$. Derivatives of the field operator will be denoted by

$$A_{(\mu)}(x) = \partial_{(\mu)} A(x) \tag{2.21}$$

where

$$\begin{aligned} (\mu) &= (\mu_1, \ldots, \mu_a) \\ \partial_{(\mu)} &= \partial_{\mu_1} \cdots \partial_{\mu_a}. \end{aligned} \tag{2.22}$$

The time ordered product of field operator derivatives is defined by

$$\begin{gathered} TA_{(\mu)_1}(x_1) \cdots A_{(\mu)_n}(x_n) = \partial^{x_1}_{(\mu)_1} \cdots \partial^{x_n}_{(\mu)_n} TA(x_1) \cdots A(x_n) \\ (\mu)_j = (\mu_{j1}, \ldots, \mu_{jn(j)}) \\ \partial^{x_j}_{(\mu)_j} = \partial^{x_j}_{\mu_{j1}} \cdots \partial^{x_j}_{\mu_{jn(j)}}, \qquad \partial^{x_j}_{\mu_{j\alpha}} = \partial/\partial(x_j)^{\mu_{j\alpha}}. \end{gathered} \tag{2.23}$$

We will also use abbreviations of the form

$$\begin{aligned} &T(A_{(\mu)_1}(x_1) \cdots A_{(\mu)_m}(x_m)\, \tilde{A}^{(n)_1}(q_1) \cdots \tilde{A}^{(n)_\nu}(q_\nu)) \\ &\quad = \partial^{x_1}_{(\mu)_1} \cdots \partial^{x_m}_{(\mu)_m} \partial^{(n)_1}_{q_1} \cdots \partial^{(n)_\nu}_{q_n} T(A(x_1) \cdots A(x_m)\, \tilde{A}(q_1) \cdots \tilde{A}(q_\nu)) \end{aligned} \tag{2.24}$$

3. Feynman Rules for Renormalized Green's Functions

In perturbation theory the Green's function

$$\langle TA(x_1) \cdots A(x_n)\rangle \tag{3.1}$$

is defined by the finite part of the Gell-Mann Low expansion (1.5) which we write as an expansion with respect to Feynman diagrams

$$\langle TA(x_1) \cdots A(x_n)\rangle = \sum_{\Gamma \in \mathscr{G}} \langle TA(x_1) \cdots A(x_n)\rangle_\Gamma \tag{3.2}$$

The sum extends over the set $\mathscr{C}_n$ of Feynman diagrams with endpoints $x_1, \ldots, x_n$ which can be built by using 4-vertices (corresponding to the A^4-term), 3-vertices (corresponding to the A^3-term) and 2-vertices (corresponding to the quadratic terms in $\mathscr{L}_{\text{int}}$). Disconnected closed loops (i.e., connected components to which no coordinates x_j are attached) are excluded.

Let $\Gamma \in \mathscr{C}_n$ be a diagram with N vertices. The endpoints carrying the coordinates $x_1, \ldots, x_n$ are called exterior[5] vertices and are denoted by $V_1, \ldots, V_n$. The remaining vertices $V_{n+1}, \ldots, V_N$ are called interior vertices. The lines connecting the vertices V_a, V_b will be denoted by $L_{ab\sigma}$ ($\sigma = 1, \ldots, \nu(ab)$). V_a, V_b are called endpoints of L_{ab} connecting an exterior vertex V_a with another vertex are called exterior lines. The exterior line L_{ab} is also denoted by L_a. Lines connecting a vertex with itself will be excluded.[6]

In momentum space the unrenormalized contribution from the connected diagram $\Gamma \in \mathscr{C}_n$ has the form

$$\langle T\tilde{A}(p_1) \cdots \tilde{A}(p_n)\rangle_\Gamma^u = \frac{\delta_\Gamma}{\mathscr{S}(\Gamma)} \lim_{\epsilon \to +0} \int \frac{dk_1}{(2\pi)^4} \cdots \frac{dk_s}{(2\pi)^4} I_\Gamma$$
$$\delta_\Gamma = (2\pi)^4 \, \delta(p_1 + \cdots + p_n) \tag{3.3}$$

The symmetry number $\mathscr{S}(\Gamma)$ is

$$\mathscr{S}(\Gamma) = \sigma 2^\alpha (3!)^\beta \tag{3.4}$$

where α is the number of double lines (i.e., pairs of vertices connected by two lines in Fig. 1), β is the number of triple lines (pairs of vertices connected by three lines in Fig. 2), and σ is the number of automorphisms of the diagram, i.e., permutations of vertices which leave the diagram unchanged.

The integrand I_Γ is constructed according to the following insertion rules. Each line $L_{ab\sigma}$ carries the internal momentum

$$l_{ab\sigma} = -l_{ba\sigma} \tag{3.5}$$

To the exterior vertices $V_1, \ldots, V_n$ the external momenta

$$p_1, \ldots, p_n \tag{3.6}$$

are assigned. For an interior vertex the external momentum is set equal to zero

$$p_j = 0 \qquad \text{if} \quad j = n + 1, \ldots, N. \tag{3.7}$$

[5] Exterior lines should be distinguished from external lines which will be introduced later in connection with subdiagrams.

[6] Contributions from diagrams containing such loops would vanish anyhow after renormalization.

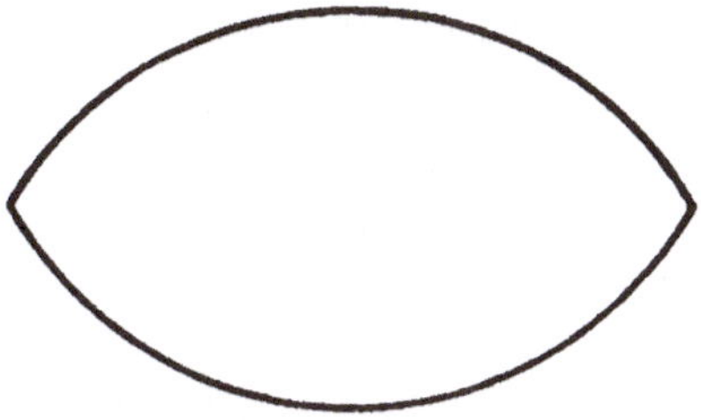

FIG. 1. Double line.

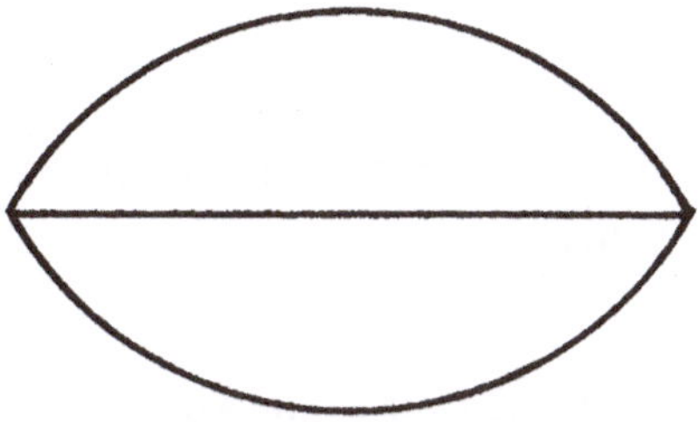

FIG. 2. Triple line.

The momenta are subject to momentum conservation at each vertex

$$\sum_{b\sigma\Gamma}^{a} l_{ab\sigma} = p_a , \qquad a = 1,..., N. \tag{3.8}$$

$\sum_{b,\sigma\Gamma}^{a}$ denotes the sum over all lines $L_{ab\sigma}$ of Γ with V_a as one of its endpoints. The internal momenta are written as linear combinations

$$l_{ab\sigma} = l_{ab\sigma}(k) + q_{ab\sigma}(p) \tag{3.9}$$

The $q_{ab\sigma}$ are linear combinations of the external momenta $p = (p_1 \cdots p_n)$ and form a particular solution of

$$\sum_{b\sigma\Gamma}^{a} q_{ab\sigma} = p_a , \qquad q_{ab\sigma} + q_{ba\sigma} = 0. \tag{3.10}$$

The $k_{ab\sigma}$ are linear combination of s independent internal momenta $k = (k_1 \cdots k_s)$ and form the general solution of the homogeneous equations

$$\sum_{b\sigma\Gamma}^{a} k_{ab\sigma} = 0, \qquad k_{ab\sigma} + k_{ba\sigma} = 0 \tag{3.11}$$

With these notations the insertion rules of I_Γ are:

$$I_\Gamma = \prod_{ab\sigma\,\Gamma} \Delta_F^{ab\sigma} \prod_{c\,\Gamma} P_c \tag{3.12}$$

with

$$\Delta_F^{ab\sigma} = i(l_{ab\sigma}^2 - m^2 + i\epsilon(l_{ab\sigma}^2 + m^2))^{-1} \tag{3.13}$$

$P_c = -i\lambda$ for a 4-vertex,
$P_c = -ig$ for a 3-vertex,
$P_c = ia + ib(l^2 - m^2)$ for a 2-vertex,
($\pm l$ is the momentum of the lines joining at the 2-vertex)
$P_c = 1$ for an exterior vertex,
and $l_{ab\sigma}$ given by (3.9).

The products $\prod_{ab\sigma\Gamma}$ and $\prod_{c\Gamma}$ extend over all lines or vertices of the diagram Γ.

The renormalized contribution of the connected diagram Γ is given by the finite part of (3.3)

$$\langle T\tilde{A}(p_1) \cdots \tilde{A}(p_n)\rangle_\Gamma = \frac{\delta_\Gamma}{\mathscr{S}(\Gamma)} \lim_{\epsilon \to +0} \int \frac{dk_1}{(2\pi)^4} \cdots \frac{dk_r}{(2\pi)^4} R_\Gamma \tag{3.14}$$

R_Γ is constructed by applying Bogoliubov's subtraction rules to the original integrand I_Γ (for details see [5]). The number of subtractions used in the definition of R_Γ is specified by the degree

$$\delta(\gamma) = 4 - r(\gamma) \tag{3.15}$$

which is assigned to every proper subdiagram γ of Γ. $r(\gamma)$ denotes the number of external lines of γ (for a precise definition see the Appendix). As is shown in the Appendix the degree (3.15) satisfies conditions which are sufficient for the convergence of (3.14). In particular $\delta(\gamma)$ is greater or equal than the dimension $d(\gamma)$ of the unrenormalized integral

$$\delta(\gamma) \geqslant d(\gamma) = 4 - r(\gamma) - v_3(\gamma) \tag{3.16}$$

with $v_3(\gamma)$ denoting the number of 3-vertices in γ.

If the diagram Γ is not connected the contribution factorizes in the usual way. Let Γ consist of the connected components $\Gamma_1, \ldots, \Gamma_c$. Some of the Γ_j may consist of a single disconnected line. Let $p_{j1}, \ldots, p_{jn_j}$ be the external momenta attached to Γ_j. Then the contribution from Γ is

$$\langle T\tilde{A}(p_1) \cdots \tilde{A}(p_n)\rangle_\Gamma = \prod_{j=1}^{c} \langle T\tilde{A}(p_{j1}) \cdots \tilde{A}(p_{jn_j})\rangle_{\Gamma_j}$$
$$= \frac{\delta_\Gamma}{\mathscr{S}(\Gamma)} \lim_{\epsilon\to+0} \int \frac{dk_1}{(2\pi)^4} \cdots \frac{dk_s}{(2\pi)^4} R_\Gamma(p_1 \cdots p_n, k_1 \cdots k_s) \tag{3.17}$$

where

$$\delta_\Gamma = \prod_{j=1}^{c} (2\pi)^4 \, \delta\left(\sum_{\alpha=1}^{n_j} p_{j\alpha}\right)$$
$$\mathscr{S}(\Gamma) = \prod_j \mathscr{S}(\Gamma_j), \qquad R_\Gamma = \prod_j R_{\Gamma_j}, \tag{3.18}$$

$k_1, \ldots, k_s$ are the independent internal momenta of all nontrivial connected components.

4. Definition of Composite Field Operators

In this section we will define composite field operators through their time-ordered functions. We consider the monomial

$$M_{\{\mu\}}\{A(x)\} = A_{(\mu)_1}(x) \cdots A_{(\mu)_m}(x) \tag{4.1}$$

with the notation

$$\begin{aligned} A_{(\mu)_j} &= \partial_{(\mu)_j} A, \\ (\mu)_j &= (\mu_{j1}, \ldots, \mu_{jm(j)}), \\ \partial_{(\mu)_j} &= \partial_{\mu_{j1}} \cdots \partial_{\mu_{jm(j)}} \qquad \partial_\mu = \partial/\partial x^\mu \\ \{\mu\} &= ((\mu)_1, \ldots, (\mu)_m). \end{aligned} \tag{4.2}$$

The dimension of the monomial (4.1) is

$$d = d(M_{\{\mu\}}) = m + \sum_{j=1}^{m} \#(\mu)_j \tag{4.3}$$

$\#s$ denotes the number of elements of a set s,

$$\#(\mu)_j = m(j).$$

We begin with the definition of the normal product

$$B_{\{u\}}(x) = N[M_{\{u\}}\{A(x)\}] \tag{4.4}$$

of minimal degree. The time ordered Green's functions of (4.4) are defined by the finite part of the expansion

$$\begin{aligned}&\langle TB_{\{u\}}(x)\, A(y_1) \cdots A(y_m)\rangle \\ &\quad = \mathrm{Fin}(\langle Te^{i\int \mathscr{L}^0_{\mathrm{int}} dz} M_{\{u\}}\{A^0(x)\}\, A^0(y_1) \cdots A^0(y_n)\rangle_0 / \langle Te^{i\int \mathscr{L}^0_{\mathrm{int}} dz}\rangle_0)\end{aligned} \tag{4.5}$$

The diagrams contributing to (4.5) contain a new vertex with the coordinate x which requires additional subtractions. The number of subtractions will be specified later. We first give the rules for constructing the unrenormalized integral of the contribution

$$\langle T\tilde{B}_{\{u\}}(p)\, \tilde{A}(q_1) \cdots \tilde{A}(q_n)\rangle^{\Delta} \tag{4.6}$$

from a given diagram Δ. $\mathscr{C}_{m+n}$ is the set of all diagrams contributing to the Green's function

$$\langle TA(x_1) \cdots A(x_m)\, A(y_1) \cdots A(y_n)\rangle.$$

Let Γ be such a diagram with coordinates $x_1, ..., x_m$; $y_1, ..., y_n$ and corresponding exterior vertices $V_1, ..., V_m$; $V_{m+1}, ..., V_{m+n}$. The interior vertices are $V_{m+n+1}, ..., V_N$. Identifying the coordinates x_j by setting $x_j = x$ we obtain a diagram contributing to (4.6). For diagrams Γ, Δ related in this way we use the notation

$$\Delta = \check{\Gamma}, \qquad \Gamma = \hat{\Delta}. \tag{4.7}$$

The new exterior vertex of Δ obtained by identifying $V_1, ..., V_m$ will be denoted by V_0. The exterior vertices of Δ are then V_0, $V_{m+1}, ..., V_{m+n}$, the interior vertices are $V_{m+n+1}, ..., V_N$. The set of all diagrams contributing to (4.6) will be denoted by $\mathscr{D}_{mn}$. (4.7) represents a one-to-one correspondence between $\mathscr{C}_{m+n}$ and $\mathscr{D}_{mn}$.

Let $\Gamma \in \mathscr{C}_{m+n}$ have the following connected components:

$$\Gamma_1, ..., \Gamma_c, \qquad \Gamma_1', ..., \Gamma_{c'}', \qquad \Gamma_1'', ..., \Gamma_{c''}''. \tag{4.8}$$

To Γ_j at least one x_i and one y_j are attached. To Γ_j' only y's are attached and to Γ_j'' only x's. With this the unrenormalized contribution from Γ becomes

$$\langle T\tilde{A}(p_1) \cdots \tilde{A}(p_m)\, \tilde{A}(q_1) \cdots \tilde{A}(q_n)\rangle_\Gamma^u = \frac{\delta_\Gamma}{\mathscr{S}(\Gamma)} \lim_{\epsilon \to +0} \int \frac{dk_1}{(2\pi)^4} \cdots \frac{dk_s}{(2\pi)^4} I_\Gamma \tag{4.9}$$

$$\delta_\Gamma = \prod_{j=1}^{c} (2\pi)^4\, \delta\Big(\sum_{\alpha=1}^{m_j} p_{j\alpha} + \sum_{\alpha=1}^{n_j} q_{j\alpha}\Big) \prod_{j=1}^{c'} (2\pi)^4\, \delta\Big(\sum_{\alpha=1}^{n_j'} q_{j\alpha}'\Big) \prod_{j=1}^{c''} (2\pi)^4\, \delta\Big(\sum_{\alpha=1}^{m_j''} p_{j\alpha}''\Big) \tag{4.10}$$

$p_{j1},\dots, p_{jm(j)}$ and $q_{j1},\dots, q_{jn(j)}$ denote the external momenta of $p_1,\dots, p_m$ or $q_1,\dots, q_n$ which belong to exterior lines of the connected component Γ_j. Likewise $q'_{j1},\dots, q'_{jn'(j)}$ denote the momenta of $q_1,\dots, q_n$ belonging to Γ_j' and $p''_{j1},\dots, p''_{jm''(j)}$ denote the external momenta of $p_1,\dots, p_m$ which belong to Γ''_j. To simplify the notation we set

$$q_1 = p_{m+1},\dots, q_n = p_{m+n}. \tag{4.11}$$

I_Γ is given by (3.12–13) with

$$p = (p_1,\dots, p_{m+n}). \tag{4.12}$$

A Fourier transformation with respect to p_j yields

$$\langle TA_{(\mu)_1}(x_1)\cdots A_{(\mu)_m}(x_m)\tilde{A}(q_1)\cdots\tilde{A}(q_n)\rangle_\Gamma^u$$
$$= \int\frac{dp_1}{(2\pi)^4}\cdots\frac{dp_m}{(2\pi)^4}\,e^{-i\Sigma p_j x_j}P_{(\mu)}\delta_\Gamma\lim_{\epsilon\to+0}\int\frac{dk_1}{(2\pi)^4}\cdots\frac{dk_s}{(2\pi)^4}\,I_\Gamma \tag{4.13}$$

with

$$P_{(\mu)} = (-i)^{\Sigma \varkappa(\mu)_j}p_{1(\mu)_1}\cdots p_{m(\mu)_m}$$
$$\{\mu\} = ((\mu)_1,\dots,(\mu)_m). \tag{4.14}$$

Introducing new variables of integration

$$p_{j\alpha} = \frac{p^{(j)}}{m_j} + K_{j\alpha},\qquad p''_{j\alpha} = K''_{j\alpha}$$
$$p^{(j)} = -\sum_{\alpha=1}^{n_j} q_{j\alpha} \tag{4.15}$$

and carrying out the integration over K_{jm_j}, $K''_{jm''_j}$

$$\langle TA_{(\mu)_1}(x_1)\cdots A_{(\mu)_m}(x_m)\tilde{A}(q_1)\cdots\tilde{A}(q_n)\rangle_\Gamma^u$$
$$= \frac{\delta_\Gamma'}{\mathscr{S}(\Gamma)}\int\prod_j\prod_{\alpha=1}^{m_j-1}\frac{dK_{j\alpha}}{(2\pi)^4}\prod_j{}''\prod_{\alpha=1}^{m_j''-1}\frac{dK''_{j\alpha}}{(2\pi)^4}\,e^{-i\Sigma p_j x_j}P_{(\mu)}\lim_{\epsilon\to+0}\int\prod_{j=1}^{s}\frac{dk_j}{(2\pi)^4}\,I_\Gamma \tag{4.16}$$

$$\delta_\Gamma' = \prod_{j=1}^{c'}(2\pi)^4\,\delta\Bigl(\sum_{\alpha=1}^{n_j'} q'_{j\alpha}\Bigr) \tag{4.17}$$

In $P_{(M)}$ and I_Γ the substitutions (4.15) are made for the external momenta. The product $\prod_j$ is restricted to values j with $m_j \geqslant 1$, $\prod_j''$ is restricted to values of j with $m_j'' \geqslant 1$. Setting $x_j = x$ and taking a Fourier transformation with respect to x we get

$$\langle T\tilde{M}_{\{u\}}(p)\, \tilde{A}(q_1) \cdots \tilde{A}(q_m)\rangle_{\Delta}^{u}$$

$$= \frac{\delta_\Delta}{\mathscr{S}(\Delta)} \lim_{\epsilon \to +0} \int \prod_j \prod_{\alpha=1}^{m_j-1} \frac{dK_{j\alpha}}{(2\pi)^4} \prod_j'' \prod_{\alpha=1}^{m_j''-1} \frac{dK_{j\alpha}''}{(2\pi)^4} \prod_{j=1}^{s} \frac{dk_j}{(2\pi)^4} I_{\Delta\{u\}} \tag{4.18}$$

where

$$\delta_\Delta = (2\pi)^4\, \delta\left(p + \sum q_{j\alpha}\right) \prod_{j=1}^{c'} (2\pi)^4\, \delta\left(\sum_{\alpha=1}^{n_j'} q_{j\alpha}'\right)$$

$$\mathscr{S}(\Delta) = \mathscr{S}(\Gamma) \tag{4.19}$$

and

$$I_{\Delta\{u\}} = SP_{\{u\}} I_\Gamma\,. \tag{4.20}$$

The substitution operator S prescribes that the substitutions (4.15) be made for momenta $p_1, \ldots, p_m$. According to (4.20) the insertion rules for the construction of I_Δ are the usual ones except that the monomial $P_{\{u\}}$ is assigned to the vertex V_0.

The renormalized Green's function (4.5) is then defined by

$$\langle TB_{\{u\}}(x)\, A(y_1) \cdots A(y_n)\rangle = \sum_{\Delta \in \mathscr{D}_{mn}} \langle TB_{\{u\}}(x)\, A(y_1) \cdots A(y_n)\rangle_\Delta \tag{4.21}$$

with the Fourier transform of

$$\langle TB_{\{u\}}(x)\, A(y_1) \cdots A(y_n)\rangle_\Delta \tag{4.22}$$

given by the finite part of (4.18), namely

$$\langle T\tilde{B}_{\{u\}}(p)\, \tilde{A}(q_1) \cdots \tilde{A}(q_n)\rangle_\Delta$$

$$= \frac{\delta_\Delta}{\mathscr{S}(\Delta)} \lim_{\epsilon \to +0} \int \prod_j \prod_{\alpha=1}^{m_j-1} \frac{dK_{j\alpha}}{(2\pi)^4} \prod_j'' \prod_{\alpha=1}^{m_j''-1} \frac{dK_{j\alpha}''}{(2\pi)^4} \prod_{j=1}^{s} \frac{dk_j}{(2\pi)^4} R_\Delta^{(d)}. \tag{4.23}$$

We still have to specify the number of subtractions to be used in constructing $R_\Delta^{(d)}$ from the unrenormalized integrand I_Δ. This is done by prescribing the degree $\delta_d(\gamma)$ of any proper subdiagram γ of Δ. If γ does not contain V_0 we define $\delta_d(\gamma)$ by (3.15). If γ contains V_0 we define $\delta_d(\gamma)$ by

$$\delta_d(\gamma) = d - r(\gamma) \tag{4.24}$$

where d is the dimension (4.3) of the operator $B^{(d)}_{\{M\}}$ and $r(\gamma)$ denotes the number of external lines of γ (for a precise definition of $r(\gamma)$ see the Appendix). As is shown in the Appendix $\delta_d(\gamma)$ satisfies the conditions (A.35) and (A.36) which are sufficient for the convergence of (4.23).

We will also need composite operators

$$B^{(a)}_{\{\mu\}}(x) = N_a[M_{\{\mu\}}\{A(x)\}] \tag{4.25}$$

which are constructed with additional subtractions. The degree a of the operator may be any integer which is greater or equal to the dimension (4.3)

$$a \geqslant d, \qquad a - d = 0, 1, 2, \ldots. \tag{4.26}$$

The Green's functions

$$\langle TB^{(a)}_{\{\mu\}}(x)\, A(y_1) \cdots A(y_n)\rangle \tag{4.27}$$

are defined similarly to (4.4), (4.21–23) but with a larger number of subtractions

$$\langle TB^{(a)}_{\{\mu\}}(x)\, A(y_1) \cdots A(y_n)\rangle = \sum_{\Delta \in \mathscr{D}_{mn}} \langle TB^{(a)}_{\{\mu\}}(x)\, A(y_1) \cdots A(y_n)\rangle_\Delta$$

$$\langle T\tilde{B}^{(a)}_{\{\mu\}}(p)\, \tilde{A}(q_1) \cdots \tilde{A}(q_n)\rangle \tag{4.28}$$

$$= \frac{\delta_\Delta}{\mathscr{S}_\Delta} \lim_{\epsilon \to +0} \int \prod_j \prod_{\alpha=1}^{n_j-1} \frac{dK_{j\alpha}}{(2\pi)^4} \prod_j \prod_{\alpha=1}^{n_j''-1} \frac{dK''_{j\alpha}}{(2\pi)^4} \prod_{j=1}^{s} \frac{dk_j}{(2\pi^4)} R^{(a)}_\Delta$$

In order to specify the number of subtractions used in constructing $R^{(a)}_\Delta$ from I_Δ we assign a degree $\delta_a(\gamma)$ to any proper subdiagram δ of Δ. If γ does not contain V_0 we set

$$\delta_a(\gamma) = \delta(\gamma) \tag{4.29}$$

If V_0 belongs to γ we set

$$\delta_a(\gamma) = a - r(\gamma) \tag{4.30}$$

According to (4.24)

$$\delta_a(\gamma) = \delta_d(\gamma) + a - d \text{ if } V_0 \text{ belongs to } \gamma. \tag{4.31}$$

Again the degree $\delta_a(\gamma)$ satisfies the conditions (A.35), (A.36) of the Appendix which are sufficient for the convergence of (4.28).

For the derivation of some normalization conditions we write down the expansion of the proper part, as defined by (2.16–17);

$$\langle TB^{(a)}_{\{\mu\}}(x)\,A(y_1)\cdots A(y_n)\rangle^{\text{prop}} = \langle T{:}A^0_{(\mu)_1}(x)\cdots A^0_{(\mu)_m}(x){:}\,A^0(y_1)\cdots A^0(y_n)\rangle^{\text{prop}}_0 + \sum_{\Phi\in\mathscr{F}_{mn}} \frac{1}{\mathscr{S}(\Phi)}\langle TB^{(a)}_{\{\mu\}}(x)\,A(y_1)\cdots A(y_n)\rangle^{\text{prop}}_\Phi \tag{4.32}$$

The first term on the right side

$$\langle T{:}A^0_{(\mu)_1}(x)\cdots A^0_{(\mu)_m}(x){:}\,A^0(y_1)\cdots A^0(y_n)\rangle^{\text{prop}}_0$$

$$= \delta_{mn}\partial^{x_1}_{(\mu)_1}\cdots\partial^{x_m}_{(\mu)_m}\sum_P \delta(x_1-y_{i_1})\cdots\delta(x_n-y_{i_n})\,|_{x_j=x} \tag{4.33}$$

represents the zero order contribution and comes from the trivial diagram consisting of the vertex V_0 only. A^0 denotes a free field, $\langle\ \rangle_0$ the free field vacuum expectation value. The sum $\sum_{\Phi\in\mathscr{F}_{mn}}$ on the right side of (4.32) extends over the set $\mathscr{F}_{mn}$ of all proper diagrams Φ which can be obtained by dropping the lines with endpoints $y_1,\ldots,y_n$ from a diagram Φ^* contributing to $\langle TB^{(a)}_{\{M\}}(x)\,A(y_1)\cdots A(y_n)\rangle$.

In momentum space the contribution from a diagram Φ is

$$\langle T\tilde{B}^{(a)}_{\{\mu\}}(p)\,\tilde{A}(q_1)\cdots\tilde{A}(q_n)\rangle^{\text{prop}}_\Phi$$

$$= \frac{\delta_\Phi}{\mathscr{S}(\Phi)}\lim_{\epsilon\to+0}\int\prod_i\prod_{\alpha=1}^{n_i-1}\frac{dK_{i\alpha}}{(2\pi)^4}\prod_j{}'\prod_{\alpha=1}^{n_j'-1}\frac{dK''_{j\alpha}}{(2\pi)^4}\prod_{j=1}^{s}\frac{dk_j}{(2\pi)^4}R^{(a)}_\Phi \tag{4.34}$$

$$\delta_\Phi = (2\pi)^4\,\delta\left(p-\sum q_j\right),\qquad \mathscr{S}(\Phi)=\mathscr{S}(\Phi^*)$$

in the notation of (4.28). The renormalized integrand $R^{(a)}_\Phi$ can be written as

$$R^{(a)}_\Phi = (1-t^\Phi)\,\bar{R}^{(a)}_\Phi \tag{4.35}$$

where t^Φ denotes the Taylor operator with respect to $q_1,\ldots,q_n$ up to and including the order $\delta_a(\Phi)$. For the definition of $\bar{R}^{(a)}_\Phi$ see [5]. Equation (4.35) implies

$$\partial^{q_1}_{(\nu)_1}\cdots\partial^{q_n}_{(\nu)_n}R^{(a)}_\Phi = 0 \tag{4.36}$$

or

$$\partial^{q_1}_{(\nu)_1}\cdots\partial^{q_n}_{(\nu)_n}\langle T\tilde{B}^{(a)}_{\{\mu\}}(0)\,\tilde{A}(q_1)\cdots\tilde{A}(q_n)\rangle^{\text{prop}}_\Phi = 0 \tag{4.37}$$

at $q_1=\cdots=q_n=0$ provided

$$\sum \#(\nu)_j \leqslant \delta_a(\Phi) = a-n. \tag{4.38}$$

Hence

$$\langle TB^{(a)}_{\{\mu\}}(0)\, \tilde{A}_{(\nu)_1}(0) \cdots \tilde{A}_{(\nu)_n}(0)\rangle^{\text{prop}} \tag{4.39}$$

$$= \langle T{:}A^0_{(\mu)_1}(0) \cdots A^0_{(\mu)_m}(0){:}\, \tilde{A}^0_{(\nu)_1}(0) \cdots \tilde{A}^0_{(\nu)_n}(0)\rangle_0^{\text{prop}} \text{ provided } n + \sum \#(\nu)_j \leqslant a.$$

$$\langle TB^{(a)}_{\{\mu\}}(0)\, \tilde{A}_{(\nu)_1}(0) \cdots \tilde{A}_{(\nu)_n}(0)\rangle^{\text{prop}} = 0$$
$$\text{if} \quad n + \sum \#(\nu)_j \leqslant a \quad \text{and} \quad \{\nu\} \not\sim \{\mu\} \tag{4.40}$$
$$\langle TB^{(a)}_{\{\mu\}}(0)\, \tilde{A}_{(\nu)_1}(0) \cdots \tilde{A}_{(\nu)_n}(0)\rangle^{\text{prop}} = i^{\Sigma \#(\mu)_j} \prod_{j=1}^{t} s_j\,! \prod_{j=1}^{m} \prod_{\alpha=0}^{3} a_{j\alpha}\,!$$
$$\text{if} \quad n + \sum \#(\nu)_j \leqslant a \quad \text{and} \quad \{\nu\} \sim \{\mu\}.$$

In these equations the following notations were used. Two sets $(\mu)_i$, $(\nu)_j$ are called equivalent

$$(\mu)_i \sim (\nu)_j \tag{4.41}$$

if $(\nu)_j$ is a permutation of $(\mu)_i$. $\{\mu\}$ and $\{\nu\}$ are called equivalent

$$\{\mu\} \sim \{\nu\} \tag{4.42}$$

if $n = m$ and if there is a permutation $((\nu)_{i_1}, \ldots, (\nu)_{i_n})$ of $\{\nu\}$ such that

$$(\mu)_j \sim (\nu)_{i_j}$$

Otherwise we write $\{\mu\} \not\sim \{\nu\}$, $s_1, \ldots, s_t$ are the numbers of equivalent $(\mu)_k$ occurring in $\{\mu\}$. $a_{j\alpha}$ is the number of α's occurring in $(\mu)_j$.

We finally show that the vacuum expectation values

$$\langle B^{(a)}_{\{\mu\}}(x)\rangle = 0 \tag{4.43}$$

vanish. The contribution from a Feynman diagram Δ is given by

$$\langle \tilde{B}^{(a)}_{\{\mu\}}(p)\rangle_\Delta = \frac{\delta(p)}{\mathscr{S}(\Delta)} \lim_{\epsilon \to +0} \int \prod_j \prod_{\alpha=1}^{n_j''-1} \frac{dK''_{j\alpha}}{(2\pi)^4} \prod_{j=1}^{s} \frac{dk_j}{(2\pi)^4}\, R^{(a)}_\Delta \tag{4.44}$$

If Δ is proper the integrand is of the form

$$R^{(a)}_\Delta = (1 - t_p{}^a)\, \bar{R}^{(a)}_\Delta = 0$$

and vanishes since $R^{(a)}$ does not depend on p. If Δ is not proper it contains at least one proper subdiagram γ with one external momentum p only and degree $b = \delta(\gamma) \geqslant 0$. Such a diagram contributes to $R_\Delta^{(a)}$ the factor

$$R_\gamma^{(b)} = (1 - t_{p\gamma}^{b})\, \bar{R}_\gamma^{(b)} = 0$$

since $\bar{R}_\gamma^{(b)}$ does not depend on p^γ.

As has been emphasized by Lowenstein [13] the time ordered Green's functions defined in this section are not uniquely determined. It frequently occurs that the same local operator $B(x)$ may be represented by different polynomials of $A(x)$

$$B(x) = \sum c_j N_{a_j}[M_j\{A(x)\}] \tag{4.45}$$

$$B(x) = \sum c_j' N_{a_j'}[M_j'\{A(x)\}] \tag{4.46}$$

M_j, M_j' denote monomials of A and its derivatives, j includes Lorentz indices. While the time ordered functions of a given monomial were uniquely prescribed the Green's function

$$\langle TB(x)\, A(y_1) \cdots A(y_n)\rangle \tag{4.47}$$

as calculated from the representation (4.45) and (4.46) may differ by δ-function terms.

5. Algebraic Identities

Let Γ be a diagram of $\mathscr{C}_{m+n}$ and $\Delta = \bar{\Gamma} \in \mathscr{D}_{mn}$. In this section we will investigate the relation between R_Γ and the integrand $R_\Delta^{(a)}$ which we used in (4.28) for defining the Green's functions of composite operators. Using the explicit formulas derived in [5, Section 3], R_Γ and $R_\Delta^{(a)}$ may be written as

$$R_\Gamma = S_\Gamma \sum_{U \in \mathscr{U}(\Gamma)} \prod_{\gamma \in U} (-t^\gamma S_\gamma)\, S_U I_\Gamma \tag{5.1}$$

$$R_\Delta^{(a)} = S_\Delta \sum_{U \in \mathscr{U}_a(\Gamma)} \prod_{\gamma \in U} (-t_a^\gamma S_\gamma)\, S_U I_\Delta \tag{5.2}$$

I_Γ and I_Δ are related by (4.20). $U(\Gamma)$ is the set of all Γ-forests, $U_a(\Delta)$ the set of all Δ-forests relative to the degree $\delta_a(\gamma)$. For the definition of the substitution operators S_γ see [5]. In addition we introduce the substitution operator S_U. If applied to a function of the momenta $l_{ab\sigma}(p, k)$ the operator S_U prescribes the substitutions

$$l_{ab\sigma}(pk) \to l_{ab\sigma}^\gamma(p^\gamma k^\gamma)$$

if $L_{ab\sigma}$ belongs to $\gamma \in U$ but not to any other $\gamma' \in U$ with $\gamma' \subset \gamma$.

Apparently

$$\mathscr{U}_a(\Delta) = \mathscr{U}(\Gamma) \cup \Phi \tag{5.3}$$

where Φ is the set of all Δ-forests with at least one element γ containing V_0. From (5.1–3)

$$R_\Delta^{(a)} = SP_{\{u\}} R_\Gamma + X \tag{5.4}$$

with

$$X = S_\Delta \sum_{U \subset \Phi} \prod_{\gamma \in U} (-t_a{}^\gamma S_\gamma)\, S_U I_\Delta \tag{5.5}$$

Our aim is to write X in a more convenient form. First we note that any two elements γ, γ' of U containing V_0 must satisfy

$$\gamma \subset \gamma' \text{ or } \gamma' \subset \gamma$$

For γ and γ' are not disjoint since they have the vertex V_0 in common. Hence among all elements $\gamma \in U$ containing V_0 there is a smallest one which we call τ. Then

$$U = U_1 \cup U_2 \cup \{\tau\} \tag{5.6}$$

U_1 is the set of all $\gamma \in U (\gamma \neq \tau)$ with $\tau \subset \gamma$ or τ and γ disjoint. U_2 is the set of all $\gamma \in U (\gamma \neq \tau)$ with $\gamma \subset \tau$. (5.6) implies

$$\prod_{\gamma \in U} (-t_a{}^\gamma S_\gamma) = \prod_{\gamma \in U_1} (-t_a{}^\gamma S_\gamma)(-t_a{}^\tau S_\tau) \prod_{\gamma \in U_2} (-t^\gamma S_\gamma) \tag{5.7}$$

With this information Eq. (5.5) becomes

$$X = \sum_{\tau \in T_a} \sum_{U_1 \in \mathscr{M}_\tau} \sum_{U_2 \in \mathscr{U}(\tau)} \prod_{\gamma \in U_1} (-t_a{}^\gamma S_\gamma)(-t_a{}^\tau S_\tau) \prod_{\gamma \in U_2} (-t^\gamma S_\gamma)\, S_U I_\Delta \tag{5.8}$$

Here T_a denotes the set of all proper subdiagrams τ of Δ which contain V_0 and have degree $\delta_a(\tau) \geqslant 0$. $\mathscr{M}_\tau$ is the set of all Δ-forests U_1 which have the property that each $\gamma \in U_1$ satisfies

$$\tau \subset \gamma \qquad \text{or } \tau \cap \gamma = \varnothing .$$

$u(\tau)$ is the set of all τ-forests. (5.4) and (5.8) represent the final form of the algebraic identities.

We finally discuss a generalization of (5.4), (5.8) which will be needed in the following section for establishing relations between normal products of different degrees. Suppose $R_\Delta^{(a)}$ and $R_\Delta^{(b)}$ are formed with different degrees δ_a and δ_b defined through (4.29–30), assuming $a > b$. We have

$$R_\Delta^{(a)} = S_\Delta \sum_{U\in\mathscr{U}_a(\Delta)} \prod_{\gamma\in U} (-t_a{}^\gamma S_\gamma)\, S_U I_\Delta \tag{5.9}$$

$$R_\Delta^{(b)} = S_\Delta \sum_{U\in\mathscr{U}_a(\Delta)} \prod_{\gamma\in U} (-t_b{}^\gamma S_\gamma)\, S_U I_\Delta \tag{5.10}$$

In (5.10) we used the convention

$$t_b{}^\gamma = 0 \qquad \text{if} \quad \delta_b(\gamma) < 0.$$

For any γ containing V_0 we split

$$t_a{}^\gamma = t_b{}^\gamma + (t_a{}^\gamma - t_b{}^\gamma) \tag{5.11}$$

while

$$t_a{}^\gamma = t_b{}^\gamma = t_{p\gamma}^{d(\gamma)} \tag{5.12}$$

holds for all γ's which do not contain V_0. Substituting (5.11) into (5.9) we obtain

$$R_\Delta^{(a)} = S_\Delta \sum_{U\in\mathscr{U}_a(\Delta)} \sum_{F\in\mathscr{F}(U)} \prod_{\gamma\in U} F_\gamma S_U I_\Delta \tag{5.13}$$

Here $\mathscr{F}(U)$ is the family of functions with the property

(i) $F_\gamma = -t_b{}^\gamma S_\gamma$ or $F_\gamma = -(t_a{}^\gamma - t_b{}^\gamma)\, S_\gamma$
if γ contains V_0,

(ii) $F_\gamma = -t_a{}^\gamma S_\gamma = -t_b{}^\gamma S_\gamma$
if γ does not contain V_0.

For any U there is a function F_0 in $\mathscr{F}(U)$ which assigns $F_\gamma \equiv -t_b{}^\gamma S_\gamma$ to any $\gamma \in U$. Taking out all terms with $F = F_0$ we find

$$R_\Delta^{(a)} = R_\Delta^{(b)} + X \tag{5.14}$$

$$X = S_\Delta \sum_{U\in\mathscr{U}'} \sum_{\substack{F\in\mathscr{F}(U)\\ F\neq F_0}} \prod_{\gamma\in U} F_\gamma S_U I_\Delta \tag{5.15}$$

$\mathscr{U}'$ is the set of all forests having at least one element γ which contains V_0. Our aim is again to write X in a convenient form. For $F \neq F_0$ there is a smallest element τ in U which contains V_0 and satisfies

$$F_\tau = -(t_a{}^\tau - t_b{}^\tau)\, S_\tau \tag{5.16}$$

For given U and F we decompose

$$U = U_1 \cup U_2 \cup \{\tau\} \tag{5.17}$$

where U_1 and U_2 are defined as below Eq. (5.6). As generalization of (5.4), (5.8) we finally obtain

$$\begin{aligned} R_\Delta^{(a)} &= R_\Delta^{(b)} + X \\ X &= \sum_{\tau\in T_a} \sum_{U_1\in\mathcal{M}_\tau} \sum_{U_2\in\mathcal{U}(\tau)} \prod_{\gamma\in U_1} (-t_a{}^\gamma S_\gamma)(-(t_a{}^\tau - t_b{}^\tau))\, S_\tau \prod_{\gamma\in U_2} (-t_b{}^\gamma S_\gamma)\, S_U I_\Delta \end{aligned} \tag{5.18}$$

T_a, $\mathcal{M}_\tau$ are defined as below Eq. (5.8).

6. Relations between Composite Operators of Different Degree

In this section we will derive some identities which relate composite operators of different degree. We consider the formal operator product

$$M_{\{\mu\}}\{A(x)\} = A_{(\mu)_1}(x) \cdots A_{(\mu)_m}(x), \qquad m \geqslant 2 \tag{6.1}$$

of dimension

$$d = m + \sum \#(\mu)_i \geqslant 2 \tag{6.2}$$

and form the composite operators

$$B_{\{\mu\}}^{(a)}(x) = N_a[M_{\{\mu\}}\{A(x)\}] \tag{6.3}$$

$$B_{\{\mu\}}^{(b)}(x) = N_b[M_{\{\mu\}}\{A(x)\}]. \tag{6.4}$$

It will always be assumed that

$$d \leqslant b < a. \tag{6.5}$$

We will compare the power series expansions of the Green's functions

$$\langle TB_{\{\mu\}}^{(a)}(x)\, A(y_1) \cdots A(y_n)\rangle = \sum_{\Delta\in\mathcal{D}_{mn}} \langle TB_{\{\mu\}}^{(a)}(x)\, A(y_1) \cdots A(y_n)\rangle_\Delta \tag{6.6}$$

$$\langle TB_{\{\mu\}}^{(b)}(x)\, A(y_1) \cdots A(y_n)\rangle = \sum_{\Delta\in\mathcal{D}_{mn}} \langle TB_{\{\mu\}}^{(b)}(x)\, A(y_1) \cdots A(y_n)\rangle_\Delta\,. \tag{6.7}$$

According to (4.28) the contributions from the diagram Δ are given by

$$\langle TB^{(a)}_{\{\mu\}}(p)\,\tilde{A}(q_1)\cdots\tilde{A}(q_n)\rangle_\Delta$$

$$= \frac{\delta_\Delta}{\mathscr{S}(\Delta)}\lim_{\epsilon\to+0}\int\prod_j\prod_{\alpha=1}^{m_j-1}\frac{dK_{j\alpha}}{(2\pi)^4}\prod_j{}''\prod_{\alpha=1}^{n_j''-1}\frac{dK''_{j\alpha}}{(2\pi)^4}\prod_{j=1}^{s}\frac{dk_j}{(2\pi)^4}R_\Delta^{(a)} \tag{6.8}$$

$$\langle TB^{(b)}_{\{\mu\}}(p)\,\tilde{A}(q_1)\cdots\tilde{A}(q_n)\rangle_\Delta$$

$$= \frac{\delta_\Delta}{\mathscr{S}(\Delta)}\lim_{\epsilon\to+0}\int\prod_j\prod_{\alpha=1}^{m_j-1}\frac{dK_{j\alpha}}{(2\pi)^4}\prod_j{}''\prod_{\alpha=1}^{n_j''-1}\frac{dK''_{j\alpha}}{(2\pi)^4}\prod_{j=1}^{s}\frac{dk_j}{(2\pi)^4}R_\Delta^{(b)}. \tag{6.9}$$

The renormalized integrands $R_\Delta^{(a)}$ and $R_\Delta^{(b)}$ are related by the identity (5.14–15). We first determine the set T_a appearing in (5.15). By definition T_a is the set of all proper subdiagrams of Δ which contain V_0 and have nonnegative degree

$$\delta_a(\tau) \geqslant 0. \tag{6.10}$$

The definition (4.24) of the degree yields the inequality

$$a - r \geqslant 0 \tag{6.11}$$

where r is the number of external lines of τ. This implies

$$r = 1,\ldots, a \tag{6.12}$$

The case $r = 0$ may be excluded since

$$(t_a{}^\tau - t_b{}^\tau)\,S_\tau\prod_{\gamma\in U_2}(-t_b{}^\gamma S_\gamma)\,S_U I_\Delta = 0$$

vanishes if τ has no external lines. For then all momenta l^γ_{abu} are independent of the external momenta p^τ. Hence T_a is the set of all proper subdiagrams of which contain V_0 and have $r = 1,\ldots, a$ external lines.

The unrenormalized integral I_Δ factorize according to

$$I_\Delta = I_{\Delta/\tau} I_\tau \qquad \text{(see Eq. (A.26))}. \tag{6.13}$$

With (6.13) the identity (5.14–15) becomes

$$\begin{aligned} R_\Delta^{(a)} &= R_\Delta^{(b)} + X \\ X &= -\sum_{\tau\subset T_a}\sum_{U_1\in\mathscr{M}_\tau}\sum_{U_2\in\mathscr{U}(\tau)}\prod_{\gamma\in U_1}(-t_a{}^\gamma S_\gamma) \\ &\quad\times S_{U_1} I_{\Delta/\tau}(t_a{}^\tau - t_b{}^\tau)\,S_\tau\prod_{\gamma\in U_2}(-t_b{}^\gamma S_\gamma)\,S_{U_2} I_\tau\,. \end{aligned} \tag{6.14}$$

In order to work out the explicit form of the differential operator $t_a^\tau - t_b^\tau$ we list all nonvanishing external momenta of τ. The external lines of τ are given by (A.2). Some of the subscripts a_j may be zero. In this case external lines of τ are attached to V_0. The nonvanishing external momenta of τ are then

$$p_0^\tau = p_0 = p$$

and

$$p_1^\tau, \ldots, p_r^\tau$$

corresponding to the external lines (A.2). The momenta $p_{m+1} = q_1, \ldots, p_{m+n} = q_n$ cannot be external to τ. For the corresponding exterior vertices $V_{m+1}, \ldots, V_{m+n}$ do not belong to τ since τ is proper. The sum of all external momenta of τ must vanish

$$p_0^\tau + \sum_{j=1}^{r} p_j^\tau = 0 \tag{6.15}$$

Since τ is connected (6.15) is the only linear relation among the external momenta. Choosing $p_1^\tau, \ldots, p_r^\tau$ as linearly independent external momenta we have

$$t_a^\tau - t_b^\tau = \sum_{\{\rho\}}{}_b^a \frac{1}{\prod \#(\rho)_j\,!} p_{1(\rho)_1}^\tau \cdots p_{r(\rho)_r}^\tau \theta^\tau \partial_{p_1^\tau}^{(\rho)_1} \cdots \partial_{p_r^\tau}^{(\rho)_r} \tag{6.16}$$

with

$$(\rho)_j = (\rho_{j1}, \ldots, \rho_{jr(j)}), \qquad \{\rho\} = ((\rho)_1, \ldots, (\rho)_r)$$
$$p_{j(\rho)_j}^\tau = p_{j\rho_1}^\tau \cdots p_{j\rho_{jr(j)}}^\tau, \qquad \partial_{p_j^\tau}^{(\rho)_j} = \partial_{p_j^\tau}^{\rho_{j1}} \cdots \partial_{p_j^\tau}^{\rho_{jr(j)}} \tag{6.17}$$
$$\partial_{p_k^\tau}^{\rho_{jk}} = \frac{\partial}{\partial (p_k^\tau)_{\rho_{jk}}}.$$

The sum $\sum_{\{\rho\}b}^{a}$ extends over all sets $\{\rho\}$ with

$$b < r + \sum_{j=1}^{r} \#(\rho)_j \leqslant a \tag{6.18}$$

The operator θ^τ prescribes that the external momenta p_j^τ be set equal to zero. Let S_μ be the first substitution operator of

$$S_\Gamma \prod_{\gamma \in U_1} (-t_a^\gamma S_\gamma)$$

to act on (6.16). That is, μ is either Γ or the smallest element of U_1 which contains τ. S_μ then effects the substitution

$$p_j^\tau \to p_j^\tau(k^\mu, p^\mu)$$

with

$$p_j^\tau(k^\mu, p^\mu) = -l^\mu_{a_j b_j \sigma_j} = -k^\mu_{a_j b_j \sigma_j}(k^\mu) - q^\mu_{a_j b_j \sigma_j}(p^\mu) \tag{6.19}$$

In (6.14) we may therefore replace

$$\cdots S_\mu \cdots p^\tau_{1(\rho)_1} \cdots p^\tau_{r(\rho)_r} \cdots$$

by

$$\cdots S_\mu \cdots (-il^\mu_{a_1 b_1 \sigma_1})_{(\mu)_1} \cdots (-il^\mu_{a_r b_r \sigma_r})_{(\rho)_r} (-i)^r \cdots.$$

We thus obtain the following factorization of X:

$$X = -\sum_{\tau \in T_a} \sum_{\{\mu\}}{}_b^a \frac{(-i)^{\Sigma \#(\rho)_j}}{\prod \#(\rho)_j\,!} R^{(a)}_{\Delta/\tau\{\rho\}} \{\partial^{(\rho)_1}_{p_1{}^\tau} \cdots \partial^{(\rho)_m}_{p_m{}^\tau} R^{(b)}_\tau\}_{p_j{}^\tau = 0} \tag{6.20}$$

$$\begin{aligned} R^{(a)}_{\Delta/\tau\{\rho\}} &= \sum_{U_1 \in \mathscr{M}_\tau} \prod_{\gamma \in U_1} (-t_a{}^\gamma S_\gamma) \prod^r l^\mu_{a_j b_j \sigma_j}(k^\mu, p^\mu)_{(\rho)_j} S_{U_1} I_{\Delta/\tau\{\rho\}} \\ &= \sum_{V \in \mathscr{F}(\Delta/\tau)} \prod_{\rho \in V} (-t_a{}^\rho S_\rho)\, S_V I_{\Delta/\tau\{\rho\}} \end{aligned} \tag{6.21}$$

$$R^{(b)}_\tau = \sum_{U \in \mathscr{F}(\tau)} \prod_{\gamma \in U} (-t_b{}^\gamma S_\gamma)\, S_U I_\tau\,. \tag{6.22}$$

In (6.21) the substitution $\rho = \gamma/\tau$ was used. For the notion of the reduced diagram Δ/τ and related definitions see the Appendix. $I_{\Delta/\tau\{\rho\}}$ is defined by

$$I_{\Delta/\tau\{\rho\}} = \prod_{j=1}^{r} l^{\Delta/\tau}_{0 b_j \sigma_j}(k, p)_{(\rho)_j} I_{\Delta/\tau} \tag{6.23}$$

Integrating (6.20) over the internal momenta, summing over all Δ and taking a Fourier transformation with respect to $p, q_1, \ldots, q_n$ we obtain

$$\begin{aligned} \langle TB^{(b)}_{\{\mu\}}(x)\, A(y_1) \cdots A(y_n)\rangle &= \langle TB^{(a)}_{\{\mu\}}(x)\, A(y_1) \cdots A(y_n)\rangle \\ &\quad + \sum_{\{\rho\}}{}_b^a \frac{(-i)^{\Sigma \#(\rho)_j}}{\prod \#(\rho)_j\,!} G^{(b)\{\rho\}}_{\{\mu\}} \langle TB^{(a)}_{\{\rho\}}(x)\, A(y_1) \cdots A(y_n)\rangle \end{aligned} \tag{6.24}$$

$$b < r + \sum_{j=1}^{r} \#(\rho)_j \leqslant a.$$

The coefficients are

$$G^{(b)\{\rho\}}_{\{\mu\}} = \sum_{\tau \in \mathscr{F}_r} \langle TB^{(b)}_{\{\mu\}}(0)\, \tilde{A}^{(\rho)_1}(0) \cdots \tilde{A}^{(\rho)_r}(0)\rangle^{\mathrm{prop}}_\tau \tag{6.25}$$

where $\mathscr{F}_r$ is the set of all nontrivial diagrams contributing to the proper part. The notation in (6.25) is similar to (2.24), i.e.,

$$\begin{aligned}&\langle TB^{(b)}_{\{\mu\}}(x)\, \tilde{A}^{(\rho)_1}(q_1) \cdots \tilde{A}^{(\rho)_r}(q_r)\rangle^{\text{prop}} \\ &\quad = \partial^{(\rho)_1}_{q_1} \cdots \partial^{(\rho)_r}_{q_r} \langle TB^{(b)}_{\{\mu\}}(x)\, \tilde{A}(q_1) \cdots \tilde{A}(q_r)\rangle^{\text{prop}}\end{aligned} \tag{6.26}$$

The contribution

$$\langle T{:}A_{0(\mu)_1}(0) \cdots A_{0(\mu)_m}(0){:}\, \tilde{A}_0^{(\rho)_1}(0) \cdots \tilde{A}_0^{(\rho)_r}(0)\rangle^{\text{prop}}$$

from the trivial diagram (with the notation of (2.24)) vanishes because

$$d = m + \sum \#(\mu)_j \leqslant b < r + \sum \#(\rho)_j$$

in (6.24), and therefore $\{\rho\} \not\sim \{\mu\}$. Hence

$$G^{(b)\{\rho\}}_{\{\mu\}} = \langle TB^{(b)}_{\{\mu\}}(0)\, \tilde{A}^{(\rho)_1}(0) \cdots \tilde{A}^{(\rho)_r}(0)\rangle^{\text{prop}} \tag{6.27}$$

for the coefficients of (6.24) with the notation (6.26). According to this the $G^{(b)\{\rho\}}_{\mu}$ are invariant tensors which can be composed of $g_{\mu\nu}$- and $\epsilon_{\mu\nu\rho\sigma}$-tensors only.

Applying the reduction technique to the variables $y_1, \ldots, y_n$ of Eq. (6.24) we obtain the operator identity

$$\begin{aligned}&B^{(b)}_{\{\mu\}}(x) = B^{(a)}_{\{\mu\}}(x) + \sum_{\{\rho\}}{}^{a}_{b} \frac{(-i)^{\sum \#(\rho)_j}}{r! \prod \#(\rho)_j\,!} G^{(b)\{\rho\}}_{\{\mu\}} B^{(a)}_{\{\rho\}}(x) \\ &\qquad b < r + \sum_{j=1}^{r} \#(\rho)_j \leqslant a\end{aligned} \tag{6.28}$$

with the coefficients given by (6.27). If the reduction technique is applied to only part of the variables y_j we obtain (6.24) in operator form

$$\begin{aligned}&TB^{(b)}_{\{\mu\}}(x)\, A(y_1) \cdots A(y_n) \\ &\quad = TB^{(a)}_{\{\mu\}}(x)\, A(y_1) \cdots A(y_n) + \sum_{\{\rho\}}{}^{a}_{b} \frac{(-i)^{\sum \#(\rho)_j}}{r! \prod \#(\rho)_j\,!} G^{(b)\{\rho\}}_{\{\mu\}} TB^{(a)}_{\{\rho\}}(x)\, A(y_1) \cdots A(y_n) \\ &\qquad b < r + \sum_{j=1}^{r} \#(\rho)_j \leqslant a.\end{aligned} \tag{6.29}$$

The coefficients $G^{(b)\{\rho\}}_{\{\mu\}}$ can also be calculated directly from (6.24) and the normalization conditions (4.40) of the normal products.

We finally derive a formula which expresses $B^{(b)}_{\{\mu\}}$ in terms of the operators $B_{\{\rho\}}$ of minimal degree. To this end we write (6.28) for $a = b + 1$

$$B^{(b)}_{\{\mu\}} = B^{(b+1)}_{\{\mu\}} + \sum_{\{\rho\}}{}^{b+1}_{b} \frac{(-i)^{\Sigma\#(\rho)_j}}{r! \prod \#(\rho)_j !} G^{(b)\{\rho\}}_{\{\mu\}} B^{(b+1)}_{\{\rho\}}. \tag{6.30}$$

The sum is restricted by

$$b < r + \sum_{j=1}^{r} \#(\rho)_j \leqslant b + 1.$$

Hence the degree $b + 1$ equals the dimension of the composite operators appearing in the sum

$$r + \sum_{j=1}^{r} \#(\rho)_j = b + 1$$

$$B^{(b+1)}_{\{\rho\}} = B_{\{\rho\}} .$$

With the abbreviation

$$G^{(b)\{\rho\}}_{\{\mu\}} = G^{\{\rho\}}_{\{\mu\}} \qquad \text{if} \quad b = r + \sum_{j=1}^{r} \#(\rho)_j - 1 \tag{6.31}$$

we find

$$B^{(a)}_{\{\mu\}} = B_{\{\mu\}} - \sum_{\{\rho\}}{}^{a}_{d} \frac{(-i)^{\Sigma\#(\rho)_j}}{r! \prod \#(\rho)_j !} G^{\{\rho\}}_{\{\mu\}} B_{\{\rho\}} . \tag{6.32}$$

Here d denotes the dimension of $B_{\{\mu\}}$

$$d = m + \sum_{j=1}^{m} \#(\mu)_j .$$

The sum is restricted by

$$d < r + \sum \#(\rho)_j \leqslant a \tag{6.33}$$

A corresponding formula holds for time ordered products

$$\langle TB^{(a)}_{\{\mu\}}(x) A(y_1) \cdots A(y_n)\rangle = \langle TB_{\{\mu\}}(x) A(y_1) \cdots A(y_n)\rangle - \sum_{\{\rho\}}{}^{a}_{d} \frac{(-i)^{\Sigma\#(\rho)_j}}{r! \prod \#(\rho)_j !} G^{\{\rho\}}_{\{\mu\}} \langle TB_{\{\rho\}}(x) A(y_1) \cdots A(y_n)\rangle \tag{6.34}$$

Many examples for the relations (6.24–34) can be found in the applications of the normal product algorithm by Lowenstein and Schroer [18, 24–26].

APPENDIX: SUBDIAGRAM AND REDUCED DIAGRAMS

The conventions concerning subdiagrams and reduced diagrams used in this paper are slightly different from those of [5]. We briefly introduce the relevant modifications.

A subdiagram γ of $\Gamma \in \mathscr{C}_n$ or $\Delta \in \mathscr{D}_{mn}$

$$\gamma \subset \Gamma \quad \text{or} \quad \gamma \subset \Delta$$

is defined by a nonempty subset of lines of Γ.[7] The vertices of γ are formed by all endpoints of lines belonging to γ. A line of Γ or Δ not belonging to γ is called an external line of γ if one or both endpoints lie in γ.

We next introduce external and internal momenta for γ. Let

$$V_{c_1}, \ldots, V_{c_f} \tag{A.1}$$

be the vertices and

$$L_{a_1 b_1 \sigma_1}, \ldots, L_{a_r b_r \sigma_r} \tag{A.2}$$

be the external lines of γ. The indices a_j, b_j are ordered such that a_j refers to the endpoint V_{a_j} which belongs to γ. If both endpoints V_a and V_b of a line $L_{ab\sigma}$ belong to γ this line is listed twice in (A.2) as $L_{ab\sigma}$ and $L_{ba\sigma}$.

As external momenta of γ we introduce

$$p_{c_1}, \ldots, p_{c_f} \tag{A.3}$$

(corresponding to each vertex (A.4) of γ) and the new variables

$$p_1^\gamma, \ldots, p_r^\gamma \tag{A.4}$$

corresponding to each external line of γ as listed in (A.2). The combined sets (A.3) and (A.4) of variables will be denoted by the symbol

$$p^\gamma = (p_{c_1}, \ldots, p_{c_f}, p_1^\gamma, \ldots, p_r^\gamma) \tag{A.5}$$

The momenta p_{c_j} associated with an interior vertex V_{c_j} of Γ or Δ vanish. Hence only those momenta p_j in (A.3) can be different from zero which belong to the exterior vertices

$$V_1, \ldots, V_n \quad \text{of} \quad \Gamma \in \mathscr{C}_n \tag{A.6}$$

[7] The empty set or a trivial diagram consisting of a single vertex and no line is not a subdiagram in the sense of this definition. On the other hand the full diagram itself is considered as a sub diagram.

or

$$V_0, V_{m+1}, \ldots, V_{m+n} \quad \text{of} \quad \Delta \in \mathscr{D}_{mn}. \tag{A.7}$$

At each vertex V_a of γ we introduce the total external momentum

$$q_a^\gamma = q_a^\gamma(p^\gamma) = p_a + \sum_\gamma^{a^{\text{ext}}} p_j^\gamma \qquad a = c_1, \ldots, c_{N(\gamma)} \tag{A.8}$$

The sum extends over all external lines $L_{ab_j\sigma_j}$ of γ which are attached to V_a. These are all lines listed in (A.2) with $a_j = a$.

The internal momenta $l^\gamma_{ab\sigma}$ are required to satisfy momentum conservation at each vertex of γ

$$\sum_\gamma^a l^\gamma_{ab\sigma} = q_a^\gamma(p^\gamma) \tag{A.9}$$

The $l^\gamma_{ab\sigma}$ will be written in the form

$$l^\gamma_{ab\sigma} = k^\gamma_{ab\sigma}(k^\gamma) + q^\gamma_{ab\sigma}(p^\gamma) \tag{A.10}$$

where the $q^\gamma_{ab\sigma}(p^\gamma)$ are linear combinations of the external momenta p^γ and form a particular solution of

$$\sum_\gamma^a q^\gamma_{ab\sigma} = q_a^\gamma(p^\gamma), \qquad q^\gamma_{ab\sigma} + q^\gamma_{ba\sigma} = 0. \tag{A.11}$$

The $k^\gamma_{ab\sigma}$ are linear combinations of $s(\gamma)$ independent internal momenta

$$k^\gamma = (k_1^\gamma, \ldots, k^\gamma_{s(\gamma)})$$

and form the general solution of

$$\sum_\gamma^a k^\gamma_{ab\sigma} = 0, \qquad k^\gamma_{ab\sigma} + k^\gamma_{ba\sigma} = 0. \tag{A.12}$$

k^γ and p^γ are expressed in terms of k and p

$$k^\gamma = k^\gamma(k, p), \qquad p^\gamma = p^\gamma(k, p) \tag{A.13}$$

by requiring

$$l^\gamma_{ab\sigma}(k^\gamma, p^\gamma) \equiv l_{ab\sigma}(k, p) \tag{A.14}$$

for each internal line of γ and

$$p^\gamma_{a_j} = -l_{ab\sigma}(k, p) \tag{A.15}$$

for each external line of γ.

Let μ be a subdiagram of $\gamma \subset \Gamma$. In an analogous way we introduce the functions

$$k^{\mu} = k_{\gamma}{}^{\mu}(k^{\gamma}, p^{\gamma}), \qquad p^{\mu} = p_{\gamma}{}^{\mu}(k^{\gamma}, p^{\gamma}) \tag{A.16}$$

by setting

$$l^{\mu}_{ab\sigma}(k^{\mu}, p^{\mu}) \equiv l^{\gamma}_{ab\sigma}(k^{\gamma}, p^{\gamma}) \tag{A.17}$$

for every line $L_{ab\sigma}$ of μ. Moreover, we require

$$p_j{}^{\mu} = -l^{\gamma}_{a_j b_j \sigma_j}(k^{\gamma}, p^{\gamma}) \tag{A.18}$$

if $L_{a_j b_j \sigma_j}$ is an external line of μ, but not of γ, and

$$p_j{}^{\mu} = p_k{}^{\gamma} \tag{A.19}$$

if $L_{a_j b_j \sigma_j} = L_{a_k b_k \sigma_k}$ is an external line of μ and γ. (The numbering (A.2) of external lines may be different for μ and γ, hence $j \neq k$ in general.)

The choice of the particular solutions $q_{ab\sigma}(p)$ of (A.11) should be such that k^{γ} depends on k only and k^{μ} on k^{γ} only in (A.13) and (A.16)

$$k = k(k^{\gamma}), \qquad k^{\mu} = k_{\gamma}{}^{\mu}(k^{\gamma}). \tag{A.20}$$

As was shown in [5] such $q_{ab\sigma}(p)$, $q^{\gamma}_{ab\sigma}(p^{\gamma})$ can always be constructed.

The unrenormalized Feynman integral associated with a subdiagram γ of $\Gamma \in \mathscr{C}_n$ or $\Delta \in \mathscr{D}_{mn}$ is given by

$$J_{\gamma} = \int dk_1{}^{\gamma} \cdots dk^{\gamma}_{s(\gamma)}\, I_{\gamma} \tag{A.21}$$

where

$$I_{\gamma} = \prod_{ab\sigma}{}^{\gamma} \Delta_{Fab\sigma} \prod_{c}{}^{\gamma} P_c \tag{A.22}$$

with the usual insertion rules. The products extend over all lines and vertices of γ. The dimension $d(\gamma)$ is defined by the dimension of the integral (A.21).

A diagram is called trivial if it consists of a single vertex and no line. A diagram is called proper if it cannot be separated in two parts by cutting a single line. Two diagrams are called disjoint if they have neither a line nor a vertex in common. Let γ be a subdiagram of $\Gamma \in \mathscr{C}_n$ or $\Delta \in \mathscr{D}_{mn}$. Let $\gamma_1, \ldots, \gamma_c$ be mutually disjoint proper subdiagrams of γ. The reduced diagram

$$\bar{\gamma} = \gamma / \gamma_1 \cdots \gamma_c \tag{A.23}$$

is then defined by reducing each γ_i in γ to a vertex. In other words, the lines of $\bar{\gamma}$

are formed by those lines of γ which do not belong to any γ_j. The vertices which are formed by the corresponding endpoints but with all vertices identified which belong to the same γ_j. The unrenormalized Feynman integral associated with (A.23) is given by

$$J_{\bar{\gamma}} = \int dk_1^{\gamma} \cdots dk_{s(\bar{\gamma})}^{\gamma} I_{\bar{\gamma}} \tag{A.24}$$

where

$$I_{\bar{\gamma}} = \prod_{abc}{}^{\bar{\gamma}} \Delta_{Fabc} \prod_{c}{}^{\bar{\gamma}} P_c \tag{A.25}$$

is formed with the usual insertion rules. The products extend over the lines and vertices of γ which do not belong to any γ_j. Accordingly the factor 1 is assigned to each reduced vertex. I_γ, $I_{\gamma/\gamma_1\cdots\gamma_c}$ and I_{γ_j} are related by

$$I_\gamma = I_{\gamma/\gamma_1\cdots\gamma_c} I_{\gamma_1} \cdots I_{\gamma_c} \tag{A.26}$$

Let γ be a subdiagram of $\Gamma \in \mathscr{C}_n$. The dimension $d(\gamma)$ of γ is defined by the dimension of the integral (A.21) which is given by

$$d(\gamma) = 4s(\gamma) + 2v_2(\gamma) - 2l(\gamma) \tag{A.27}$$

Here $s(\gamma)$ is the number of independent internal momenta of γ, $v_2(\gamma)$ the number of 2-vertices, $l(\gamma)$ the number of lines belonging to γ. The dimension $d(\bar{\gamma})$ of the reduced diagram, i.e., the dimension of the corresponding integral (A.24), is given by

$$d(\bar{\gamma}) = 4s(\bar{\gamma}) + 2v_2'(\bar{\gamma}) - 2l(\bar{\gamma}) \tag{A.28}$$

where $v_2'(\bar{\gamma})$ is the number of original 2-vertices belonging to $\bar{\gamma}$ (i.e., not counting those 2-vertices which were obtained by reducing proper self-diagrams of γ). The relation

$$d(\gamma) = d(\bar{\gamma}) + \sum_{a=1}^{c} d(\gamma_a) \tag{A.29}$$

then follows from (A.27–28). This is obvious in the absence of 2-vertices in γ. Moreover, the insertion of 2-vertices into the lines of γ does not change (A.29) since the degree of a factor $A + B(l^2 - m^2)$ is cancelled by the degree of an additional Feynman denominator. Hence (A.29) holds in general.

We next derive two conditions for the degree $\delta(\gamma)$ defined by (3.15). (A.24) may also be written as

$$d(\gamma) = 4 - r(\gamma) - v_3(\gamma). \tag{A.30}$$

Here $r(\gamma)$ is the number of external lines of γ with those external lines counted twice for which both endpoints lie in γ. $v_3(\gamma)$ denotes the number of 3-vertices in γ. From (3.15) and (A.27) we obtain

$$\delta(\gamma) = d(\gamma) + v_3(\gamma) \tag{A.31}$$

which implies

$$\delta(\gamma) \geqslant d(\gamma). \tag{A.32}$$

Moreover, (A.29), (A.31) and

$$v_3(\gamma) \geqslant \sum_{j=1}^{c} v_3(\gamma_j) \tag{A.33}$$

imply

$$\delta(\gamma) \geqslant d(\bar{\gamma}) + \sum_{j=1}^{c} d(\gamma_j). \tag{A.34}$$

The conditions (A.32) and (A.34) guarantee that the final renormalized integral (3.14) is absolutely convergent and approaches a distribution in the limit $\epsilon \to +0$.[8]

We finally derive the corresponding formulas for the contribution from a diagram $\Delta \in \mathscr{D}_{mn}$ to the Green's function (4.28) of the operator $B^{(a)}_{\{\mu\}}$. Let γ be a subdiagram of Δ. We have to verify that the degree function $\delta_a(\gamma)$ as defined by (4.31) satisfies the conditions

$$\delta_a(\gamma) \geqslant d(\gamma) \tag{A.35}$$

$$\delta_a(\gamma) \geqslant d(\bar{\gamma}) + \sum_{j=1}^{c} \delta_a(\gamma_j) \tag{A.36}$$

for any reduced diagram (A.23). Only the case where γ contains the vertex V_0 need be checked. In this case the dimension of γ is given by

$$d(\gamma) = \sum \#(\mu)_i + 4s(\gamma) + 2v_2(\gamma) - 2l(\gamma) \tag{A.37}$$

which may also be written as

$$d(\gamma) = d - r(\gamma) - v_3(\gamma). \tag{A.38}$$

d is the dimension (4.3) of the operator $B^{(a)}_{\{\mu\}}$, $s(\gamma)$ the number of independent internal momenta of γ, $v_2(\gamma)$ the number of 2-vertices (excluding V_0), $v_3(\gamma)$ the number

[8] See ref. [5]. Apart from minor changes the convergence proof is not affected by the modifications of this paper.

of 3-vertices (excluding V_0), $l(\gamma)$ the number of lines belonging to γ. (4.30) and (A.38) imply

$$\delta_a(\gamma) = d(\gamma) + v_3(\gamma) + a - d \tag{A.39}$$

yielding (A.35).

In order to check (A.36) we note that (A.37) implies

$$d(\gamma) = d(\bar{\gamma}) + \sum_{j=1}^{c} d(\gamma_j) \tag{A.40}$$

in the absence of 2-vertices. Since the insertion of a 2-vertex does not change the dimension of a diagram Eq. (A.40) holds in general. Assuming that none of the subdiagrams $\gamma_1, \ldots, \gamma_c$ contains V_0 we have

$$\delta_a(\gamma_j) = \delta(\gamma_j). \tag{A.41}$$

Then (A.31), (A.39), (A.41), and

$$v_3(\gamma) \geqslant \sum_{j=1}^{c} v_3(\gamma_j) \tag{A.42}$$

imply (A.36). Next we assume that γ_j contains V_0. Since $\gamma_1, \ldots, \gamma_c$ are disjoint only γ_j contains V_0. Hence

$$\delta_a(\gamma_j) = d(\gamma_j) + v_3(\gamma_j) + a - d$$
$$\delta_a(\gamma_i) = d(\gamma_i) + v_3(\gamma_i) \qquad \text{if} \quad i \neq j.$$

Combined with (A.39), (A.42) these relations imply (A.36).

References

1. M. Gell-Mann and F. Low, *Phys. Rev.* **84** (1951), 350.
2. N. N. Bogoliubov and D. V. Shirkov, "Introduction to the Theory of Quantized Fields," Interscience, New York, 1960.
3. K. Hepp, *Comm. Math. Phys.* **6** (1967), 161.
4. E. Speer, Generalized Feynman Amplitudes, *in* "Annals of Mathematics Studies," No. 62, Princeton Univ. Press, Princeton, 1969.
5. W. Zimmermann, *Comm. Math. Phys.* **15** (1969), 208.
6. H. Epstein and V. Glaser, *in* "Statistical Mechanics and Quantum Field Theory," Les Houches 1970, Gordon and Breach, New York, in press; preprint CERN, No. Th 1344 and 1400 (1971).
7. R. Stora, Lecture Notes, Les Houches Summer School, 1971.
8. F. Dyson, *Phys. Rev.* **75** (1949), 1736.

9. A. SALAM, *Phys. Rev.* **82** (1951), 217; **84** (1951), 426.
10. O. STEINMANN, "Perturbation Expansions in Axiomatic Field Theory, Theorem 2.1, p. 16, Springer–Verlag, Heidelberg, 1971.
11. G. WICK, *Phys. Rev.* **80** (1950), 268.
12. K. WILSON, unpublished.
13. J. VALATIN, *Proc. Roy. Soc. A* **225** (1954), 535; **226** (1954), 254.
14. K. WILSON, On products of quantum field operators at short distances, Cornell Report, 1964.
15. R. BRANDT, *Ann. Phys.* **44** (1967), 221; *Ann. Phys.* **52** (1969), 122 and *Fortschr. Physik.* **19** (1970), 249.
16. W. ZIMMERMANN, *Comm. Math. Phys.* **6** (1967), 161; **10** (1968), 325.
17. W. ZIMMERMANN, in "Lectures on Elementary Particles and Quantum Field Theory," 1970 Brandeis Summer Institute in Theoretical Physics (S. Deser, M. Grisaru, and H. Pendleton, Eds.), Chap. 4, MIT Press, Cambridge, Mass., 1971.
18. R. LOWENSTEIN, *Phys. Rev.* **D4** (1971), 2281.
19. H. ROUET AND R. STORA, CNRS preprints (1972).
20. K. WILSON, A finite energy-momentum tensor was first constructed in perturbation theory, unpublished.
21. C. CALLAN, S. COLEMAN, AND R. JACKIW, *Ann. Phys. (N.Y.)* **59** (1970), 42.
22. K. SYMANZIK, private communication.
23. C. CALLAN, *Phys. Rev. D* **2** (1970), 1951; K. SYMANZIK, *Comm. Math. Phys.* **18** (1970), 227.
24. B. SCHROER, *Lett. Nuovo Cimento* **2** (1971), 867.
25. J. LOWENSTEIN, *Commun. Math. Phys.* **24** (1971), 1.
26. J. LOWENSTEIN AND B. SCHROER, *Phys. Rev. D* **6** (1972), 1553; and to be published.
27. Y. LAM, *Phys. Rev. D* **6** (1972), 2145, 2161; and to be published.
28. M. GOMES AND J. LOWENSTEIN, *Nucl. Phys. B* **45** (1972), 252.
29. M. GOMES AND J. LOWENSTEIN, *Nucl. Phys. B* (September, 1972).

Printed by the St Catherine Press Ltd., Tempelhof 37, Bruges, Belgium.

Reprinted from ANNALS OF PHYSICS
All Rights Reserved by Academic Press, New York and London

Vol. 77, No. 1-2, May 1, 1973
Printed in Belgium

Normal Products and the Short Distance Expansion in the Perturbation Theory of Renormalizable Interactions*†

WOLFHART ZIMMERMANN

Department of Physics, New York University, New York, N.Y. 10003

Received May 23, 1972

Normal products are defined in perturbation theory for renormalizable interactions of a scalar field. Various identities involving normal products are derived. As an application Wilson's asymptotic short distance expansion is verified in perturbation theory for a product of two field operators.

1. INTRODUCTION

In a previous paper composite operators

$$N_a[M_{\{\mu\}}\{A(x)\}] \tag{1.1}$$

were defined in perturbation theory for the model of a scalar field A with mixed A^3, A^4-coupling [1]. $M_{\{\mu\}}$ denotes the formal monomial

$$M_{\{\mu\}}\{A(x)\} = A_{(\mu)_1}(x) \cdots A_{(\mu)_m}(x) \tag{1.2}$$

with the notation

$$\begin{aligned} &A_{(\mu)_j} = \partial_{(\mu)_j} A_j \\ &(\mu)_j = (\mu_{j1}, \dots, \mu_{jm(j)}), \qquad m(j) \geqslant 0 \\ &\partial_{(\mu)_j} = \partial_{\mu_{j1}} \cdots \partial_{\mu_{jm(j)}}, \qquad \partial_{(\mu)_j} = 1 \quad \text{if} \quad m(j) = 0 \\ &\{\mu\} = ((\mu)_1, \dots, (\mu)_m), \qquad \partial_\mu = \partial/\partial x^\mu. \end{aligned} \tag{1.3}$$

a may be any integer greater or equal than the dimension of the monomial

$$a \geqslant d(M_{\{\mu\}}) \tag{1.4}$$

* This work was supported in part by the National Science Foundation Grant No. GP-25609.
† A preliminary version of the material contained in this paper appeared in W. Zimmermann, Renormalization and Composite Field Operators, in "Lectures on Elementary Particles and Quantum Field Theory," 1970 Brandeis Summer Institute in Theoretical Physics, Vol. I (S. Deser, M. Grisaru, and H. Pendleton, Eds.) MIT Press, Cambridge, Mass. (1971).

P. Breitenlohner and D. Maison (Eds.): Proceedings 1998, LNP 558, pp. 278–309, 2000.

$d(M_{(\mu)})$ is the dimension of $M_{(\mu)}$ in the naive sense, where the dimension 1 is assigned to the field A and the derivative ∂_μ, i.e.,

$$d(M_{(\mu)}\{A(x)\}) = m + \sum_{j=1}^{m} m(j) \tag{1.5}$$

Composite operators for which the degree a equals the dimension (1.5) are called composite operators of minimal degree and denoted by

$$N[M_{(\mu)}\{A(x)\}] = N_a[M_{(\mu)}\{A(x)\}] \qquad \text{with} \qquad a = d(M_{(\mu)}). \tag{1.6}$$

In this paper (1.1) will be generalized to a definition of a normal product

$$N_a[A_{(\mu)_1}(x_1) \cdots A_{(\mu)_m}(x_m)] \tag{1.7}$$

which depends on different coordinates $x_1, \ldots, x_m$. a may be any integer. For $a < 0$ the operator (1.7) equals the corresponding time ordered product

$$N_a[A_{(\mu)_1}(x_1) \cdots A_{(\mu)_m}(x_m)] = \partial^{x_1}_{(\mu)_1} \cdots \partial^{x_m}_{(\mu)_m} TA(x_1) \cdots A(x_m)$$
$$\text{if} \quad a = -1, -2, \ldots . \tag{1.8}$$

If a is greater than or equal to the dimension of the monomial the normal product (1.7) approaches the composite operator (1.1) in the limit $x_j \to x$.

For the special case of two operators the normal product

$$N_a[A_{(\mu)_1}(x_1)\, A_{(\mu)_2}(x_2)] \tag{1.9}$$

will be defined in two different ways. In Section 2 the operator (1.9) is defined by a linear combination

$$N_a[A_{(\mu)_1}(x_1)\, A_{(\mu)_2}(x_2)] = \partial^{x_1}_{(\mu)_1} \partial^{x_2}_{(\mu)_2} TA(x_1)\, A(x_2) + \sum_{j=1}^{N} G_j(\xi)\, B_j(x)$$
$$x = (x_1 + x_2)/2, \qquad \xi = (x_1 - x_2)/2 \tag{1.10}$$

of the corresponding time ordered product and composite operators $B_j(x)$ of minimal degree $d \leqslant a$. The coefficients G_j are explicitly given in terms of Green's functions. An alternative method of defining normal products is developed in Section 3. There the Green's functions

$$\langle TN_a[A_{(\mu)_1}(x_1) \cdots A_{(\mu)_m}(x_m)]\, A(y_1) \cdots A(y_n)\rangle \tag{1.11}$$

are defined in perturbation theory by a simple generalization of the corresponding definition of

$$\langle TN_a[A_{(\mu)_1}(x) \cdots A_{(\mu)_m}(x)]\, A(y_1) \cdots A(y_n)\rangle$$

which was given in [1], (Ref. [1] is referred to as I throughout). For the case of $m = 2$ the complete equivalence of both definitions is established in Section 5.

A generalization of (1.10) to the case of more than two operators is possible, but will not be given in the present paper.

The normal product (1.9) satisfies

$$\partial_\mu{}^\varepsilon N_a[A_{(\mu)_1}(x_1)\, A_{(\mu)_2}(x_2)] = N_a \partial_\mu{}^\varepsilon [A_{(\mu)_1}(x_1)\, A_{(\mu)_2}(x_2)] \tag{1.12}$$

and Lowenstein's rule [2]

$$\partial_\mu{}^\alpha N_a[A_{(\mu)_1}(x_1)\, A_{(\mu)_2}(x_2)] = N_{a+1} \partial_\mu{}^\alpha [A_{(\mu)_1}(x_1)\, A_{(\mu)_2}(x_2)] \tag{1.13}$$

The normal product formalism is applied to Wilson's short distance expansion for a product of two operators [3]. In Section 4 the principal part of the Wilson expansion

$$\begin{gathered} TA(x_1)\, A(x_2) = \sum f_j(\xi)\, B_j(x) + R(x_1 x_2) \\ x = (x_1 + x_2)/2, \qquad \xi = (x_1 - x_2)/2, \qquad \lim_{\xi \to 0} R = 0 \end{gathered} \tag{1.14}$$

is derived in perturbation theory, confirming earlier results by Brandt [4][1]. On the right side the sum extends over all composite operators (1.6) of minimal degree and dimension

$$d \leqslant 2,$$

including the identity. The index j includes possible Lorentz indices. Explicit expressions for the coefficients f_j in terms of Green's functions are found. The remainder is obtained in the form

$$R(x_1 x_2) = N_2[A(x_1)\, A(x_2)] - N_2[A(x)^2] \tag{1.15}$$

which vanishes in the limit $x_j \to x$.

In Section 6 the asymptotic form of the Wilson expansion

$$\partial^{\alpha_1}_{(\mu)_1} \partial^{\alpha_2}_{(\mu)_2} TA(x_1)\, A(x_2) = \sum f_j(\xi)\, B_j(x) + R_n(x_1 x_2) \tag{1.16}$$

is proved to be valid in perturbation theory[2]. The sum extends over all composite operators (1.6) of minimal degree and dimension

$$d \leqslant n. \tag{1.17}$$

[1] The form (1.14–15) of the principal part was applied by K. Symanzik in ref. [5] to the asymptotic behavior of vertex functions at large momenta. The paper also contains another derivation of (1.14–15) which makes use of the 2-particle structure of Green's functions.

[2] Another proof of the asymptotic form of the short distance expansion has been given by K. Wilson [6], using a different method.

The coefficients f_j are expressed in terms of Green's functions. The remainder R_n is related to the normal product by

$$\begin{gathered}R_n(x_1x_2) = N_n[A_{(\mu)_1}(x_1)\,A_{(\mu)_2}(x_2)] - t_\xi{}^c N_n[A_{(\mu)_1}(x+\xi)\,A_{(\mu)_2}(x-\xi)] \\ x_1 = x+\xi\,, \qquad x_2 = x-\xi \\ (\mu)_j = \mu_{j1}\cdots\mu_{jm(j)} \\ c = n-2-m(1)-m(2).\end{gathered} \tag{1.18}$$

Applied to a function F of ξ the expression t^cF denotes the Taylor series of F at $\xi = 0$ up to and including terms of order c. The remainder (1.18) vanishes at least like $(\xi^2)^{c/2}$ for $\xi \to 0$.

2. Definition of Normal Products

As a generalization of the Wick product we introduce a family of normal products

$$N_a[A_{(\mu)_1}(x_1)\,A_{(\mu)_2}(x_2)] \tag{2.1}$$

where a may be any integer. If a is negative we set[3]

$$N_a[A_{(\mu)_1}(x_1)\,A_{(\mu)_2}(x_2)] = TA_{(\mu)_1}(x_1)\,A_{(\mu)_2}(x_2), \qquad a<0. \tag{2.2}$$

N_0 is identical with Wick's : : -product

$$\begin{aligned}N_0[A_{(\mu)_1}(x_1)\,A_{(\mu)_2}(x_2)] &= {:}A_{(\mu)_1}(x_1)\,A_{(\mu)_2}(x_2){:} \\ &= TA_{(\mu)_1}(x_1)\,A_{(\mu)_2}(x_2) - \langle TA_{(\mu)_1}(x_1)\,A_{(\mu)_2}(x_2)\rangle\cdot 1\end{aligned} \tag{2.3}$$

For positive a we define

$$\begin{gathered}N_a[A_{(\mu)_1}(x_1)\,A_{(\mu)_2}(x_2)] = TA_{(\mu)_1}(x_1)\,A_{(\mu)_2}(x_2) - \langle TA_{(\mu)_1}(\xi)\,A_{(\mu)_2}(-\xi)\rangle\,1 \\ -\sum_{\{\rho\}}^{a}\frac{(-i)^{\Sigma\#(\rho)_j}}{r!\prod\#(\rho)_j\,!}\,G^{\{\rho\}}_{\{\mu\}}(\xi)\,B_{(\rho)}(x) \\ x=(x_1+x_2)/2, \qquad \xi=(x_1-x_2)/2\end{gathered} \tag{2.4}$$

[3] The time ordered product of derivatives of field operators is defined by
$$\langle TA_{(\nu)_1}(x_1)\cdots A_{(\nu)_n}(x_n)\rangle = \partial^{x_1}_{(\nu)_1}\cdots\partial^{x_n}_{(\nu)_n}\langle TA(x_1)\cdots A(x_n)\rangle.$$

The sum extends over all sets

$$\{\rho\} = ((\rho)_1, \ldots, (\rho)_r)$$
$$(\rho)_j = (\rho_{j1}, \ldots, \rho_{jn(j)}) \quad \#n(j) = 0, 1, \ldots$$

with

$$0 < r + \sum_{j=1}^{r} \#(\rho)_j \leqslant a \tag{2.5}$$

$\#(\rho)_j$ denotes the number $n(j)$ of elements in $(\rho)_j$. $B_{\{\rho\}}(x)$ is the composite operator

$$B_{\{\rho\}}(x) = N[A_{(\rho)_1}(x) \cdots A_{(\rho)_r}(x)] \tag{2.6}$$

of minimal degree

$$d = r + \sum_{j=1}^{r} \#(\rho)_j .$$

The coefficients $G^{\{\rho\}}_{\{\mu\}}$ will be given later. According to (2.4) the normal product is a linear combination of the time ordered product and local operators depending on $x = (x_1 + x_2)/2$. This suggests a natural definition of the time ordered product

$$T(N_a[A_{(\mu)_1}(x_1)\, A_{(\mu)_2}(x_2)]\, A(y_1) \cdots A(y_n)). \tag{2.7}$$

In order to form (2.7) we multiply (2.4) by the operators $A(y_1), \ldots, A(y_n)$ and apply the time ordering separately to every term on the right side. We thus arrive at the following formula for the time ordered products (2.7)

$$\begin{aligned} &TN_a[A_{(\mu)_1}(x_1)\, A_{(\mu)_2}(x_2)]\, A(y_1) \cdots A(y_n) \\ &\quad = TA_{(\mu)_1}(x_1)\, A_{(\mu)_2}(x_2)\, A(y_1) \cdots A(y_n) \\ &\quad\quad - \langle TA_{(\mu)_1}(\xi)\, A_{(\mu)_2}(-\xi)\rangle\, TA(y_1) \cdots A(y_n) \\ &\quad\quad - \sum_{\{\rho\}}^{a} \frac{(-i)^{\Sigma \#(\rho)_j}}{r! \prod \#(\rho)_j !}\, G^{\{\rho\}}_{\{\mu\}}(\xi)\, TB_{\{\rho\}}(x)\, A(y_1) \cdots A(y_n). \end{aligned} \tag{2.8}$$

We finally define the coefficients $G^{\{\rho\}}_{\{\mu\}}(\xi)$ by

$$G^{\{\rho\}}_{\{\mu\}}(\xi) = \langle TN_b[A_{(\mu)_1}(\xi)\, A_{(\mu)_2}(-\xi)]\, \tilde{A}^{(\rho)_1}(0) \cdots \tilde{A}^{(\rho)_r}(0)\rangle^{\text{prop}} \quad \text{if} \quad r \neq 2 \tag{2.9}$$

and

$$\begin{aligned} G^{\{\rho\}}_{\{\mu\}}(\xi) = {} & \langle TN_b[A_{(\mu)_1}(\xi)\, A_{(\mu)_2}(-\xi)]\, \tilde{A}^{(\rho)_1}(0)\, \tilde{A}^{(\rho)_2}(0)\rangle^{\text{prop}} \\ & - i^{\#(\rho)_1+\#(\rho)_2}(-1)^{\#(\mu)_2+\#(\rho)_2}\, \partial_{(\mu)_1}\xi^{(\rho)_1}\partial_{(\mu)_2}\xi^{(\rho)_2} \\ & - i^{\#(\rho)_1+\#(\rho)_2}(-1)^{\#(\mu)_2+\#(\rho)_1}\, \partial_{(\mu)_1}\xi^{(\rho)_2}\partial_{(\mu)_2}\xi^{(\rho)_1} \end{aligned} \tag{2.10}$$

$$\xi^{(\rho)_1} = \xi^{\rho_{11}} \cdots \xi^{\rho_{1n(1)}}, \qquad \xi^{(\rho)_2} = \xi^{\rho_{21}} \cdots \xi^{\rho_{2n(2)}} \qquad \text{if} \quad r = 2.$$

b denotes the integer

$$b = r + \sum \#(\rho)_j - 1.$$

(For the definition of the proper part $\langle\ \rangle^{\text{prop}}$ see I, Section 2].) The notation used in (2.9–10) is

$$\langle TO\tilde{A}^{(\rho)_1}(q_1) \cdots \tilde{A}^{(\rho)_r}(q_r)\rangle^{\text{prop}} = \partial^{(\rho)_1}_{q_1} \cdots \partial^{(\rho)_r}_{q_r} \langle TO\tilde{A}(q_1) \cdots \tilde{A}(q_r)\rangle^{\text{prop}} \tag{2.11}$$

for the proper part of a time ordered function involving the operator O. In the formula (2.4) for the normal product N_a all coefficients $G^{\{\rho\}}_{\{\mu\}}$ are determined recursively. For every term of the sum satisfies

$$b = r + \sum \#(\rho)_j - 1 \leqslant a - 1 \tag{2.12}$$

so that all coefficients $G^{\{\rho\}}_{\{\mu\}}$ are given by normal products N_b with $b < a$.

According to (2.8) reduction formulas hold in their usual form for the time ordered product (2.7). In particular, the matrix elements of (2.4) are uniquely determined by the vacuum expectation values of (2.7).

The normal product (2.4) satisfies the differentiation rules

$$\partial_\nu{}^\xi N_a[A_{(\mu)_1}(x + \xi)\, A_{(\mu)_2}(x - \xi)] = N_a \partial_\nu{}^\xi [A_{(\mu)_1}(x + \xi)\, A_{(\mu)_2}(x - \xi)] \tag{2.13}$$

$$\partial_\nu{}^x N_a[A_{(\mu)_1}(x + \xi)\, A_{(\mu)_2}(x - \xi)] = N_{a+1} \partial_\nu{}^x [A_{(\mu)_1}(x + \xi)\, A_{(\mu)_2}(x - \xi)] \tag{2.14}$$

(2.13) follows immediately from the definition (2.4), (2.9–10) of the normal product. (2.14) was derived by Lowenstein [2] in perturbation theory. It can also be derived by using the definition (2.4), (2.9–10) and Lowenstein's relation for composite operators

$$\partial_\nu{}^x N[A_{(\rho)_1}(x) \cdots A_{(\rho)_r}(x)] = N\partial_\nu{}^x [A_{(\rho)_1}(x) \cdots A_{(\rho)_r}(x)].$$

Repeated application of (2.13–14) yields

$$\partial^{\xi}_{(\nu)} N_a[A_{(\mu)_1}(x+\xi)\, A_{(\mu)_2}(x-\xi)] = N_a \partial^{\xi}_{(\nu)}[A_{(\mu)_1}(x+\xi)\, A_{(\mu)_2}(x-\xi)] \tag{2.15}$$

$$\partial^{x}_{(\nu)} N_a[A_{(\mu)_1}(x+\xi)\, A_{(\mu)_2}(x-\xi)] = N_b \partial^{x}_{(\nu)}[A_{(\mu)_1}(x+\xi)\, A_{(\mu)_2}(x-\xi)]$$

$$b = a + \#(\nu). \tag{2.16}$$

Using (2.15) and (2.16) any derivative of a normal product with respect to x_1 and x_2 may be written as a linear combination of normal products.

For time ordered products we have similar differentiation rules

$$\partial^{\xi}_{(\nu)} TN_a[A_{(\mu)_1}(x+\xi)\, A_{(\mu)_2}(x-\xi)]\, A(y_1) \cdots A(y_n)$$
$$= TN_a \partial^{\xi}_{(\nu)} N_a[A_{(\mu)_1}(x+\xi)\, A_{(\mu)_2}(x-\xi)]\, A(y_1) \cdots A(y_n) \tag{2.17}$$

$$\partial^{x}_{(\nu)} TN_a[A_{(\mu)_1}(x+\xi)\, A_{(\mu)_2}(x-\xi)]\, A(y_1) \cdots A(y_n)$$
$$= TN_b \partial^{x}_{(\nu)}[A_{(\mu)_1}(x+\xi)\, A_{(\mu)_2}(x-\xi)]\, A(y_1) \cdots A(y_n) \tag{2.18}$$
$$b = a + \#(\nu).$$

Normalization conditions for the normal products (2.1) can easily be worked out. We first consider the Green's function

$$Q^{(\nu)_1 \cdots (\nu)_n}_{(\mu)_1 (\mu)_2}(x_1 x_2) = \langle TN_a[A_{(\mu)_1}(x_1)_, A_{(\mu)_2}(x_2)]\, \tilde{A}^{(\nu)_1}(0) \cdots \tilde{A}^{(\nu)_n}(0)\rangle^{\text{prop}} \tag{2.19}$$

for $x_1 = \xi$, $x_2 = -\xi$ with the notation (2.11). Taking the vacuum expectation value of (2.8) and setting $x = 0$ we find

$$Q^{(\nu)_1 \cdots (\nu)_n}_{(\mu)_1 (\mu)_2}(\xi, -\xi) = 0 \tag{2.20}$$

if $\sum \#(\nu)_i \leqslant a$ and $n \neq 2$,

$$Q^{(\nu)_1 (\nu)_2}_{(\mu)_1 (\mu)_2}(\xi, -\xi) = i^{\#(\nu)_1 + \#(\nu)_2}(-1)^{\#(\mu)_2 + \#(\nu)_2}\, \partial_{(\mu)_1}\xi^{(\nu)_1} \partial_{(\mu)_2}\xi^{(\nu)_2}$$
$$+\, i^{\#(\nu)_1 + \#(\nu)_2}(-1)^{\#(\mu)_2 + \#(\nu)_1}\, \partial_{(\mu)_1}\xi^{(\nu)_2} \partial_{(\mu)_2}\xi^{(\nu)_1} \qquad \text{if} \quad n = 2. \tag{2.21}$$

In deriving these relations the normalization conditions (see [1, Eq. 4.40]) for the composite operators of minimal degree were used.

Next we consider (2.19) for arbitrary values

$$x_1 = x + \xi, \qquad x_2 = x - \xi$$

By translation invariance (2.19) is a polynomial in x. Likewise

$$P^{(\nu)_1(\nu)_2}_{(\mu)_1(\mu)_2}(x_1x_2) = i^{\#(\nu)_1+\#(\nu)_2}\partial^{x_1}_{(\mu)_1}\partial^{x_2}_{(\mu)_2}x_1^{(\nu)_1}x_2^{(\nu)_2} + i^{\#(\nu)_1+\#(\nu)_2}\partial^{x_1}_{(\mu)_1}\partial^{x_2}_{(\mu)_2}x_2^{(\nu)_1}x_1^{(\nu)_2}$$

is a polynomial in x. Differentiating (2.19) with respect to x and using (2.18), (2.20–21) we find that $Q^{(\nu)_1\cdots(\nu)_n}_{(\mu)_1(\mu)_2}$ and $\delta_{2n}P^{(\nu)_1(\nu)_2}_{(\mu)_1(\mu)_2}$ coincide at $x = 0$ with all derivatives. Hence the normalization condition

$$\langle TN_a[A_{(\mu)_1}(x_1)\,A_{(\mu)_2}(x_2)]\,\tilde{A}^{(\nu)_1}(0)\cdots\tilde{A}^{(\nu)_r}(0)\rangle^{\text{prop}}$$

$$= \delta_{n2}i^{\#(\nu)_1+\#(\nu)_2}\partial^{x_1}_{(\mu)_1}\partial^{x_2}_{(\mu)_2}x_1^{(\nu)_1}x_2^{(\nu)_2} + \delta_{n2}i^{\#(\nu)_1+\#(\nu)_2}\partial^{x_1}_{(\mu)_1}\partial^{x_2}_{(\mu)_2}x_2^{(\nu)_1}x_1^{(\nu)_2}, \quad (2.22)$$

$$\sum \#(\nu)_j \leqslant a,$$

is valid everywhere, with the notation (2.11).

As examples for the definition of normal products we work out $N_1[A(x_1)\,A(x_2)]$ and $N_2[A(x_1)\,A(x_2)]$. It is convenient to use the notation $G(\xi)$ for the coefficient of $A(x)$, $G_s(\xi)$ for the coefficient of

$$B_s(x) = N[A^s(x)], \quad (2.23)$$

$G^{\rho_1\cdots\rho_n}(\xi)$ for the coefficient of $A_{\rho_1\cdots\rho_n}(x)$, and $G_s^{(\rho)_1\cdots(\rho)_r}(\xi)$ for the coefficient of

$$B_{s(\rho)_1\cdots(\rho)_r}(x) = N[A^s(x)\,A_{(\rho)_1}(x)\cdots A_{(\rho)_r}(x)] \quad (2.24)$$

With this notation we have

$$N_1[A(x_1)\,A(x_2)] = TA(x_1)\,A(x_2) - \langle TA(\xi)\,A(-\xi)\rangle\,1 - G(\xi)\,A(x), \quad (2.25)$$

$$G(\xi) = \langle T:A(\xi)\,A(-\xi):\,\tilde{A}(0)\rangle^{\text{prop}}, \quad (2.26)$$

$$N_2[A(x_1)\,A(x_2)] = TA(x_1)\,A(x_2) - \langle TA(\xi)\,A(-\xi)\rangle\,1 - G(\xi)\,A(x) + iG^{\rho}(\xi)\,A_{\rho}(x) - \tfrac{1}{2}G_2(\xi)\,N[A^2(x)], \quad (2.27)$$

$$G^{\rho}(\xi) = \langle TN_1[A(\xi)\,A(-\xi)]\,\tilde{A}^{\rho}(0)\rangle^{\text{prop}} = \langle T:A(\xi)\,A(-\xi):\,\tilde{A}^{\rho}(0)\rangle^{\text{prop}}, \quad (2.28)$$

$$G_2(\xi) = \langle TN_1[A(\xi)\,A(-\xi)]\,\tilde{A}(0)\,\tilde{A}(0)\rangle^{\text{prop}} - 2 = \langle T:A(\xi)\,A(-\xi):\,\tilde{A}(0)\,\tilde{A}(0)\rangle^{\text{prop}} - 2. \quad (2.29)$$

In (2.28) and (2.29) the definition (2.25) of the N_1-product was used with

$$\langle A(x)\rangle = 0$$

and

$$\langle TA(0)\, \tilde{A}^{\rho}(0)\rangle^{\text{prop}} = 0,$$
$$\langle TA(0)\, \tilde{A}(0)\, \tilde{A}(0)\rangle^{\text{prop}} = 0.$$

3. Green's Functions of Normal Products

The time ordered Green's functions of the composite operator

$$N_a\{A_{(\mu)_1}(x) \cdots A_{(\mu)_m}(x)\},$$
$$a \geqslant d = m + \sum \#(\mu)_j\,, \tag{3.1}$$

were defined in [1] by a formal power series with respect to the coupling constants. In this section we will extend this definition by constructing a set of Green's functions

$$\langle TN_a[A_{(\mu)_1}(x_1) \cdots A_{(\mu)_m}(x_m)]\, A(y_1) \cdots A(y_n)\rangle \tag{3.2}$$

in perturbation theory which in lim $x_j \to x$ approach the Green's functions

$$\langle TN_a[A_{(\mu)_1}(x) \cdots A_{(\mu)_m}(x)]\, A(y_1) \cdots A(y_n)\rangle \tag{3.3}$$

of (3.1), provided $a \geqslant d$. For $m = 2$ the functions thus defined will turn out to be identical with the vacuum expectation values of the time ordered products (2.7). The equivalence of the two definitions will be established in Section 5.

The power series of (3.1) will be set up in the form

$$\langle TN_a[A_{(\mu)_1}(x_1) \cdots A_{(\mu)_m}(x_m)]\, A(y_1) \cdots A(y_n)\rangle$$
$$= \sum_{\Gamma \in \mathscr{C}_{m+n}} \langle TN_a[A_{(\mu)_1}(x_1) \cdots A_{(\mu)_m}(x_m)]\, A(y_1) \cdots A(y_n)\rangle_{\Gamma} \tag{3.4}$$

A corresponding expansion

$$\langle TN_a[A_{(\mu)_1}(x) \cdots A_{(\mu)_m}(x)]\, A(y_1) \cdots A(y_n)\rangle$$
$$= \sum_{\Delta \in \mathscr{D}_{mn}} \langle TN_a[A_{(\mu)_1}(x) \cdots A_{(\mu)_m}(x)]\, A(y_1) \cdots A(y_n)\rangle_{\Delta} \tag{3.5}$$

for (3.2) was given in [1]. The diagrams $\Gamma \in \mathscr{C}_{m+n}$ and $\Delta \in \mathscr{D}_{mn}$ are related by the one-to-one correspondence

$$\Delta = \check{\Gamma}, \qquad \Gamma = \hat{\Delta}. \tag{3.6}$$

The terms of the expansion (3.5) were defined in [1, Eq. 4.28]. A Fourier transformation with respect to x yields

$$\langle TB_{\{\mu\}}^{(a)}(x)\, \tilde{A}(q_1) \cdots \tilde{A}(q_n)\rangle_\Delta$$

$$= \frac{\delta_\Delta'}{\mathscr{S}(\Delta)} e^{-ipx} \lim_{\epsilon\to+0} \int \prod_j \prod_{\alpha=1}^{n_j-1} \frac{dK_{j\alpha}}{(2\pi)^4} \prod_j{}'' \prod_{\alpha=1}^{n_j''-1} \frac{dK_{j\alpha}''}{(2\pi)^4} \prod_{j=1}^{s} \frac{dk_j}{(2\pi)^4} R_\Delta^{(a)}, \tag{3.7}$$

$$\delta_\Delta' = \prod_{j=1}^{c'} (2\pi)^4\, \delta\Big(\sum_{\alpha=1}^{n_j'} q_{j\alpha}'\Big), \qquad p = -\sum_{j=1}^{n} q_j .$$

This suggests defining

$$\langle TN_a[A_{(\mu)_1}(x_1) \cdots A_{(\mu)_m}(x_m)]\, \tilde{A}(q_1) \cdots \tilde{A}(q_n)\rangle_\Gamma$$

$$= \frac{\delta_\Gamma'}{\mathscr{S}(\Gamma)} \int \prod_j \prod_{\alpha=1}^{n_j-1} \frac{dK_{j\alpha}}{(2\pi)^4} \prod_j{}'' \prod_{\alpha=1}^{n_j''-1} \frac{dK_{j\alpha}''}{(2\pi)^4} e^{-i\Sigma p_j x_j} \lim_{\epsilon\to+0} \int \prod_{j=1} \frac{dk_j}{(2\pi)^4} R_\Delta^{(a)} \tag{3.8}$$

$$\delta_\Gamma' = \delta_\Delta', \qquad \mathscr{S}(\Gamma) = \mathscr{S}(\Delta),$$

which, for $a \geqslant m + \sum \#(\mu)_j$, approaches (3.7) in a suitable limit $x_j \to x$. However, due to the presence of the ϵ-limit it is not easy to study the limit $x_j \to x$ rigorously. The limit should be taken such that all quantities

$$(x_i - x_j)/(|(x_i - x_j)^2|)^{1/2} \tag{3.9}$$

stay bounded. Otherwise the limit may diverge since (3.8) in general has singularities on the light cone. Thus

$$\lim_{x_j\to x} \langle TN_a[A_{(\mu)_1}(x_1) \cdots A_{(\mu)_m}(x_m)]\, A(y_1) \cdots A(y_n)\rangle$$

$$= \langle TN_a[A_{(\mu)_1}(x) \cdots A_{(\mu)_m}(x)]\, A(y_1) \cdots A(y_n)\rangle$$

$$\text{for } (x_i - x_j)/(|(x_i - x_j)^2|)^{1/2} \text{ bounded and } a \geqslant m + \sum \#(\mu)_j . \tag{3.10}$$

If a is negative the degree $\delta_a(\gamma)$ of any proper subdiagram γ containing V_0 is negative. Consequently no subtractions for any such diagram are made, therefore

$$\langle TN_a[A_{(u)_1}(x_1) \cdots A_{(u)_m}(x_m)]\, A(y_1) \cdots A(y_n)\rangle$$
$$= \langle TA_{(u)_1}(x_1) \cdots A_{(u)_m}(x_m)\, A(y_1) \cdots A(y_n)\rangle \qquad \text{if} \quad a = -1, -2, \ldots . \tag{3.11}$$

If $a = 0$ only those proper subdiagrams containing V_0 require subtractions which have no external lines. A connected diagram $\Delta \in \mathscr{D}_{mn}$ contains such subdiagrams iff Δ itself has no exterior vertices except V_0 (see Fig. 1). In this case

$$R_\Delta = (1 - t_\Delta)\, \bar{R}_\Delta = 0, \tag{3.12}$$

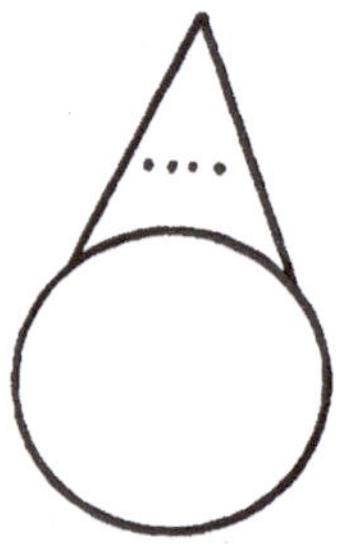

FIG. 1. Subdiagram of Δ which has no external lines.

since $\bar{R}_\Delta$ does not depend on the external momentum. Hence

$$\langle TN_0[A_{(u)_1}(x_1) \cdots A_{(u)_m}(x_m)]\, A(y_1) \cdots A(y_n)\rangle$$
$$= \langle TA_{(u)_1}(x_1) \cdots A_{(u)_m}(x_m)\, A(y_1) \cdots A(y_n)\rangle$$
$$- \langle TA_{(u)_1}(x_1) \cdots A_{(u)_m}(x_m)\rangle \langle TA(y_1) \cdots A(y_n)\rangle. \tag{3.13}$$

We finally give a generalization of the normalization conditions I.4.40. The proper part of (3.4) as defined by Eq. (I.2.17) has the expansion

$$\langle TN_a[A_{(u)_1}(x_1) \cdots A_{(u)_m}(x_m)]\, A(y_1) \cdots A(y_n)\rangle^{\text{prop}}$$
$$= \langle T{:}A^0_{(u)_1}(x_1) \cdots A^0_{(u)_m}(x_m){:}\, A^0(y_1) \cdots A^0(y_n)\rangle_0^{\text{prop}}$$
$$+ \sum_{\Pi \in \mathscr{P}} \frac{1}{\mathscr{S}(\pi)} \langle TN_a[A_{(u)_1}(x_1) \cdots]\, A(y_1) \cdots\rangle_\Pi^{\text{prop}} \tag{3.14}$$

The sum extends over the set $\mathscr{P}$ of all diagrams Π which are related to the diagrams Φ of the class $\mathscr{F}_{mn}$ (as defined below Eq. (I.4.33) by

$$\Pi = \hat{\Phi}, \qquad \Phi = \hat{\Pi}$$

Applying the differential operator $\partial^{q_1}_{(\nu)_1} \cdots \partial^{q_n}_{(\nu)_n}$ and using (I.4.36) we obtain

$$\partial^{q_1}_{(\nu)_1} \cdots \partial^{q_n}_{(\nu)_n} \langle TN_a[A_{(\mu)_1}(x_1) \cdots A_{(\mu)_m}(x_m)\, \tilde{A}(q_1) \cdots \tilde{A}(q_n)\rangle_{\Pi}^{\mathrm{prop}} = 0 \quad \text{if} \quad q_j = 0$$
$$\text{if} \quad n + \sum \#(\nu)_j \leqslant a \tag{3.15}$$

for any diagram $\Pi \in \mathscr{P}$. Hence the relations

$$\langle TN_a[A_{(\mu)_1}(x_1) \cdots A_{(\mu)_m}(x_m)]\, \tilde{A}^{(\nu)_1}(0) \cdots \tilde{A}^{(\nu)_n}(0)\rangle^{\mathrm{prop}} \tag{3.16}$$
$$= \langle T{:}A^0_{(\mu)_1}(x_1) \cdots A^0_{(\mu)_m}(x_m){:}\, \tilde{A}^{(\nu)_1}(0) \cdots \tilde{A}^{(\nu)_n}(0)\rangle_0^{\mathrm{prop}} \quad \text{for} \quad n + \sum \#(\nu)_j \leqslant a,$$

follow as normalization conditions for the Green's functions of normal products, with the notation (I.2.24) and (2.11). Using (I.2.20) the right side of (3.16) can be worked out as follows

$$\langle T{:}A^0_{(\mu)_1}(x_1) \cdots A^0_{(\mu)_m}(x_m){:}\, \tilde{A}^{(\nu)_1}(0) \cdots \tilde{A}^{(\nu)_n}(0)\rangle^{\mathrm{prop}}$$
$$= \delta_{mn} \sum_P i^{\Sigma \#(\nu)_j} \partial^{\alpha_1}_{(\mu)_1} \cdots \partial^{\alpha_m}_{(\mu)_m} x^{(\nu)_1}_{\alpha_1} \cdots x^{(\nu)_m}_{\alpha_m} \tag{3.17}$$

The sum extends over all permutations

$$P = \begin{pmatrix} \alpha_1 & \cdots & \alpha_m \\ 1 & \cdots & m \end{pmatrix}$$

For $m = 2$, Eqs. (3.16–17) are in agreement with the normalization condition (2.22).

4. Principal Part of Wilson's Short Distance Expansion

According to Wilson's hypothesis the short distance expansion of $TA(x_1)\, A(x_2)$ has the principal part

$$TA(x + \xi)\, A(x - \xi) = E_0(\xi)\, 1 + E_1(\xi)\, A(x) - iE_2{}^{\mu}(\xi)\, \partial_\mu A(x)$$
$$+ \tfrac{1}{2} E_3(\xi)\, N[A(x)^2] + R(x, \xi) \tag{4.1}$$

where the remainder vanishes as $\xi \to 0$ with $\xi^\mu/(|\xi^2|)^{1/2}$ bounded [3], Wilson's dimensional rules imply the following bounds on the dimensions of the coefficients

$$\begin{aligned} d(E_0) &\leqslant 2, \qquad d(E_1) \leqslant 1 \\ d(E_2{}^\mu) &\leqslant 0, \qquad d(E_3) \leqslant 0. \end{aligned} \tag{4.2}$$

Here the following definition of the dimension $d(f)$ of a function $f(\xi)$ is used

$$\lim_{\rho\to 0} \rho^s t(\rho\eta) = 0 \qquad \text{if} \quad s > d(t)$$

$$\lim_{\rho\to 0} \rho^s t(\rho\eta) = \infty \qquad \text{if} \quad s < d(t)$$

for any fixed vector η with $\eta^2 \neq 0$.

It was first shown by R. Brandt that these statements hold to all orders of renormalized perturbation theory [4]. An alternative proof will be given here as a first illustration of the combinatorial methods developed in [1]. The generalization to the product of field operator derivatives is not discussed in this section since this will follow from the complete short distance expansion of $TA(x_1)\, A(x_2)$ by differentiation.

We will first check Wilson's hypothesis for Green's functions in the form

$$\begin{aligned} &\langle TA(x+\xi)\, A(x-\xi)\, A(y_1) \cdots A(y_n)\rangle \\ &\quad = E_0(\xi)\langle TA(y_1) \cdots A(y_n)\rangle + E_1(\xi)\langle TA(x)\, A(y_1) \cdots A(y_n)\rangle \\ &\quad\quad - iE_2{}^\mu(\xi)\, \partial_\mu\langle TA(x)\, A(y_1) \cdots A(y_n)\rangle \\ &\quad\quad + \tfrac{1}{2}E_3(\xi)\langle TN[A(x)^2]\, A(y_1) \cdots A(y_n)\rangle \\ &\quad\quad + r(x\xi y_1 \cdots y_n). \end{aligned} \tag{4.3}$$

We begin by comparing the Green's functions

$$\langle T\!:\!A(x_1)\, A(x_2)\!:\, A(y_1) \cdots A(y_n)\rangle \tag{4.4}$$

$$\langle TN[A(x)^2]\, A(y_1) \cdots A(y_n)\rangle \tag{4.5}$$

$$\langle TN_2[A(x_1)\, A(x_2)]\, A(y_1) \cdots A(y_n)\rangle \tag{4.6}$$

in perturbation theory. For (4.6) we will use the definitions (3.5), (3.8) of the preceding section. It will be shown later in this section that the Green's functions thus defined are identical with the time ordered function of the normal product $N_2[A(x_1)\, A(x_2)]$ as defined by (2.27).

We first give convenient expressions for the contributions of a diagram $\Gamma \in \mathscr{C}_{n+2}$ or $\Delta \in \mathscr{D}_{2,n}$ to (4.4–6). In the notation of [1] $\mathscr{C}_{n+2}$ is defined as the set of all diagrams contributing to $\langle TA(x_1)\, A(x_2)\, A(y_1) \cdots A(y_n)\rangle$. We distinguish the classes $\mathscr{A}_1$, $\mathscr{A}_2$, $\mathscr{A}_3$ of diagrams in $\mathscr{C}_{2n}$ which are defined as follows.

$\mathscr{A}_1$ is the set of diagrams $\Gamma \in \mathscr{C}_{n+2}$ for which x_1, x_2 and at least one coordinate y_j are attached to the same connected component of Γ (Fig. 2).

$\mathscr{A}_2$ is the set of diagrams $\Gamma \in \mathscr{C}_{n+2}$ for which x_1 and x_2 belong to different connected components of Γ (Fig. 3).

$\mathscr{A}_3$ is the set of diagrams $\Gamma \in \mathscr{C}_{n+2}$ for which x_1 and x_2 belong to the same connected component of Γ to which no coordinate y_j is attached (Fig. 4).

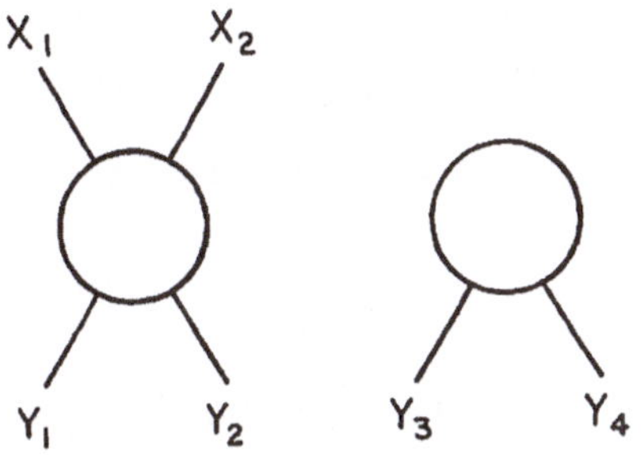

FIG. 2. Example of a diagram belonging to $\mathscr{A}_1$.

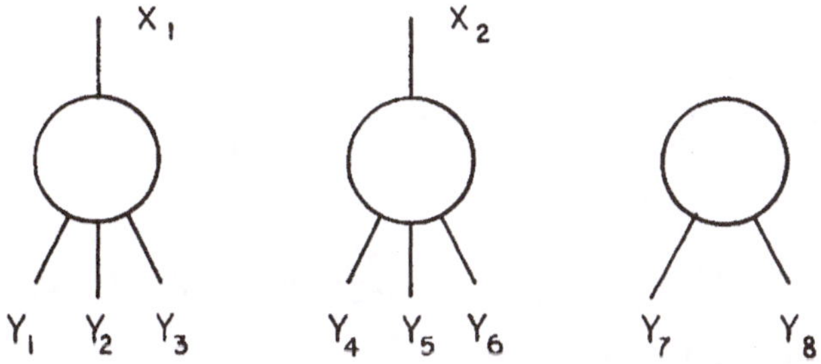

FIG. 3. Example of a diagram belonging to $\mathscr{A}_2$.

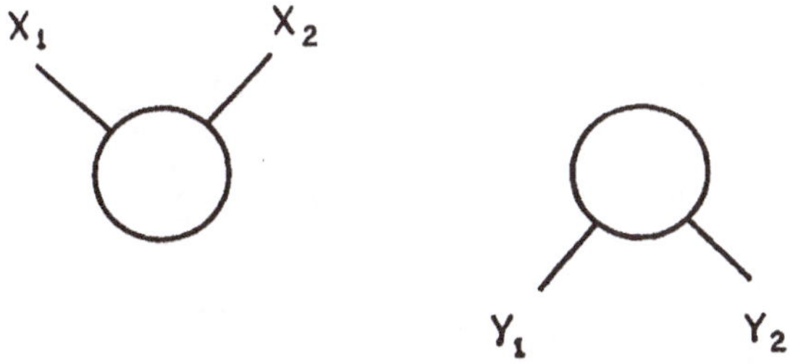

FIG. 4. Example of a diagram belonging to $\mathscr{A}_3$.

As for the notation of the connected components we follow the convention of Eq. (I.4.8). If $\Gamma \in \mathscr{A}_1$, the connected component containing x_1 and x_2 is denoted by Γ_1. If $\Gamma \in \mathscr{A}_2$ the connected component containing x_j is called Γ_j. If $\Gamma \in \mathscr{A}_3$ the connected component containing x_1, x_2 is denoted by Γ_1''. Accordingly the connected components of Γ are

$$\begin{aligned} &\Gamma_1, \Gamma_1', \ldots, \Gamma_{c'}' && \text{if} \quad \Gamma \in \mathscr{A}_1 \\ &\Gamma_1, \Gamma_2, \Gamma_1', \ldots, \Gamma_{c'}' && \text{if} \quad \Gamma \in \mathscr{A}_2 \\ &\Gamma_1'', \Gamma_1', \ldots, \Gamma_{c'}' && \text{if} \quad \Gamma \in \mathscr{A}_3 \end{aligned}$$

where the Γ_j' are the connected components to which only coordinates y_j are attached.

Apparently

$$\mathscr{C}_{n+2} = \mathscr{A}_1 \cup \mathscr{A}_2 \cup \mathscr{A}_3$$

The set of diagrams obtained from $\mathscr{A}_j$ by identifying $x_1 = x_2 = x$ will be denoted by $\mathscr{B}_j$.

$$\mathscr{D}_{2,n} = \mathscr{B}_1 \cup \mathscr{B}_2 \cup \mathscr{B}_3 .$$

The contributions from a diagram $\Gamma \in \mathscr{C}_{n+2}$ or $\Delta_{2,n}$ have the form[4]

$$\langle T{:}A(x_1)\, A(x_2){:}\, \tilde{A}(q_1) \cdots \tilde{A}(q_n)\rangle_\Gamma \tag{4.7}$$
$$= \frac{\delta_\Gamma'}{\mathscr{S}(\Gamma)} \int \frac{dK}{(2\pi)^4}\, e^{-i(p_1x_1+p_2x_2)} \lim_{\epsilon\to+0} \int \frac{dk_1}{(2\pi)^4} \cdots \frac{dk_s}{(2\pi)^4} R_\Gamma \qquad \text{if} \quad \Gamma \in \mathscr{A}_1 ,$$

$$\langle TN[A(x)^2]\, \tilde{A}(q_1) \cdots \tilde{A}(q_n)\rangle_\Delta$$
$$= \frac{\delta_\Delta'}{\mathscr{S}(\Delta)}\, e^{-ipx} \lim_{\epsilon\to+0} \int \frac{dK}{(2\pi)^4} \frac{dk_1}{(2\pi)^4} \cdots \frac{dk_s}{(2\pi)^4} R_\Delta \qquad \text{if} \quad \Delta \in \mathscr{B}_1 \text{ or } \mathscr{B}_3 , \tag{4.8}$$

$$\langle TN_2[A(x_1)\, A(x_2)]\, \tilde{A}(q_1) \cdots \tilde{A}(q_n)\rangle_\Gamma \tag{4.9}$$
$$= \frac{\delta_\Gamma'}{\mathscr{S}(\Gamma)} \int \frac{dK}{(2\pi)^4}\, e^{-i(p_1x_1+p_2x_2)} \lim_{\epsilon\to+0} \int \frac{dk_1}{(2\pi)^4} \cdots \frac{dk_s}{(2\pi)^4} R_\Delta \qquad \text{if} \quad \Gamma \in \mathscr{A}_1 \text{ or } \mathscr{A}_3$$

where

$$\Delta = \bar{\Gamma}, \qquad \delta_\Gamma' = \delta_\Delta' = \sum_{j=1}^{c'} (2\pi)^4\, \delta\Big(\sum_{\alpha=1}^{n_j'} q_{j\alpha}'\Big)$$
$$p_1 = (p/2) + K, \qquad p_2 = (p/2) - K, \qquad p = -\sum_{j=1}^{n} q_j , \qquad \mathscr{S}(\Gamma) = \mathscr{S}(\Delta). \tag{4.10}$$

$q_{j1}', \ldots, q_{jn_j'}'$ denote the external momenta of the connected component.

[4] For integrands R_Γ it is always understood that the substitutions $p_1 = p/2 + K$, $p_2 = p/2 - K$ are made. Hence R_Γ corresponds to SR_Γ in the notation of *I*.

For $\Gamma \in \mathscr{A}_2, \mathscr{A}_3$ we have

$$\langle TN_2[A(x_1)\,A(x_2)]\,A(y_1)\cdots A(y_n)\rangle_\Gamma \tag{4.11}$$
$$= \langle T{:}A(x_1)\,A(x_2){:}\,A(y_1)\cdots A(y_n)\rangle_\Gamma$$
$$= \frac{\delta_\Gamma}{\mathscr{S}(\Gamma)}\,e^{-i(p_1+p_2)x-i(p_1-p_2)\xi}\lim_{\epsilon\to+0}\int\frac{dk_1}{(2\pi)^4}\cdots\frac{dk_s}{(2\pi)^4}\,R_\Gamma \qquad \text{if}\quad \Gamma\in\mathscr{A}_2,$$

$$\langle TN_2[A(x_1)\,A(x_2)]\,A(y_1)\cdots A(y_n)\rangle_\Gamma$$
$$= \langle T{:}A(x_1)\,A(x_2){:}\,A(y_1)\cdots A(y_n)\rangle_\Gamma = 0 \qquad \text{if}\quad \Gamma\in\mathscr{A}_3. \tag{4.12}$$

In (4.11) it was used that any subdiagram of $\Delta = \bar{\Gamma}$ containing V_0 is one-particle-reducible (see Fig. 5). If $\Gamma \in \mathscr{A}_3$ the integrand R_Δ of (4.9) contains the factor

$$R_{a\Delta_1''} = (1 - t_p^{\,q})\,\bar{R}_{a\Delta_1''} = 0, \qquad \Delta_1'' = \bar{\Gamma}_1'' \tag{4.13}$$

which vanishes since $\bar{R}_{a\Delta_1''}$ does not depend on p. Therefore (4.12) holds.

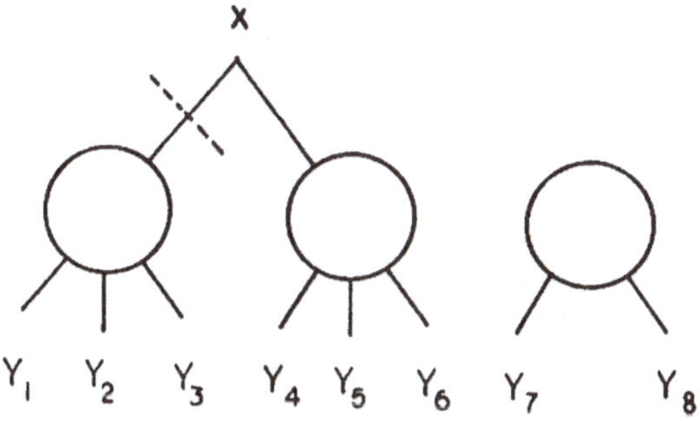

FIG. 5. One-particle-cut of the diagram $\Delta = \bar{\Gamma}$ with Γ as shown in Fig. 3.

Because of (4.11–12) we may restrict ourselves to the nontrivial case $\Gamma \in \mathscr{A}_1$ or $\Delta = \bar{\Gamma} \in \mathscr{B}_1$ for the comparison of the Green's functions (4.7) and (4.9). In general R_Γ and R_Δ are related by (I.5.4–5), i.e.

$$R_\Delta = R_\Gamma + X \tag{4.14}$$

$$X = S_\Delta \sum_{\tau\in T}\ \sum_{U_1\in\mathscr{M}_\tau}\ \sum_{U_2\in\mathscr{F}(\sigma)}\ \prod_{\gamma\in U_1}(-t^\gamma S_\gamma)(-t^\tau S_\tau)\prod_{\gamma\in U_2}(-t^\gamma S_\gamma)\,S_U I_\Delta. \tag{4.15}$$

We will study the expression X in some detail. First we determine all possible diagrams τ to be summed over in (4.15). The set T was introduced as the set of all proper subdiagrams τ of which contain V and have nonnegative degree. The definition (I.4.24) of the degree $\delta_2(\tau)$ yields the inequality

$$0 \leqslant \delta_2(\tau) = 2 - r \tag{4.16}$$

where r is the number of external lines of τ. This implies $r = 0, 1, 2$. $r = 0$ is not possible since in any $\Gamma \in \mathscr{A}_1$ x_1, x_2 are connected to at least one y_j. Hence only diagrams τ with $r = 1, 2$ are eligible. Accordingly the set T is given by

$$T = T_1 \cup T_2 \tag{4.17}$$

where T_j is the set of diagrams τ with $r = j$. The corresponding contributions to (4.15) are denoted by X_1 and X_2

$$X = X_1 + X_2 \tag{4.18}$$

$$X_1 = S_\Delta \sum_{\tau \in T_1} \sum_{U_1 \in \mathscr{M}_\tau} \sum_{U_2 \in \mathscr{F}(\sigma)} \prod_{\gamma \in U_1} (-t^\gamma S_\gamma) \prod_{\gamma \in U_2} (-t^\gamma S_\gamma) S_U I_\Delta \tag{4.19}$$

$$X_2 = S_\Delta \sum_{\tau \in T_2} \sum_{U_1 \in \mathscr{M}_\tau} \sum_{U_2 \in \mathscr{F}(\sigma)} \prod_{\gamma \in U_1} (-t^\gamma S_\gamma)(-t^\tau S_\tau) \prod_{\gamma \in U_2} (-t^\gamma S_\gamma) S_U I_\Delta . \tag{4.20}$$

The diagram σ is defined by

$$\sigma = \hat{\tau}, \tag{4.21}$$

i.e., τ is obtained from σ by identifying $x_1 = x_2$. The set of all such σ is called S. (4.21) is a one-to-one correspondence between S and T. The unrenormalized integrands belonging to σ and τ are identical:

$$I_\tau = I_\sigma . \tag{4.22}$$

For any subdiagram τ of Δ containing V_0 the unrenormalized Feynman integrand factorizes

$$I_\Delta = I_{\Delta/\tau} I_\tau = I_{\Delta/\tau} I_\sigma . \tag{4.23}$$

It will be seen that X can be written as a sum of factorizing terms. We begin with a discussion of X_2. In this case

$$\delta(\tau) = 2 - m = 0. \tag{4.24}$$

Therefore, the Taylor operator becomes

$$t^\tau = t_{q_\tau}^{\delta(\tau)} = t_{q_\tau}^0 \tag{4.25}$$

and merely prescribes that the external momenta of τ be set equal to zero.

We rearrange the products occurring in (4.20) by using (4.23), (4.25), and the fact that $I_{\Delta/\tau}$ does not depend on the external momenta q_τ of the diagram τ

$$\begin{aligned} S_\Delta \prod_{\gamma\in U_1} (-t^\gamma S_\gamma)(-t^\tau S_\tau) \prod_{\gamma\in U_2} (-t^\gamma S_\gamma)\, S_U I_U \\ = S_\Delta \prod_{\gamma\in U_1} (-t^\gamma S_\gamma)(-t^\tau S_\tau) \prod_{\gamma\in U_2} (-t^\gamma S_\gamma)\, S_{U_1} I_{\Delta/\tau} S_{U_2} I_\sigma \\ = -S_\Delta \prod_{\gamma\in U_1} (-t^\gamma S_\gamma)\, S_{U_1} I_{\Delta/\tau} t^0_{q_\tau} \Big\{ S_\tau \prod_{\gamma\in U_2} (-t^\gamma S_\gamma)\, S_{U_2} I_\sigma \Big\}. \end{aligned} \tag{4.26}$$

With this X_2 becomes

$$X_2 = -\sum_{\tau\in T_2} X^\tau_{21} X^\tau_{22} \tag{4.27}$$

where

$$X^\tau_{21} = S_\Delta \sum_{U_1\in\mathcal{M}_\tau} \prod_{\gamma\in U_1} (-t^\gamma S_\gamma)\, S_{U_1} I_{\Delta/\tau} \tag{4.28}$$

$$X^\tau_{22} = t^0 \Big\{ S_\sigma \sum_{U_2\in\mathcal{F}(\sigma)} \prod_{\gamma\in U_2} (-t^\gamma S_\gamma)\, S_{U_2} I_\sigma \Big\}. \tag{4.29}$$

We recall that the sum $\sum_{U_1\in\mathcal{M}_\tau}$ extends over all forests of renormalization parts γ of Γ which satisfy $\gamma \supset \tau$ or are disjoint. Accordingly (4.28) may be replaced by the sum over all forests U' of Δ/τ with elements $\gamma' = \gamma/\tau$. Hence

$$X^\tau_{21} = S_{\Delta/\tau} \sum_{U'\in\mathcal{F}(\Delta/\tau)} \prod_{\gamma'\in U'} (-t^{\gamma'} S_{\gamma'})\, S_{U'} I_{\Delta/\tau} = R_{\Delta/\tau}, \tag{4.30}$$

$$t^\gamma = t^{\gamma'}, \qquad q^\gamma = q^{\gamma'}, \qquad \delta_2(\gamma) = \delta_2(\gamma') \qquad \text{if} \quad \gamma' = \gamma/\tau.$$

Thus X_{21} represents the renormalized integrand $R^{(2)}_{\Delta/\tau}$ of the reduced diagram Δ/τ with respect to the degree δ_2. On the other hand,

$$S_\sigma \sum_{U\in\mathcal{F}(\sigma)} \prod_{\gamma\in U} (-t^\gamma S_\gamma)\, I_\sigma = R_\sigma \tag{4.31}$$

represents the renormalized integrand belonging to the diagram σ. Combining (4.27) with (4.28–30) we obtain

$$X_2 = -\sum_{\tau\in T} R^{(2)}_{\Delta/\tau} \mathring{R}_\sigma \tag{4.32}$$

$$\mathring{R}_\sigma = R_\sigma \big|_{q_\tau=0} \tag{4.33}$$

Hence X_2 is a sum of terms which factorize into the renormalized integrand of the reduced diagram Δ/τ and the renormalized integrand of σ.

In case of X_1 we have

$$\delta(\tau) = 2 - r = 1 \tag{4.34}$$

There is only one external momentum for each τ which we denote by q. Hence the Taylor operator becomes

$$t^\tau = t_q^1 = 1 + \sum_\mu q_\mu \frac{\partial}{\partial q_\mu} \tag{4.35}$$

with q_μ set equal to zero afterwards. We thus obtain the following factorization of X_1

$$X_1 = -\sum_{\tau\in T_1} R_{\Delta/\tau} \mathring{R}_\sigma - \sum_{\tau\in T_1} \sum_\mu (-i)\, R_{\mu\Delta/\tau}\{\partial_q{}^\mu R_\tau\}_{q=0} \tag{4.36}$$

where

$$R_{\Delta/\tau} = S_{\Delta/\tau} \sum_{U\in\mathscr{M}_\tau} \prod_{\gamma\in U_1} (-t^\gamma S_\gamma)\, I_{\Delta/\tau} \tag{4.37}$$

is a contribution to the Green's function $\langle TA(x)\, \tilde{A}(q_1) \cdots \tilde{A}(q_n)\rangle$ and

$$R_{\mu\Delta/\tau} = S_{\Delta/\tau} \sum_{U\in\mathscr{M}_\tau} \prod_{\gamma\in U_1} (-t^\gamma S_\gamma)\{(-iq_\mu)\, I_{\Delta/\tau}\} \tag{4.38}$$

is a contribution to the Green's function

$$\langle TA_\mu(x)\, \tilde{A}(q_1) \cdots \tilde{A}(q_m)\rangle$$

Combining (4.18–20) with (4.32–33) and (4.36) we find

$$R_\Gamma = R_\Delta^{(2)} + \sum_{\tau\in T_1} R^{\Delta/\tau} \mathring{R}_\sigma + \sum_{\tau\in T_1} \sum_\mu (-i)\, R_\mu^{\Delta/\tau}\{\partial_q{}^\mu R^\tau\}_{q=0} + \sum_{\tau\in T_2} R_{\Delta/\tau}^{(2)} \mathring{R}_\sigma\,, \tag{4.39}$$

Multiplying by

$$\delta(K_1 + K_2)\, e^{-i(p_1x_1+p_2x_2)} = \delta(K_1 + K_2)\, e^{-ipx} e^{-i(K_1-K_2)\xi}$$

with

$$x_1 = x + \xi, \qquad x_2 = x - \xi, \qquad p = p_1 + p_2$$

and integrating over K_i and k_j we obtain

$$\begin{aligned}&\langle T{:}A(x_1)\,A(x_2){:}\,\tilde{A}(q_1)\cdots\tilde{A}(q_m)\rangle_\Gamma \qquad (4.40)\\ &\quad=\langle TN\{A(x_1)\,A(x_2)\}\,\tilde{A}(q_1)\cdots\tilde{A}(q_m)\rangle_\Gamma\\ &\qquad+\sum_{\tau\in T_1}\langle TA(x)\,\tilde{A}(q_1)\cdots\tilde{A}(q_m)\rangle_{\Delta/\tau}\langle TA(\xi)\,A(-\xi)\,\tilde{A}(0)\rangle_\sigma\\ &\qquad\sum_{\tau\in T_2}(-i)\langle TA_\mu(x)\,\tilde{A}(q_1)\cdots\tilde{A}(q_m)\rangle_{\Delta/\tau}\langle TA(\xi)\,A(-\xi)\,\tilde{A}^\mu(0)\rangle_\sigma\\ &\qquad+\sum_{\tau\in T_3}\langle TN[A(x)^2]\,\tilde{A}(q_1)\cdots\tilde{A}(q_m)\rangle_{\Delta/\tau}\langle TA(\xi)\,A(-\xi)\,\tilde{A}(0)\,\tilde{A}(0)\rangle_\sigma\end{aligned}$$

for any $\Gamma\in\mathscr{A}_1$. Summing over all $\Gamma\in\mathscr{A}_1$ and taking the Fourier transform with respect to $q_1,\ldots,q_n$ we obtain

$$\begin{aligned}&\sum_{\Gamma\in\mathscr{A}_1}\langle T{:}A(x_1)\,A(x_2){:}\,A(y_1)\cdots A(y_n)\rangle_\Gamma\\ &\quad=\sum_{\Gamma\in\mathscr{A}_1}\langle N_2[A(x_1)\,A(x_2)]\,A(y_1)\cdots A(y_n)\rangle_\Gamma\\ &\qquad+E_1(\xi)\sum_{\Gamma\in\mathscr{C}'_{1n}}\langle TA(x)\,A(y_1)\cdots A(y_n)\rangle_\Gamma\\ &\qquad-iE_2{}^\mu(\xi)\sum_{\Gamma\in\mathscr{C}'_{1n}}\langle T\partial_\mu A(x)\,A(y_1)\cdots A(y_n)\rangle_\Gamma\\ &\qquad+\tfrac{1}{2}\hat{E}_3(\xi)\sum_{\Delta\in\mathscr{D}'_{2n}}\langle TN[A(x)^2]\,A(y_1)\cdots A(y_n)\rangle_\Delta \qquad (4.41)\end{aligned}$$

with

$$E_1(\xi)=\sum_{\sigma\in\mathscr{F}_{21}}\langle T{:}A(\xi)\,A(-\xi){:}\,\tilde{A}(0)\rangle_\sigma \qquad (4.42)$$

$$E_2{}^\mu(\xi)=\sum_{\sigma\in\mathscr{F}_{21}}\langle T{:}A(\xi)\,A(-\xi){:}\,\tilde{A}^\mu(0)\rangle_\sigma \qquad (4.43)$$

$$\hat{E}_3(\xi)=\sum_{\sigma\in\mathscr{F}_{22}}\langle T{:}A(\xi)\,A(-\xi){:}\,\tilde{A}(0)\,\tilde{A}(0)\rangle_\sigma \qquad (4.44)$$

$\mathscr{C}'_{1n}$ and $\mathscr{D}'_{2n}$ denote the set of all diagrams $\Gamma\in\mathscr{C}_{n+1}$ or $\Delta\in\mathscr{D}_{2n}$ for which x is connected to at least one y_j. $\mathscr{F}_{mn}$ is the set of all proper nontrivial diagrams with coordinates $x_1,\ldots,x_m$, $y_1,\ldots,y_n$ as defined below Eq. (I.4.33).

For diagrams in which x is not connected to any y_j we have

$$\begin{aligned}\langle TA(x)\,A(y_1)\cdots A(y_n)\rangle_\Gamma&=0\\ \langle T[A(x)^2]\,A(y_1)\cdots A(y_n)\rangle_\Delta&=0\end{aligned} \qquad (4.45)$$

because of (I.4.43). With (4.11–12) and (4.45) Eq. (4.41) yields

$$\begin{aligned}\langle T{:}A(x_1)\,A(x_2){:}\,A(y_1)\cdots A(y_n)\rangle \\ = \langle TN_2[A(x_1)\,A(x_2)]\,A(y_1)\cdots A(y_n)\rangle \\ + E_1(\xi)\langle TA(x)\,A(y_1)\cdots A(y_n)\rangle \\ - iE_2{}^{\mu}(\xi)\,\partial_\mu{}^{x}\langle TA(x)\,A(y_1)\cdots A(y_n)\rangle \\ + \tfrac{1}{2}\hat{E}_3(\xi)\langle TN[A(x)^2]\,A(y_1)\cdots A(y_n)\rangle.\end{aligned} \tag{4.46}$$

The expansions (4.42–44) represent the Green's functions

$$E_1(\xi) = \langle T{:}A(\xi)\,A(-\xi){:}\,\tilde{A}(0)\rangle^{\mathrm{prop}} \tag{4.47}$$

$$E_2{}^{\mu}(\xi) = \langle T{:}A(\xi)\,A(-\xi){:}\,\tilde{A}\,(0)\rangle^{\mathrm{prop}} \tag{4.48}$$

$$\begin{aligned}\hat{E}_3(\xi) &= \langle T{:}A(\xi)\,A(-\xi){:}\,\tilde{A}(0)\,\tilde{A}(0)\rangle^{\mathrm{prop}} \\ &\quad - \langle T{:}A_0(\xi)\,A_0(-\xi){:}\,\tilde{A}_0(0)\,\tilde{A}_0(0)\rangle_0^{\mathrm{prop}}\end{aligned} \tag{4.49}$$

In (4.49) the contribution from the trivial diagram was subtracted which equals

$$\langle T{:}A_0(\xi)\,A_0(-\xi){:}\,\tilde{A}_0(0)\,\tilde{A}_0(0)\rangle_0^{\mathrm{prop}} = 2 \tag{4.50}$$

Introducing the remainder

$$\begin{aligned}r(x\xi y_1\cdots y_n) &= \langle TN_2[A(x_1)\,A(x_2)]\,A(y_1)\cdots A(y_n)\rangle \\ &\quad - \langle TN_2[A(x)^2]\,A(y_1)\cdots A(y_n)\rangle\end{aligned} \tag{4.51}$$

we can write (4.46) in the equivalent form (4.3) with (4.47–48) and

$$E_0(\xi) = \langle TA(\xi)\,A(-\xi)\rangle \tag{4.52}$$

$$E_3(\xi) = \langle T{:}A(\xi)\,A(-\xi){:}\,\tilde{A}(0)\,\tilde{A}(0)\rangle^{\mathrm{prop}} \tag{4.53}$$

The remainder (4.51) vanishes as $\xi \to 0$ for $\xi^{\mu}/(|\,\xi^2\,|)^{1/2}$ bounded

$$\lim_{\xi\to 0} r(x\xi y_1\cdots y_n) = 0 \tag{4.54}$$

This completes the proof of the short distance expansion (4.3) of the Green's function.

We next prove Wilson's short distance expansion in the operator form (4.1). A comparison of (4.46–49) with (2.27–29) shows that the Green's functions

$$\langle TN_2[A(x_1)\,A(x_2)]\,A(y_1)\cdots A(y_n)\rangle \tag{4.55}$$

as defined in perturbation theory (Section 3) are identical with the time ordered functions of the operator $N_2[A(x_1)\,A(x_2)]$ given by (2.27). Since the matrix elements of $N_2[A(x_1)\,A(x_2)]$ are related to (4.55) by reduction formulas the relation

$$\lim_{\xi\to 0} N_2[A(x+\xi)\,A(x-\xi)] = N_2[A(x)^2] \tag{4.56}$$

follows from (4.51) and (4.54). Hence the definition (2.27–29) of the normal product represents the Wilson expansion (4.1) in operator form with the coefficients (4.47–48), (4.52–53) and the remainder

$$R(x\xi) = N_2[A(x+\xi)\,A(x-\xi)] - N_2[A(x)^2] \tag{4.57}$$

which vanishes in the limit

$$\lim_{\xi\to 0} R(x,\xi) = 0. \tag{4.58}$$

Similarly we obtain

$$\begin{aligned} &TA(x+\xi)\,A(x-\xi)\,A(y_1)\cdots A(y_n)\\ &\quad= E_0(\xi)\,TA(y_1)\cdots A(y_n) + E_1(\xi)\,TA(x)\,A(y_1)\cdots A(y_n)\\ &\qquad - iE_2{}^{\mu}(\xi)\,\partial_\mu{}^{x}TA(x)\,A(y_1)\cdots A(y_n)\\ &\qquad + \tfrac{1}{2}E_3(\xi)\,TN[A(x)^2]\,A(y_1)\cdots A(y_n)\\ &\qquad + R(x\xi y_1\cdots y_n) \end{aligned} \tag{4.59}$$

with

$$\lim_{\xi\to 0} R(x\xi y_1\cdots y_n) = 0. \tag{4.60}$$

We finally check the dimension of the coefficients E_0, E_1, $E_2{}^{\mu}$ and E_3 given by (4.47–48), (4.52–53). Let Γ be a diagram contributing to E_0. The dimension of any subdiagram γ of $\Delta = \bar{\Gamma}$ which contains V_0 is

$$d(\gamma) \leqslant 2 \qquad \text{(see Eq. (I.A.38)).}$$

Accordingly E_0 is at most quadratically divergent for $\xi \to 0$

$$d(E_0) \leqslant 2 \tag{4.61}$$

Let Γ contribute to E_2 or $E_3{}^{\mu}$, A subdiagram γ of $\Delta = \bar{\Gamma}$ which contains V_0 has the dimension

$$d(\gamma) \leqslant 0$$

as follows from Eq. (I.A.38) and the structure of the Feynman diagram for a mixed A^3, A^4-coupling. Hence

$$d(E_1) \leqslant 0 \tag{4.62}$$

and

$$d(E_2{}^\mu) \leqslant -1 \tag{4.63}$$

since $E_2{}^\mu$ is of the form

$$E_2{}^\mu = \xi^\mu E_2(\xi^2)$$

Similarly one finds

$$d(E_3) \leqslant 0 \tag{4.64}$$

It can be checked in low orders of λ and g that the indicated singularities are actually present. Hence Wilson's rule is satisfied with

$$\left.\begin{array}{ll} d(E_0) = 2, & d(E_1) = 0 \\ d(E_2{}^\mu) = -1, & d(E_3) = 0 \end{array}\right\} \quad \text{if} \quad g, \lambda \neq 0. \tag{4.65}$$

For pure A^4-coupling, i.e., $g = 0$, one has

$$E_1 \equiv 0, \qquad E_2{}^\mu \equiv 0 \qquad \text{if} \quad g = 0 \tag{4.66}$$

and

$$d(E_0) = 2, \qquad d(E_3) = 0 \qquad \text{if} \quad \lambda \neq 0, \qquad g = 0 \tag{4.67}$$

5. Relations Between Normal Products of Different Degree

In this section we will derive some identities which relate normal products of different degree. To this end we compare the power series expansions of the Green's functions

$$\langle TN_a[A_{(\mu)_1}(x_1)\, A_{(\mu)_2}(x_2)]\, A(y_1) \cdots A(y_n)\rangle \tag{5.1}$$

$$\langle TN_b[A_{(\mu)_1}(x_1)\, A_{(\mu)_2}(x_2)]\, A(y_1) \cdots A(y_n)\rangle \tag{5.2}$$

assuming $a > b \geqslant 0$. The Equation (4.46) of the preceding section represents the special case $a = 2$, $b = 0$ since

$$\langle TN_0[A(x_1)\, A(x_2)]\, A(y_1) \cdots A(y_n)\rangle = \langle T{:}A(x_1)\, A(x_2){:}\, A(y_1) \cdots A(y_n)\rangle$$

according to (3.13).

The contributions from a diagram $\Gamma \in \mathscr{A}_2$ or $\mathscr{A}_3$ to (5.1) and (5.2) are identical according to the following generalization of (4.11–12)

$$\begin{aligned}
&\langle TN_a[A_{(\mu)_1}(x_1)\, A_{(\mu)_2}(x_2)]\, A(y_1) \cdots A(y_n)\rangle_\Gamma \\
&\quad = \langle TN_b[A_{(\mu)_1}(x_1)\, A_{(\mu)_2}(x_2)]\, A(y_1) \cdots A(y_n)\rangle_\Gamma \qquad \text{if} \quad \Gamma \in \mathscr{A}_1 \text{ or } \mathscr{A}_3 .
\end{aligned} \tag{5.3}$$

For $\Gamma \in \mathscr{A}_3$ both contributions vanish.

For contributions from a diagram $\Gamma \in \mathscr{A}_1$ are given by

$$\begin{aligned}
&\langle TN_a[A_{(\mu)_1}(x_1)\, A_{(\mu)_2}(x_2)]\, \tilde{A}(q_1) \cdots \tilde{A}(q_n)\rangle_\Gamma \\
&\quad = \frac{\delta_\Gamma{}'}{\mathscr{S}(\Gamma)} \int \frac{dK}{(2\pi)^4}\, e^{-i(p_1x_1+p_2x_2)} \lim_{\epsilon\to+0} \int \frac{dk_1}{(2\pi)^4} \cdots \frac{dk_s}{(2\pi)^4}\, R_\Delta^{(a)} \\
&\langle TN_b[A_{(\mu)_1}(x_1)\, A_{(\mu)_2}(x_2)]\, \tilde{A}(q_1) \cdots \tilde{A}(q_n)\rangle_\Gamma \\
&\quad = \frac{\delta_\Gamma{}'}{\mathscr{S}(\Gamma)} \int \frac{dK}{(2\pi)^4}\, e^{-i(p_1x_1+p_2x_2)} \lim_{\epsilon\to+0} \int \frac{dk_1}{(2\pi)^4} \cdots \frac{dk_s}{(2\pi)^4}\, R_\Delta^{(b)}
\end{aligned} \tag{5.4}$$

with the notation (4.10).

The renormalized integrands $R^{(a)}$ and $R^{(b)}$ are related by the identity (I.6.14) with X given by (I.6.20–23). Multiplying (I.6.14) by

$$\delta(K_1 + K_2)\, e^{-i(p_1x_1+p_2x_2)} = \delta(K_1 + K_2)\, e^{-ipx} e^{-i(K_1-K_2)\xi}, \tag{5.5}$$

taking a Fourier transformation with respect to $q_1, \ldots, q_n$, integrating over the internal momenta and summing over all Γ we obtain

$$\begin{aligned}
&\langle TN_b[A_{(\mu)_1}(x+\xi)\, A_{(\mu)_2}(x-\xi)]\, A(y_1) \cdots A(y_n)\rangle \\
&\quad = \langle TN_a[A_{(\mu)_1}(x+\xi)\, A_{(\mu)_2}(x-\xi)]\, A(y_1) \cdots A(y_n)\rangle \\
&\qquad + \sum_{\{\rho\}}{}_b^a \frac{(-)^{\Sigma \#(\rho)_j}}{r! \prod \#(\rho)_j !}\, G_{\{\mu\}}^{(b)\{\rho\}}(\xi) \langle TB_{\{\rho\}}^{(a)}(x)\, A(y_1) \cdots A(y_n)\rangle
\end{aligned} \tag{5.6}$$

With the notation (2.11) the coefficients $G_{\{\mu\}}^{(b)\{\rho\}}$ are given by

$$G_{\{\mu\}}^{(b)\{\rho\}}(\xi) = \sum_{\tau\in\mathscr{F}_r} \langle TN_b[A_{(\mu)_1}(\xi)\, A_{(\mu)_2}(-\xi)]\, \tilde{A}^{(\rho)_1}(0) \cdots \tilde{A}^{(\rho)_r}(0)\rangle_\tau^{\text{prop}} \tag{5.7}$$

where $\mathscr{F}_r$ is the set of all nontrivial diagrams contributing to the proper part. Hence

$$\begin{aligned}
G_{\{\mu\}}^{(b)\{\rho\}}(\xi) &= \langle TN_b[A_{(\mu)_1}(\xi)\, A_{(\mu)_2}(-\xi)]\, \tilde{A}^{(\rho)_1}(0) \cdots \tilde{A}^{(\rho)_r}(0)\rangle^{\text{prop}} \\
&\quad - \langle T{:}A^0_{(\mu)_1}(\xi)\, A^0_{(\mu)_2}(-\xi){:}\, \tilde{A}_0^{(\rho)_1}(0) \cdots \tilde{A}_0^{(\rho)_r}(0)\rangle_0^{\text{prop}}
\end{aligned} \tag{5.8}$$

or explicitly

$$G^{(b)\{\rho\}}_{\{\mu\}}(\xi) = \langle TN_b[A_{(\mu)_1}(\xi)\, A_{(\mu)_2}(-\xi)]\, \tilde{A}^{(\ )_1}(0) \cdots \tilde{A}^{(\rho)_r}(0)\rangle^{\text{prop}} \tag{5.9}$$

if $r \neq 2$ and

$$\begin{aligned} G^{(b)(\rho)_1(\rho)_2}_{\{\mu\}}(\xi) &= \langle TN_b[A_{(\mu)_1}(\xi)\, A_{(\mu)_2}(-\xi)]\, \tilde{A}^{(\rho)_1}(0)\, \tilde{A}^{(\rho)_2}(0)\rangle^{\text{prop}} \\ &\quad - i^{\#(\rho)_1+\#(\rho)_2}(-1)^{\#(\rho)_2+\#(\mu)_2}\, \partial_{(\mu)_1}\xi^{(\rho)_1}\partial_{(\mu)_2}\xi^{(\rho)_2} \\ &\quad - i^{\#(\rho)_1+\#(\rho)_2}(-1)^{\#(\rho)_1+\#(\mu)_2}\, \partial_{(\mu)_1}\xi^{(\rho)_2}\partial_{(\mu)_2}\xi^{(\rho)_1} \end{aligned} \tag{5.10}$$

if $r = 2$.

Equation (5.6) is particularly interesting in the case of $a = b + 1$. Then the sum is restricted by

$$r + \sum \#(\rho)_j = a = b + 1 \tag{5.11}$$

so that only normal products of minimal degree occur on the right side. With the notation

$$G^{(b)\{\rho\}}_{\{\mu\}} = G^{\{\rho\}}_{\{\mu\}} \qquad \text{if} \quad b = r + \sum \#(\rho)_j - 1 \tag{5.12}$$

$$B^{(a)}_{\{\rho\}} = B_{\{\rho\}} \qquad \text{if} \quad a = r + \sum \#(\rho)_j\,, \tag{5.13}$$

Eq. (5.6) takes the form

$$\begin{aligned} &\langle TN_b[A_{(\mu)_1}(x_1)\, A_{(\mu)_2}(x_2)]\, A(y_1) \cdots A(y_n)\rangle \\ &\quad = \langle TN_{b+1}[A_{(\mu)_1}(x_1)\, A_{(\mu)_2}(x_2)]\, A(y_1) \cdots A(y_n)\rangle \\ &\quad + \sum_{\{\rho\}}{}^{b+1}_{b} \frac{(-)^{\sum\#(\rho)_j}}{r!\prod \#(\rho)_j\,!}\, G^{\{\rho\}}_{\{\mu\}}(\xi)\langle TB_{\{\rho\}}(x)\, A(y_1) \cdots A(y_n)\rangle. \end{aligned} \tag{5.14}$$

Solving this recursion formula we obtain

$$\begin{aligned} &\langle TA_{(\mu)_1}(x_1)\, A_{(\mu)_2}(x_2)\, A(y_1) \cdots A(y_n)\rangle \\ &\quad = \langle TA_{(\mu)_1}(\xi)\, A_{(\mu)_2}(-\xi)\rangle\langle TA(y_1) \cdots A(y_n)\rangle \\ &\quad + \sum_{\{\rho\}}{}^{a}_{0} \frac{(-)^{\sum\#(\rho)_j}}{r!\prod \#(\rho)_j\,!}\, G^{\{\rho\}}_{\{\mu\}}(\xi)\langle TB_{\{\rho\}}(x)\, A(y_1) \cdots A(y_n)\rangle \\ &\quad + \langle TN_a[A_{(\mu)_1}(x_1)\, A_{(\mu)_2}(x_2)]\, A(y_1) \cdots A(y_n)\rangle. \end{aligned} \tag{5.15}$$

The coefficients $G^{\{\rho\}}_{\{\mu\}}$ are given recursively by (5.9–10) with

$$b = r + \sum \#(\rho)_s - 1 \tag{5.16}$$

Comparing (5.9–10), (5.15–16) with (2.8), (2.10–11) we find complete equivalence of the two definitions of

$$\langle TN_a[A_{(\mu)_1}(x_1)\, A_{(\mu)_2}(x_2)]\, A(y_1) \cdots A(y_n)\rangle. \tag{5.17}$$

The Green's function (5.17) as defined by the power series expansions (3.4), (3.8) is identical with the corresponding time ordered function of the normal product (2.4).

This equivalence may be used to prove the existence of derivatives

$$\partial^{\xi}_{(\nu)} TN_a[A_{(\mu)_1}(x+\xi)\, A_{(\mu)_2}(x-\xi)]\, A(y_1) \cdots A(y_n) \tag{5.18}$$

According to (3.10) the limit of the right side of (2.17) exists for

$$\xi \to 0, \qquad \xi\mu/(|\, \xi^2\, |)^{1/2} \text{ bounded} \tag{5.19}$$

provided

$$2 + \#(\nu) + \#(\mu)_1 + \#(\mu)_2 \leqslant a. \tag{5.20}$$

Therefore, the limit (5.19) exists for all derivatives (5.18) which are of order

$$\#(\nu) \leqslant a - d. \tag{5.21}$$

This property of the normal products will be used in the following section for proving Wilson's asymptotic form of the short distance expansion.

6. Asymptotic Form of Wilson's Short Distance Expansion

The defining equation (2.4) of the normal product already has the form of a Wilson expansion

$$\begin{aligned} TA_{(\mu)_1}(x_1)\, A_{(\mu)_2}(x_2) &= \langle TA_{(\mu)_1}(\xi)\, A_{(\mu)_2}(-\xi)\rangle\, 1 \\ &\quad + \sum^a \frac{(-i)^{\sum \#(\rho)_s}}{r!\prod \#(\rho)_s\,!}\, G^{\{\rho\}}_{(\mu)_1(\mu)_2}(\xi)\, B_{(\rho)}(x) \\ &\quad + N_a[A_{(\mu)_1}(x_1)\, A_{(\mu)_2}(x_2)], \end{aligned} \tag{6.1}$$

$$x_1 = x + \xi, \qquad x_2 = x - \xi.$$

However, (6.1) is certainly not asymptotic since the remainder

$$N_a[A_{(\mu)_1}(x_1)\, A_{(\mu)_2}(x_2)]$$

approaches

$$B^{(a)}_{(\mu)_1(\mu)_2}(x) \qquad \text{for} \quad \xi \to 0$$

which in general does not vanish. In order to modify (6.1) we introduce a new kind of normal product

$$M_a[A_{(\mu)_1}(x_1)\, A_{(\mu)_2}(x_2)]$$

which will vanish for $x_j \to x$ with a sufficient number of derivatives provided the degree a is chosen large enough. We define

$$\begin{aligned} M_a[A_{(\mu)_1}(x_1)\, A_{(\mu)_2}(x_2)] &= (1 - t_\xi{}^c)\, N_a[A_{(\mu)_1}(x_1)\, A_{(\mu)_2}(x_2)], \\ x_1 = x + \xi, &\qquad x_2 = x - \xi, \\ c = a - 2 &- \#(\mu)_1 - \#(\mu)_2\,. \end{aligned} \tag{6.2}$$

This is well-defined since all ξ-derivatives involved exist according to (see (2.15))

$$\partial^\xi_{(\nu)} N_a[A_{(\mu)_1}(x_1)\, A_{(\mu)_2}(x_2)] = N_a \partial^\xi_{(\nu)}[A_{(\mu)_1}(x_1)\, A_{(\mu)_2}(x_2)] \tag{6.3}$$

which stays finite for $\xi \to 0$ provided

$$\#(\nu) \leqslant c. \tag{6.4}$$

Some examples of M-products of $A(x_1)\, A(x_2)$ are

$$M_{-1}[A(x_1)\, A(x_2)] = N_{-1}[A(x_1)\, A(x_2)] = TA(x_1)\, A(x_2) \tag{6.5}$$

$$M_0[A(x_1)\, A(x_2)] = N_0[A(x_1)\, A(x_2)] = {:}A(x_1)\, A(x_2){:} \tag{6.6}$$

$$M_1[A(x_1)\, A(x_2)] = N_1[A(x_1)\, A(x_2)] \tag{6.7}$$

$$M_2[A(x_1)\, A(x_2)] = N_2[A(x_1)\, A(x_2)] - N_2[A(x)^2] \tag{6.8}$$

$$M_3[A(x_1)\, A(x_2)] = N_3[A(x_1)\, A(x_2)] - N_3[A(x)^2] \tag{6.9}$$

$$\begin{aligned} M_4[A(x_1)\, A(x_2)] = N_4[A(x_1)\, A(x_2)] &- N_4[A(x)^2] \\ &- \xi^\mu \xi^\nu N_4[A_{\mu\nu}(x)\, A(x)] \\ &+ \xi^\mu \xi^\nu N_4[A_\mu(x)\, A_\nu(x)] \end{aligned} \tag{6.10}$$

In (6.9–10) we used

$$\partial_\nu^\xi N_a[A(x+\xi)\,A(x-\xi)]|_{\xi=0} = N_a\partial_\nu^\xi[A(x+\xi)\,A(x-\xi)]|_{\xi=0}$$
$$= N_a[A_\nu(x)\,A(x)]$$
$$- N_a[A(x)\,A_\nu(x)] = 0 \tag{6.11}$$

$$\partial_\mu^\xi\partial_\nu^\xi N_a[A(x+\xi)\,A(x-\xi)]|_{\xi=0}$$
$$= N_a\partial_\mu^\xi\partial_\nu^\xi[A(x+\xi)\,A(x-\xi)]|_{\xi=0}$$
$$= 2N_a[A_{\mu\nu}(x)\,A(x)] - 2N_a[A_\mu(x)\,A_\nu(x)]. \tag{6.12}$$

In order to discuss the behavior for $\xi \to 0$ we set

$$\xi = \rho\eta \tag{6.13}$$

and write

$$M_a[A_{(\mu)_1}(x_1)\,A_{(\mu)_2}(x_2)] = (1 - t_\rho{}^c)\,N_a[A_{(\mu)_1}(x_1)\,A_{(\mu)_2}(x_2)]$$
$$= (1 - t_\rho^{c-1})\,N_a[A_{(\mu)_1}(x_1)\,A_{(\mu)_2}(x_2)]$$
$$- \frac{\rho^c}{c!}\frac{\partial^c}{\partial\rho^c}\,N_a[A_{(\mu)_1}(x_1)\,A_{(\mu)_2}(x_2)]\Big|_{\rho=0} \tag{6.14}$$
$$x_1 = x + \rho\eta, \qquad x_2 = x - \rho\eta.$$

In perturbation theory the matrix element

$$(\Phi,\,N_a[A_{(\mu)_1}(x_1)\,A_{(\mu)_2}(x_2)]\psi) \tag{6.15}$$

between suitable state vectors exists as a continuous function of ρ (including $\rho = 0$) provided

$$a \geqslant 2 + \#(\mu)_1 + \#(\mu)_2 \qquad \text{and} \qquad \eta^2 \neq 0. \tag{6.16}$$

Likewise

$$\left(\Phi,\,N_a\frac{\partial^n}{\partial\rho^n}\,[A_{(\mu)_1}(x_1)\,A_{(\mu)_2}(x_2)]\psi\right)$$
$$x_1 = x + \rho\eta, \qquad x_2 = x - \rho\eta \tag{6.17}$$

is continuous in ρ provided

$$a \geqslant 2 + n + \#(\mu)_1 + \#(\mu)_2 \qquad \text{or} \qquad n \leqslant c. \tag{6.18}$$

Equation (6.3) implies

$$\frac{\partial^n}{\partial\rho^n}(\Phi, N_a[A_{(\mu)_1}(x_1)\, A_{(\mu)_2}(x_2)]\psi) = \left(\Phi, N_a \frac{\partial^n}{\partial\rho^n}[A_{(\mu)_1}(x_1)\, A_{(\mu)_2}(x_2)]\psi\right) \tag{6.19}$$

Hence (6.15) is continuous in ρ with all ρ-derivatives up to and including the order c. Hence

$$\begin{aligned}(1 - t_\rho^{c-1})&(\Phi, N_a[A_{(\mu)_1}(x_1)\, A_{(\mu)_2}(x_2)]\psi)\\ &= \frac{\rho^c}{c!}\frac{\partial^c}{\partial\sigma^c}(\Phi, N_a[A_{(\mu)_1}(x+\sigma\eta)\, A_{(\mu)_2}(x-\sigma\eta)]\psi)\Big|_{\sigma=\theta\rho}\end{aligned} \tag{6.20}$$

for some value θ with

$$0 \leqslant \theta \leqslant 1$$

(6.14) thus becomes

$$\begin{aligned}M_a&[A_{(\mu)_1}(x_1)\, A_{(\mu)_2}(x_2)]\\ &= \frac{\rho^c}{c!}\left(\Phi, N_a \frac{\partial^c}{\partial\sigma^c}[A_{(\mu)_1}(x+\sigma\eta)\, A_{(\mu)_2}(x-\sigma\eta)]\psi\right)\Big|_{\sigma=\theta\rho}\\ &\quad - \frac{\rho^c}{c!}\left(\Phi, N_a \frac{\partial^c}{\partial\sigma^c}[A_{(\mu)_1}(x+\sigma\eta)\, A_{(\mu)_2}(x-\sigma\eta)]\psi\right)\Big|_{\sigma=0}.\end{aligned} \tag{6.21}$$

The behavior of the M-products for $\xi \to 0$ is therefore given by

$$\lim_{\rho\to 0}\frac{M_a[A_{(\mu)_1}(x_1)\, A_{(\mu)_2}(x_2)]}{\rho^c} = 0 \tag{6.22}$$

provided

$$c = a - 2 - \#(\mu)_1 - \#(\mu)_2 \geqslant 0. \tag{6.23}$$

We will now derive an asymptotic form of the short distance expansion with the M-product as remainder. First we write (6.1) as recursion formula

$$\begin{aligned}N_b[A_{(\mu)_1}(x_1)\, A_{(\mu)_2}(x_2)] = \textstyle\sum_b^{b+1} &\frac{(-i)^{\Sigma\#(\rho)_j}}{r!\prod \#(\rho)_j\,!} G^{(\rho)}_{(\mu)_1(\mu)_2}(\xi)\, B_{(\rho)}(x)\\ &+ N_{b+1}[A_{(\mu)_1}(x_1)\, A_{(\mu)_2}(x_2)].\end{aligned} \tag{6.24}$$

Application of $(1 - t_\xi^{c})$ with

$$c = b - 2 - \#(\mu)_1 - \#(\mu)_2$$

to (6.24) yields

$$M_b[A_{(\mu)_1}(x_1)\, A_{(\mu)_2}(x_2)] = \sum_b^{b+1} \frac{(-i)^{\Sigma \#(\rho)_j}}{r!\prod \#(\rho)_j\,!} (1 - t_\xi^{c})\, G^{\{\rho\}}_{(\mu)_1(\mu)_2}(\xi)\, B_{(\rho)}(x)$$

$$+ M_{b+1}[A_{(\mu)_1}(x_1)\, A_{(\mu)_2}(x_2)]$$

$$+ \frac{\xi^{(\nu)}}{(c+1)!} N_{b+1}\partial^{\xi}_{(\nu)}[A_{(\mu)_1}(x_1)\, A_{(\mu)_2}(x_2)]\Big|_{\xi=0}$$

$$(\nu) = (\nu_1 ,\ldots, \nu_{c+1}). \tag{6.25}$$

Introducing

$$H^{\{\rho\}}_{(\mu)_1(\mu)_2}(\xi) = \langle TM_b[A_{(\mu)_1}(\xi)\, A_{(\mu)_2}(-\xi)]\, \tilde{A}^{(\rho)_1}(0) \cdots \tilde{A}^{(\rho)_r}(0)\rangle^{\text{prop}}$$

$$b = r + \sum \#(\rho)_j - 1 \tag{6.26}$$

(notation (2.11)) we obtain

$$(1 - t_\xi^{c})\, G^{\{\rho\}}_{(\mu)_1(\mu)_2}(\xi) = H^{\{\rho\}}_{(\mu)_1(\mu)_2}(\xi) \qquad \text{if} \quad r \neq 2 \tag{6.27}$$

and

$$(1 - t_\xi^{c})\, G^{(\rho)_1(\rho)_2}_{(\mu)_1(\mu)_2}(\xi) = H^{(\rho)_1(\rho)_2}_{(\mu)_1(\mu)_2}(\xi) - i^{\#(\rho)_1+\#(\rho)_2}(-1)^{\#(\mu)_2+\#(\rho)_2}\, \partial_{(\mu)_1}\xi^{(\rho)_1}\partial_{(\mu)_2}\xi^{(\rho)_2}$$

$$- i^{\#(\rho)_1+\#(\rho)_2}(-1)^{\#(\mu)_2+\#(\rho)_1}\, \partial_{(\mu)_1}\xi^{(\rho)_2}\partial_{(\mu)_2}\xi^{(\rho)_1}. \tag{6.28}$$

(6.25) then takes the form

$$M_b[A_{(\mu)_1}(x_1)\, A_{(\mu)_2}(x_2)] = \sum_{\substack{b\\(\rho)}}^{b+1} \frac{(-i)^{\Sigma \#(\rho)_j}}{r!\prod \#(\rho)_j\,!} H^{\{\rho\}}_{(\mu)_1(\mu)_2}(\xi)\, B_{(\rho)}(x)$$

$$+ M_{b+1}[A_{(\mu)_1}(x_1)\, A_{(\mu)_2}(x_2)]. \tag{6.29}$$

In deriving (6.29) we used the fact that the last term on the right side of (6.25) cancels the contributions from the polynomial expression in (6.28) because

$$\frac{\xi^{(\nu)}}{(c+1)!} N_{b+1} \partial^{\xi}_{(\nu)} [A_{(\mu)_1}(x_1) A_{(\mu)_2}(x_2)] \Big|_{\xi=0}$$

$$= \sum_{(\nu)_1 (\nu)_2} (-1)^{\#(\nu)_2} \frac{\xi^{(\nu)_1} \xi^{(\nu)_2}}{\#(\nu)_1! \, \#(\nu)_2!} N[\partial_{(\nu)_1} A_{(\mu)_1}(x) \, \partial_{(\nu)_2} A_{(\mu)_2}(x)]$$

$$= \sum_{(\rho)_1 (\rho)_2}{}_{b}^{b+1} (-1)^{\#(\rho)_2 + \#(\mu)_2} \frac{\partial_{(\mu)_1} \xi^{(\rho)_1} \partial_{(\mu)_2} \xi^{(\rho)_2}}{\#(\rho)_1! \, \#(\rho)_2!} B_{(\rho)_1 (\rho)_2}(x)$$

$$= \sum_{(\rho)_1 (\rho)_2}{}_{b}^{b+1} (-1)^{\#(\rho)_2 + \#(\mu)_2} \frac{\partial_{(\mu)_1} \xi^{(\rho)_1} \partial_{(\mu)_2} \xi^{(\rho)_2}}{2! \, \#(\rho)_1! \, \#(\rho)_2!} B_{(\rho)_1 (\rho)_2}(x)$$

$$+ \sum_{(\rho)_1 (\rho)_2}{}_{b}^{b+1} (-1)^{\#(\rho)_1 + \#(\mu)_2} \frac{\partial_{(\mu)_1} \xi^{(\rho)_2} \partial_{(\mu)_2} \xi^{(\rho)_1}}{2! \, \#(\rho)_1! \, \#(\rho)_2!} B_{(\rho)_1 (\rho)_2}(x). \tag{6.30}$$

The sum $\sum_{(\nu)_1 (\nu)_2}$ extends over all sets

$$(\nu)_1 = (\nu_{11}, \ldots), \qquad (\nu)_2 = (\nu_{21}, \ldots)$$

with

$$\#(\nu)_1 + \#(\nu)_2 = C + 1$$

(6.29) may equivalently be written as

$$TA_{(\mu)_1}(x_1) A_{(\mu)_2}(x_2) = \langle TA_{(\mu)_1}(\xi) A_{(\mu)_2}(-\xi) \rangle \cdot 1$$

$$+ \sum_{\{\rho\}}^{a} \frac{(-i)^{\sum \#(\rho)_j}}{r! \prod \#(\rho)_j!} H^{\{\rho\}}_{\{\mu\}}(\xi) + M_a[A_{(\mu)_1}(x_1) A_{(\mu)_2}(x_2)]. \tag{6.31}$$

By (6.22) the remainder vanishes stronger than any power for $\xi \to 0$, provided a is chosen large enough. Moreover,

$$\lim_{\rho \to 0} \frac{H^{\{\rho\}}_{\{\mu\}}(\rho\eta)}{\rho^n} = 0 \tag{6.32}$$

for

$$n = r - 3 + \sum \#(\rho)_j - \#(\mu)_1 - \#(\mu)_2 \geq 0.$$

Accordingly (6.31) represents Wilson's short distance expansion in asymptotic form.

References

1. W. Zimmermann, (1972), in press.
2. J. Lowenstein, *Phys. Rev. D* 4 (1971), 2281.
3. K. Wilson, Cornell Report (1964), unpublished; and *Phys. Rev.* 179 (1969), 1499.
4. R. Brandt, *Ann. Phys.* 44 (1967), 221; 52 (1969), 122; and *Fortschritte der Physik* 19 (1970), 249.
5. K. Symanzik, *Comm. Math. Phys.* 23 (1971), 49.
6. K. Wilson, private communication.

Printed by the St Catherine Press Ltd., Tempelhof 37, Bruges, Belgium.

Commun. math. Phys. 44, 73—86 (1975)

The Power Counting Theorem for Feynman Integrals with Massless Propagators

J. H. Lowenstein

Physics Department, New York University, New York, N. Y., USA

W. Zimmermann

Max-Planck-Institut für Physik und Astrophysik, München, Federal Republic of Germany

Received March 27, 1975

Abstract. Dyson's power counting theorem is extended to the case where some of the mass parameters vanish. Weinberg's ultraviolet convergence conditions are supplemented by infrared convergence conditions which combined are sufficient for the convergence of Feynman integrals.

1. Introduction

In the theory of renormalization Dyson's power counting theorem plays a decisive part [1–3]. The contribution of a proper Feynman diagram to a Green's function has the form

$$J=\int dk R(k,p)$$

$$R=\frac{P}{\prod_{j=1}^{n}(l_j^2-m_j^2+i\varepsilon(\vec{l}_j^2+m_j^2))^{n_j}} \tag{1.1}$$

where

$$\begin{aligned} &k=(k_1\ldots k_m), && p=(p_1\ldots p_N),\\ &k_j=(k_{j0}k_{j1}k_{j2}k_{j3}), && p_j=(p_{j0}p_{j1}p_{j2}p_{j3}),\\ &dk=dk_1\ldots dk_m, && dk_j\ldots dk_{j0}dk_{j1}dk_{j2}dk_{j3},\\ &m_j\geqq 0, \quad n_j>0. \end{aligned} \tag{1.2}$$

k_j and p_j are Minkowski vectors with the metric $(+1,-1,-1,-1)$. The vectors l_j are linear combinations

$$l_j=K_j(k)+P_j(p) \tag{1.3}$$

of the vectors $k_1,\ldots,k_m$ and $p_1,\ldots,p_N$ with $K_j\neq 0$. P is a polynomial in the components of k and p. The denominator of R is the common denominator of the unrenormalized integrand and the subtraction terms.

If all masses are non-zero Weinberg's version of the power counting theorem can be used to prove that the integral (1.1) is absolutely convergent provided the renormalized integrand R has been constructed according to Bogoliubov's subtraction rules [3, 4]. It can further be shown that the limit $\varepsilon\to+0$ exists as a covariant tempered distribution.

So far the power counting theorem has only been stated for non-vanishing masses. In the present paper Weinberg's ultraviolet convergence conditions are

P. Breitenlohner and D. Maison (Eds.): Proceedings 1998, LNP 558, pp. 310–323, 2000.

supplemented by infrared convergence conditions which will be shown to be sufficient for the convergence of integrals (1.1). The limit $\varepsilon \to +0$ [5] as well as the application to field theoretic models [6–9] are discussed in separate papers.

Our results are consistent with recent work by Bergère and Lam, as well as by Trute and Pohlmeyer, on the asymptotic behavior of parametrized Feynman integrals for small mass values [10–12].

Some general definitions are given in Section 2. Section 3 contains the statement and proof of the power counting theorem. The concept of reduced integrals, which is useful for some application of the theorem, is introduced in Section 4.

2. General Definitions

We consider integrals of the form (1.1). L denotes the space of the linear forms

$$l = \sum_{j=1}^{n} a_j k_j + \sum_{j=1}^{N} b_j p_j \tag{2.1}$$

which will be interpreted as inhomogeneous linear forms in the integration variables $k_1, \ldots, k_m$. Elements of L are called linerarly (in)dependent if their homogeneous parts (in k) are linearly (in)dependent. A set of elements in L is called a basis of L if their homogeneous parts form a basis for the space of the homogeneous forms in k.

We observe that always for an absolutely convergent integral (1.1) a basis

$$l_{j_1}, \ldots, l_{j_m} \tag{2.2}$$

exists consisting of linear forms which occur in the denominators of (1.1). Otherwise there would be at most $m' < m$ linearly independent forms

$$l_{j_1}, \ldots, l_{j_{m'}} \tag{2.3}$$

with the remaining l_j being linear combinations of vectors (2.3) and p_j. Extending (2.3) to a basis

$$l_{j_1}, \ldots, l_{j_{m'}}, w_1, \ldots, w_c \qquad m' + c = m$$

of L with Jacobian one (relative to $k_1, \ldots, k_m$) we find

$$J = \int dl_{j_1} \ldots dl_{j_{m'}} dw_1 \ldots dw_c R$$

with the divergent subintegral

$$\int dw_1 \ldots dw_c R = \frac{1}{\prod (l_j^2 - m_j^2 + \ldots)^{n_j}} \int dw_1 \ldots dw_c P .$$

Therefore, a basis (2.2) of L must always exist if the integral (1.1) is to be absolutely convergent.

For the formulation of the power counting theorem we will need certain subintegrals which we set up as follows. Let

$$u_1 = l_{i_1}, \ldots, u_a = l_{i_a}, v_1 = l_{j_1}, \ldots, v_b = l_{j_b} \tag{2.4}$$

be a basis of L with Jacobian one (relative to $k_1, \ldots, k_m$). Using (2.4) as new integration variables for (1.1) we obtain

$$\begin{aligned} &J = \int du\, dv\, R , \\ &u = (u_1 \ldots u_a), \qquad v = (v_1 \ldots v_b) , \\ &du = du_1 \ldots du_a , \qquad dv = dv_1 \ldots dv_b \end{aligned} \tag{2.5}$$

where P and l_j are expressed in terms of u, v and p through

$$k=k(u,v,p).$$

We consider a hyperplane H defined by the condition that the linear forms

$$v_1=l_{j_1},\ldots,v_b=l_{j_b}$$

have constant values. The subintegral of (1.1) along H is then given by

$$J(H)=\int du R. \tag{2.6}$$

We distinguish two different definitions for the dimension of a subintegral (2.6).

The upper dimension $\overline{\dim}$ refers to the behavior for large values of the integration variables. The lower dimension $\underline{\dim}$ refers to the behavior for small values of the integration variables. We define

$$\overline{\dim} J(H)=\overline{\deg}_u R+4a, \tag{2.7}$$

$$\underline{\dim} J(H)=\underline{\deg}_u R+4a. \tag{2.8}$$

The upper degree $\overline{\deg}_u$ (or lower degree $\underline{\deg}_u$) denotes the leading power of ϱ in the limit $\varrho\to\infty$ (or $\varrho\to 0$) if $u_j=\varrho\hat{u}_j$ is substituted into R. More precisely,

$$\bar{v}=\overline{\deg}_u R, \quad \underline{v}=\underline{\deg}_u R, \tag{2.9}$$

if

$$\lim_{\varrho\to\infty}\frac{R}{\varrho^{\bar{v}}}\neq 0,\infty, \quad \lim_{\varrho\to 0}\frac{R}{\varrho^{\underline{v}}}\neq 0,\infty \tag{2.10}$$

for almost all values of $u_1,\ldots,u_a$ and the remaining parameters $v_1,\ldots,v_b,p_1,\ldots,p_N$.

We quote some rules for the upper and lower degree. Let $N, D, F, F_1,\ldots,F_r$ be complex-valued functions of real four-vectors $u_1,\ldots,u_a, v_1,\ldots,v_b, p_1,\ldots,p_N$ to which the definitions $\overline{\deg}_u$ and $\underline{\deg}_u$ may be applied. Then the following rules hold

$$\overline{\deg}_u F^n=n\overline{\deg}_u F, \tag{2.11}$$

$$\underline{\deg}_u F^n=n\underline{\deg}_u F, \tag{2.12}$$

$$\overline{\deg}_u \frac{N}{D}=\overline{\deg}_u N-\overline{\deg}_u D, \tag{2.13}$$

$$\underline{\deg}_u \frac{N}{D}=\underline{\deg}_u N-\underline{\deg}_u D, \tag{2.14}$$

$$\overline{\deg}_u \prod_{j=1}^{r} F_j=\sum_{j=1}^{r}\overline{\deg}_u F_j, \tag{2.15}$$

$$\underline{\deg}_u \prod_{j=1}^{r} F_j=\sum_{j=1}^{r}\underline{\deg}_u F_j, \tag{2.16}$$

$$\overline{\deg}_u \sum_{j=1}^{r} F_j\leqq \max_j\{\overline{\deg}_u F_j\}, \tag{2.17}$$

$$\underline{\deg}_u \sum_{j=1}^{r} F_j\geqq \min_j\{\underline{\deg}_u F_j\}. \tag{2.18}$$

Let F be a polynomial of $u=(u_1,\ldots,u_a)$, $v=(v_1,\ldots,v_b)$ and $p=(p_1,\ldots,p_N)$ with vectors u_i, v_j, p_r. Then we may write

$$F=\sum_\alpha Q_\alpha M_\alpha, \quad Q_\alpha \not\equiv 0 \tag{2.19}$$

where M_α are independent monomials in u and Q_α are polynomials in v, p which are not identically zero. The upper and lower degrees of F are given by

$$\overline{\deg}_u F=\max_\alpha\{\deg M_\alpha\}, \tag{2.20}$$

$$\overline{\deg}_u F=\min_\alpha\{\deg M_\alpha\}. \tag{2.21}$$

3. Convergence Theorem

In this section the power counting theorem will be formulated for integrals J of type (1.1) assuming that a basis (2.2) of L can be formed. Weinberg's hypothesis of the power counting theorem may be stated as follows:

Ultraviolet Convergence Condition. *The inequality*

$$\overline{\dim} J(H)=\overline{\deg}_u R+4a<0 \tag{3.1}$$

holds for any basis (2.4) *and for any hyperplane H defined by constant values of* $v_1,\ldots,v_b$.

In particular, the upper dimension of the full integral J should be negative. Weinberg's condition (3.1) is sufficient for the absolute convergence of J provided all masses are different from zero

$$m_j>0, \quad j=1,\ldots,n.$$

In the general case we propose in addition the following

Infrared Convergence Condition. *The inequality*

$$\underline{\dim} J(H)=\underline{\deg}_u R+4a>0 \tag{3.2}$$

holds for any basis (2.4) *satisfying*

$$m_{i_1}=\ldots=m_{i_a}=0 \tag{3.3}$$

and for any hyperplane H defined by constant values of $v_1,\ldots,v_b$.

The ultraviolet and infrared convergence conditions combined form the hypothesis of the

Power Counting Theorem. *Let J be an integral of the form* (1.1) *for which a basis* (2.2) *of L can be formed. J is absolutely convergent if the ultraviolet convergence condition* (3.1) *and the infrared convergence condition* (3.2–3) *hold.*

Due to the inequality

$$\frac{l_0^2+\vec{l}^2+m^2}{|l_0^2-\vec{l}^2-m^2+i\varepsilon(\vec{l}^2+m^2)|}\leqq\left(1+\frac{4}{\varepsilon^2}\right)^{\frac{1}{2}}$$

the absolute convergence of (1.1) is implied by the absolute convergence of the corresponding Euclidean integral

$$J=\int dk R(k,p)$$

$$R=\frac{P}{\prod_{j=1}^{n}(l_j^2+m_j^2)^{n_j}} \tag{3.4}$$

where now

$$l_j^2=l_{j0}^2+\vec{l}_j^2 .$$

Therefore, we may restrict ourselves to proving the absolute convergence of (3.4) under the conditions (3.1–3).

We begin proving a lemma on the infrared convergence of certain integrals which are homogeneous in the integration variables.

Lemma. *Consider integrals of the form*

$$F=\int_{u_i^2\leq 1} du_1\ldots du_a\frac{M}{\prod_j (U_j^2)^{n_j}} \tag{3.5}$$

where the U_j are linear combinations of the Euclidean four-vectors $u_1,\ldots,u_a$ and M is a monomial in the components of $u_1,\ldots,u_a$. M may be factorized as

$$M=\prod_{i=1}^{a} M_i \tag{3.6}$$

where M_i is a monomial of u_i. For any subset

$$u_{i_1},\ldots,u_{i_c} \tag{3.7}$$

of the integration variables we form the integral

$$F_{i_1\ldots i_c}=\int_{u_i^2\leq 1} du_{i_1}\ldots du_{i_c}\frac{M_{i_1}\ldots M_{i_c}}{\prod_j^{i_1\ldots i_c}(U_j^2)^{n_j}} \tag{3.8}$$

where the product $\prod_j^{i_1\ldots i_c}$ extends over all U_j which are linear combinations of vectors (3.7) only. The integrals (3.8) are called sections of (3.5).

The statement is that the integral (3.5) is absolutely convergent if the dimension $d_{i_1\ldots i_c}$ of each section (3.8) is positive:

$$\dim F_{i_1\ldots i_c}=d_{i_1\ldots i_c}>0 . \tag{3.9}$$

This condition includes the dimension of the full integral which we denote by d,

$$d=\dim F=d_{1\ldots a}>0 .$$

Proof. We decompose the integral (3.5) into

$$F=\sum_P F_P$$

$$F_P=\int_{u_{i_1}^2\leq\ldots\leq u_{i_a}^2\leq 1} du_1\ldots du_a\frac{M}{\prod_j (U_j^2)^{n_j}} \tag{3.10}$$

with the sum extending over all permutations

$$P=\begin{pmatrix}1\dots a\\ i_1\dots i_a\end{pmatrix}.$$

We will check the convergence of each term F_P. In order to simplify the notation we rename the integration variables and monomials by

$$w_1=u_{i_1},\ \dots, w_a=u_{i_a},$$
$$N_1=M_{i_1},\dots,N_a=M_{i_a}.$$

Moreover, we denote the momenta U_j and exponents n_j of the denominators by

$$W_{11},\dots,W_{21},\dots,\dots,W_{a1},\dots$$
$$n_{11},\dots,n_{21},\dots,\dots,n_{a1},\dots$$

such that each W_{ij} is a linear combination of $w_1,\dots,w_i$ with non-vanishing coefficient of w_i,

$$W_{ij}=\sum_{i'=1}^{i} c_{iji'}w_{i'}, \qquad c_{iji}\neq 0.$$

In this notation F_P may be written in the form

$$F_P=\int_{w_1^2\leq 1} dw_1\frac{N_1}{\prod_j (W_{1j}^2)^{n_{1j}}}\int_{w_1^2\leq w_2^2\leq 1} dw_2\frac{N_2}{\prod_j (W_{2j}^2)^{n_{2j}}}\cdots\int_{w_{a-1}^2\leq w_a^2\leq 1} dw_a\frac{N_a}{\prod_j (W_{aj}^2)^{n_{aj}}}. \tag{3.11}$$

According to the hypothesis of the Lemma the dimension d_c of each section

$$F_c=F_{i_1\dots i_c}=\int_{w_1^2\leq 1} dw_1\frac{N_1}{\prod_j (W_{1j}^2)^{n_{1j}}}\cdots\int_{w_c^2\leq 1} dw_c\frac{N_c}{\prod_j (W_{cj}^2)^{n_{cj}}} \tag{3.12}$$

is positive,

$$\dim F_c=d_c>0.$$

d_c satisfies the recursion formula

$$d_c=4+\deg N_c-\deg\prod_j (W_{cj}^2)^{n_{cj}}+d_{c-1}. \tag{3.13}$$

We choose a number δ with

$$d_c>\delta>0 \quad \text{for} \quad c=1,\dots,a \tag{3.14}$$

and form the integral

$$G=\int_{w_1^2\leq 1} dw_1\frac{|N_1|}{\prod_j (W_{1j}^2)^{n_{1j}}}\int_{w_1^2\leq w_2^2\leq 1} dw_2\frac{|N_2|}{\prod_j (W_{2j}^2)^{n_{2j}}}\cdots\int_{w_{a-1}^2\leq w_a^2\leq 1} dw_a\frac{|N_a|}{|w_a|^{d-\delta}\prod_j (W_{aj}^2)^{n_{aj}}} \tag{3.15}$$

Since $|w_a|\leq 1$ and $d-\delta=d_a-\delta>0$ the integral F_P is majorized by G,

$$|F_P|\leq G. \tag{3.16}$$

We will prove the convergence of G by recursively estimating the integrals

$$G_c = \int\limits_{w_c^2 \leq w_{c+1}^2 \leq 1} dw_{c+1} \frac{|N_c|}{\prod_j (W_{c+1,j}^2)^{n_{c+1,j}}} \cdots \int\limits_{w_{a-1}^2 \leq w_a^2 \leq 1} dw_a \frac{|N_a|}{|w_a|^{d-\delta} \prod_j (W_{aj}^2)^{n_{aj}}} . \quad (3.17)$$

The dimension of

$$G_{a-1} = \int\limits_{w_{a-1}^2 \leq w_a^2 \leq 1} dw_a \frac{|N_a|}{|w_a|^{d-\delta} \prod_j (W_{aj}^2)^{n_{aj}}}$$

is [see Eq. (3.13)]

$$\dim G_{a-1} = 4 + \deg N_a - \deg \prod_j (W_{aj}^2)^{n_{aj}} + \delta - d .$$

Now, by a change of integration variable,

$$G_{a-1} = \frac{1}{|u_{a-1}|^{\delta_{a-1}-d}} \int\limits_{1 \leq w_a^2 \leq \frac{1}{w_{a-1}^2}} dw_a \frac{|N_a|}{|w_a|^{d-\delta} \prod_j W_{aj}^2} .$$

In the last line the limit $1/|u_{a-1}^2| \to \infty$ could be performed since the dimension of the integral is negative. Hence

$$G_{a-1} \leq \frac{\gamma_{a-1}}{|u_{a-1}|^{d_{a-1}-\delta}}$$

where γ_{a-1} is a constant.

Repeating this argument recursively we obtain

$$G_c \leq \frac{\gamma_c}{|w_c|^{d_c-\delta}}$$

by Eq. (3.13–14). Finally

$$G \leq \gamma \int\limits_{w_1^2 \leq 1} dw_1 \frac{|N_1|}{|w_1|^{d_1-\delta} \prod_j (W_{1j}^2)^{n_{1j}}} .$$

The integral on the right hand side exists since its dimension

$$\dim F_1 - d_1 + \delta = \delta > 0$$

is positive. By (3.16) each term of the decomposition (3.10) is absolutely convergent which implies the absolute convergence of (3.5). This completes the proof of the lemma.

We now turn to the

Proof of the Power Counting Theorem. Let S_0 be the set of all momenta l_j with $m_j = 0$. Let S be any subset

$$S \subsetneqq S_0$$

T denotes the complementary set

$$T = S_0 \setminus S .$$

We require that with a momentum l_j the set S should contain any l_i which satisfies

$$l_i^2 \equiv l_j^2, \quad m_i = 0.$$

We decompose the integral (3.4) into

$$J = \sum_S A_S \tag{3.18}$$

where

$$A_S = \int_{\substack{l_j^2 \leqq r^2 \text{ in } S \\ l_j^2 \geqq r^2 \text{ in } T}} dk \frac{P}{\prod_j (l_j^2 + m_j^2)^{n_j}}. \tag{3.19}$$

For studying A_S we select momentum vectors

$$u_1 = l_{i_1}, \ldots, u_a = l_{i_a} \tag{3.20}$$

in S which form a basis of S. Then $l_j \in S$ is a linear combination of $u_1, \ldots, p_1, \ldots,$

$$l_i = U_i + Q_i$$

$$U_i = \sum_{j=1}^{a} c_{ij} u_j, \quad Q_i = \sum_{j=1}^{N} d_{ij} p_j.$$

We say that S or the integral A_S has zero external momenta if $Q_i = 0$ for all $l_i \in S$.

For r small enough the term A_S vanishes unless all external momenta vanish. For the proof we observe that $u_\alpha^2 \leqq r^2$ since the u_α occur among the $l_i \in S$. The U_j are of the form

$$U_j = \sum_{\alpha=1}^{a} \eta_\alpha u_\alpha$$

where $|\eta_\alpha| \leqq \eta$ with η being characteristic number of the integral. Now

$$|Q_j| \leqq |l_j| + |U_j| \leqq (1+\eta) r$$

implies

$$r \geqq \frac{|Q_j|}{1+\eta} \quad \text{for any} \quad Q_j,$$

if the domain of integration is not empty. If at least one $Q_j \neq 0$ we may choose r such that

$$0 < r < \frac{|Q_j|}{1+\eta}. \tag{3.21}$$

But then the domain of integration is empty and $A_S = 0$. Hence for r small enough we find

$$J = \sum_S A_S \tag{3.22}$$

where S is restricted to those subsets for which $Q_j = 0$ for any $l_j \in S$.

In each integral A_S we introduce new variables of integration as follows. By adding suitable vectors

$$v_1 = l_{j_1}, \ldots, v_b = l_{j_b}, \quad a + b = m, \tag{3.23}$$

we extend (3.20) to a basis

$$u_1, \ldots, u_a, v_1, \ldots, v_b \tag{3.24}$$

of L with Jacobian one (relative to $k_1,\ldots,k_m$). Then each $l_j \in S$ is a linear combination of $u_1,\ldots,u_a$. The remaining l_j are linear combinations of $u_1,\ldots,u_a$, $v_1,\ldots,v_b$, $p_1,\ldots,p_N$. We next write the numerator P as a polynomial in u

$$\begin{aligned} &P = \sum C_{S\alpha} M_\alpha\,, \\ &M_\alpha = \prod_{i=1}^{a} M_{i\alpha_i}\,, \qquad \alpha = (\alpha_1,\ldots,\alpha_a)\,, \\ &M_{i\alpha_i} = u_{i0}^{\alpha_{i0}} u_{i1}^{\alpha_{i1}} u_{i2}^{\alpha_{i2}} u_{i3}^{\alpha_{i3}}\,, \qquad \alpha_i = (\alpha_{i0},\ldots,\alpha_{i3})\,. \end{aligned} \tag{3.25}$$

with the coefficients being polynomials in $v_1,\ldots,v_b$, $p_1,\ldots,p_N$. Then

$$\begin{aligned} &A_S = \sum_\alpha A_{S\alpha}\,, \\ &A_{S\alpha} = \int_{l_j^2 \leqq r^2 \text{ in } S} du \frac{M_\alpha}{\prod_S (l_j^2)^{n_j}} \int dv \frac{C_{S\alpha}}{\prod_T (l_j^2)^{n_j} \prod_U (l_j^2 + m_j^2)^{n_j}}\,, \\ &U = (l_1,\ldots,l_n) \setminus S_0\,. \end{aligned} \tag{3.26}$$

We now estimate the v-integrals

$$\int_{l_i^2 \geqq r^2 \text{ in } T} dv \frac{C_{S\alpha}}{\prod_T (l_j^2)^{n_j} \prod_U (l_j^2 + m_j^2)^{n_j}}\,. \tag{3.27}$$

To this end we consider the integral

$$\int dv \frac{P}{\prod_T (l_j^2 + M^2)^{n_j} \prod_U (l_j^2 + m_j^2)^{n_j}} \tag{3.28}$$

with $M > 0$. In (3.28) all masses are different from zero. Because of the ultraviolet convergence conditions the integral (3.28) is absolutely convergent. Each l_j in (3.28) is of the form

$$l_j = V_j(v) + R_j(u, p)\,.$$

Using

$$\frac{l_j^2 + m^2}{V_j^2 + m^2} \leqq 1 + \frac{|l_j - V_j|}{m} + \frac{|l_j - V_j|^2}{m^2}$$

we find

$$\begin{aligned} &\int dv \frac{|P|}{\prod_T (V_j^2 + M^2)^{n_j} \prod_U (V_j^2 + M^2)^{n_j}} \\ &\qquad \leqq c \int dv \frac{|P|}{\prod_T (l_j^2 + M^2)^{n_j} \prod_U (l_j^2 + M^2)^{n_j}}\,. \end{aligned}$$

Hence

$$\int dv \frac{P}{\prod_T (V_j^2 + M^2)^{n_j} \prod_U (V_j^2 + m_j^2)^{n_j}}$$

is absolutely convergent. The denominator does not depend on u while the numerator is a polynomial in u. Applying Lemma 3 of Ref. [13] we find that

$$\int dv \frac{C_\alpha(v,p)}{\prod_T (V_j^2+M^2)^{n_j} \prod_U (V_j^2+m_j^2)^{n_j}}$$

is absolutely convergent. With this we can estimate (3.27):

$$\int_{l_j^2 \geqq r^2} dv \frac{|C_{S\alpha}|}{\prod_T (l_j^2)^{n_j} \prod_U (l_j^2+m_j^2)^{n_j}}$$

$$\leqq \left(1+\frac{M^2}{r^2}\right)^r \int dv \frac{|C_\alpha|}{\prod_T (l_j^2+M^2)^{n_j} \prod_U (l_j^2+M^2)^{n_j}}$$

$$\left(\text{using} \frac{l_j^2+M^2}{l_j^2} \leqq 1+\frac{M^2}{r^2}\right)$$

$$\leqq d \int dv \frac{|C_\alpha|}{\prod_T (V_j^2+M^2)^{n_j} \prod_U (V_j^2+m_j^2)^{n_j}}$$

using

$$\frac{V_j^2+M_j^2}{l_j^2+M_j^2} \leqq 1+\frac{|l_j-V_j|}{M_j}+\frac{|l_j-V_j|^2}{M_j^2}$$

The integral of the last line only depends on p and the masses. Hence

$$|J| \leqq \sum_{S\alpha} D_{S\alpha} \int_{l_j^2 \leqq r^2} du \frac{|M_\alpha|}{\prod_S (l_j^2)^{n(j)}} . \tag{3.29}$$

The integrals on the right hand side can further be estimated by

$$\int_{l_j^2 \leqq r^2 \text{ in } S} du \frac{|M_\alpha|}{\prod_S (l_j^2)^{n_j}} = r^d \int_{l_j^2 \leqq 1 \text{ in } S} du \frac{|M_\alpha|}{\prod_S (l_j^2)^{n_j}}$$

$$\leqq r^d \int_{u_j^2 \geqq 1} du \frac{|M_\alpha|}{\prod_S (l_j^2)^{n_j}}$$

where d is the dimension of the integral. Thus

$$|J| \leqq \sum_{S\alpha} C_{S\alpha} \int_{u_j^2 \leqq 1} du \frac{|M_\alpha|}{\prod_S (l_j^2)^{n_j}} \tag{3.30}$$

with the sum restricted to those M_α which occur in (3.25) with non-vanishing coefficient. According to the Lemma (page 8 of this paper) we have convergence of the integrals on the right hand side if the dimension of any section is positive. In order to check the dimension d of the full integral

$$I = \int_{u_j^2 \leqq 1} du \frac{M_\alpha}{\prod_S (l_j^2)^{n_j}} \tag{3.31}$$

we form the subintegral

$$J(H) = \int du_1 \ldots du_a \frac{P}{\prod (l_j^2+m_j^2)^{n_j}} \tag{3.32}$$

of (3.4) along a hyperplane H defined by constant values of $v_1,\ldots,v_b$. By the hypothesis (3.2–3) the lower dimension δ of $J(H)$ is positive,

$$\delta=4a+\underline{\deg}_u P-\underline{\deg}_u \prod_j (l_j^2+m_j^2)^{n_j}>0\,. \tag{3.33}$$

This implies

$$\begin{aligned}0<\delta&=4a+\underline{\deg}_u P-\deg\textstyle\prod_S(l_j^2)^{n_j}-\underline{\deg}_u\prod_T(l_j^2)^{n_j}\\ &\leqq 4a+\underline{\deg}_u P-\deg\textstyle\prod_S(l_j^2)^{n_j}\\ &\leqq 4a+\deg M_\alpha-\deg\textstyle\prod_S(l_j^2)^{n_j}=d\,.\end{aligned}$$

Hence the dimension of (3.32) is positive.

We further have to verify that the dimension $d_{i_1\ldots i_c}$ of each section

$$I_{i_1\ldots i_c}=\int du_{i_1}\ldots du_{i_c}\frac{M_{\alpha i_1\ldots i_c}}{\prod_{i_1\ldots i_c}(l_j^2)^{n_j}} \tag{3.34}$$

is positive. Here $M_{\alpha i_1\ldots i_c}$ is the restriction of the product (3.25) to factors depending on $u_{i_1}\ldots u_{i_{\alpha'}}$. $\prod_{i_1\ldots i_c}$ denotes the product over all factors for which $l_j\in S$ is a linear combination of the vectors $u_{i_1}\ldots u_{i_c}$ only. Useful information is obtained by comparing the expansion

$$P=\sum C_\alpha M_\alpha \tag{3.35}$$

of P with respect to monomials M_α in $u_1,\ldots,u_a$ with the expansion

$$P=\sum C'_{\alpha'}M'_{\alpha'} \tag{3.36}$$

with respect to independent monomials $M'_{\alpha'}$ in $u_{i_1}\ldots u_{i_c}$ only. We know that $M_{\alpha i_1\ldots i_c}$ occurs as a factor of at least one monomial M_α with $C_\alpha\neq 0$. Since the monomials M_α are linearly independent the factor $C'_{\alpha'}$ of $M'_{\alpha'}=M_{\alpha i_1\ldots i_c}$ in (3.36) must also be different from zero. This implies the inequality

$$\underline{\deg}_{u'}P\leqq\deg M_{\alpha i_1\ldots i_c} \tag{3.37}$$

which will be crucial for the proof of the theorem. $\underline{\deg}_{u'}$ denotes the lower degree with respect to the variables $u'=(u_{i_1},\ldots,u_{i_c})$. We now form the subintegral

$$J(H')=\int du_{i_1}\ldots du_{i_c}\frac{P}{\prod_j(l_j^2+m_j^2)^{n_j}} \tag{3.38}$$

along a hyperplance H' defined by constant values of $v_1,\ldots,v_b$ and the momenta u_j which do not belong to u'. The lower dimension δ' of (3.38) is positive by hypothesis (3.2–3),

$$\begin{aligned}0<\delta'&=4c+\underline{\deg}_{u'}P-\underline{\deg}_{u'}\textstyle\prod_j(l_j^2+m_j^2)^{n_j}\,,\\ &\leqq 4c+\underline{\deg}_{u'}P-\deg\textstyle\prod_{i_1\ldots i_c}(l_j^2)^{n_j}\,.\end{aligned}$$

With (3.37)

$$\begin{aligned}d_{i_1\ldots i_c}&=4c+\deg M_{\alpha i_1\ldots i_c}-\deg\textstyle\prod_{i_1\ldots i_c}(l_j^2)^{n_j}\\ &\geqq 4c+\underline{\deg}_{u'}P-\deg\textstyle\prod_{i_1\ldots i_c}(l_j^2)^{n_j}\end{aligned}$$

follows. Hence the dimension $d_{i_1 \ldots i_c}$ of (3.38) is positive. According to the lemma each integral on the right hand side of (3.30) converges. This completes the proof of the theorem.

4. Reduced Integrals

In this section we discuss integrals of the form

$$I = \int dk \prod_{j=1}^{n} \Delta_j(l_j) \tag{4.1}$$

in the notation of (1.1) and

$$\Delta_j(l_j) = \frac{M_j}{(l_j^2 - m_j^2 + i\varepsilon(\vec{l}_j^2 + m_j^2))^{n_j}}, \qquad n_j > 0, \tag{4.2}$$

where M_j is a monomial in k and p. For integrals of this type we introduce the concept of the reduced integral. Let

$$S = (l_{j_1}, \ldots, l_{j_c}) \tag{4.3}$$

be any subset of the momenta $l_1, \ldots, l_n$. From the elements of S we select a basis, i.e. we choose linearly independent forms $u_1, \ldots, u_a$ of L such that each $l_j \in S$ is a linear combination of $u_1, \ldots, u_a$ and $p_1, \ldots, p_N$. With respect to S we form the reduced integral

$$\begin{aligned} &I_{\mathrm{red}}(S) \propto \int du_1 \ldots du_a \prod\nolimits_S \Delta_j(l_j) \\ &l_j = l_j(u, p), \qquad u = (u_1, \ldots, u_a) \end{aligned} \tag{4.4}$$

where the product $\prod_S$ extends over the $l_j \in S$ only. The reduced integral (4.4) is defined up to a factor which depends on the chosen basis.

Of special interest are reduced integrals of vanishing masses and vanishing external momenta, i.e.

$$m_j = 0 \quad \text{if} \quad l_j \in S,$$

$$l_j = l_j(u), \quad \text{independent of } p, \quad \text{if} \quad l_j \in S.$$

In this case each factor $\Delta_j(l_j)$ occurring in the reduced integral is homogeneous in u.

In case that (4.1) represents an unrenormalized Feynman integral the reduced integrals have a simple graphical interpretation: $I_{\mathrm{red}}(S)$ is the Feynman integral which corresponds to the reduced diagram $S' = S/T$ where all lines of T have been contracted to a point.

With the concept of the reduced integral we can give an equivalent formulation of the infrared convergence condition for integrals of type (4.1)[1]. Consider a basis

$$u_1 = l_{i_1}, \ldots, u_a = l_{i_a}, \qquad v_1 = l_{j_1}, \ldots, v_b = l_{j_b} \tag{4.5}$$

of L with

$$m_{i_1} = \ldots = m_{i_a} = 0. \tag{4.6}$$

[1] This formulation was used by Mack [14] to study infrared convergence of integrals like (4.1) in the context of conformally in variant theorems.

For any such basis the infrared convergence condition reads

$$\underline{\deg}_u \prod \Delta_j(l_j) + 4a > 0 . \tag{4.7}$$

We now form the reduced integral

$$I_{red}(S) \propto \int du \prod_S \Delta_j(l_j) \tag{4.8}$$

with respect to the set S of all momenta l_j with $m_j=0$ and $l_j=0$ at $u=0$. Then for any $l_j \notin S$ we have

$$\left.\begin{array}{l} m_j \neq 0 \\ \text{or} \quad l_j \neq 0 \quad \text{at} \quad u=0 \end{array}\right\} \quad \text{if} \quad l_j \notin S .$$

Therefore,

$$\underline{\deg}_u \prod \Delta_j(l_j) = \deg \prod_S \Delta_j(l_j) \tag{4.9}$$

and

$$\underline{\deg}_u \prod \Delta_j(l_j) + 4a = \dim I_{red}(S) . \tag{4.10}$$

Hence an equivalent formulation of the infrared convergence condition for the integral (4.1) is

$$\dim I_{red}(S) > 0 \tag{4.11}$$

for any set S of momenta l_j with

$$\left.\begin{array}{l} m_j = 0 \\ \text{and} \quad l_j = 0 \quad \text{at} \quad u=0 \end{array}\right\} \quad \text{if} \quad l_j \in S . \tag{4.12}$$

With this result, we are able to formulate the infrared convergence condition for an integral of the type [notation of (1.1) and (4.1)]

$$\int dk Q \prod_{j=1}^{n} \Delta_j(l_j) \tag{4.13}$$

where Q is a polynomial in k and p, in terms of a power counting criterion involving the formal integral

$$\int dk \prod_{S_0} \Delta_j(l_j) \tag{4.14}$$

where the product is restricted to the set S_0 of momenta with $m_j=0$. In particular, we have the following

Corollary to the Power Counting Theorem. *The integral* (4.13) *is absolutely convergent if the ultraviolet convergence condition* (3.1) *holds and if any reduced integral of* (4.14) *with vanishing external momenta has positive dimension.*

Proof.

$$\underline{\deg}_u Q \prod_{j=1}^{n} \Delta_j(l_j) = \underline{\deg}_u Q + \underline{\deg}_u \prod_{j=1}^{n} \Delta_j(l_j)$$

$$\geq \underline{\deg}_u \prod_{j=1}^{n} \Delta_j(l_j) .$$

Hence the infrared convergence condition of (4.13) is implied by that of (4.1). Any reduced integral of (4.1) with (4.12) is also a reduced integral of (4.14). This completes the proof.

References

1. Dyson, F. J.: Phys. Rev. **75**, 1736 (1949)
2. Weinberg, S.: Phys. Rev. **118**, 838 (1960)
3. Bogoliubov, N. N., Shirkov, D. V.: Introduction to the theory of quantized fields. New York: Interscience 1959
4. Gomes, M., Lowenstein, J., Zimmermann, W.: Commun. math. Phys. **39**, 81 (1974)
5. Lowenstein, J.: in preparation
6. Lowenstein, J., Zimmermann, W.: to be published
7. Clark, T.: to be published
8. Becchi, C.: in preparation
9. Lowenstein, J.: in preparation
10. Bergère, M., Lam, P.: Commun. math. Phys. **39**, 1 (1974) and to be published
11. Trute, H.: DESY preprint 74/44 (1974)
12. Pohlmeyer, K.: DESY preprint 74/36 (1974)
13. Hahn, Y., Zimmermann, W.: Commun. math. Phys. **10**, 330 (1968)
14. Mack, G.: In scale and conformal symmetry in hadron physics, ed. R. Gatto, New York: J. Wiley 1973

Communicated by K. Symanzik

J. H. Lowenstein
Physics Department
New York University
New York, N. Y., USA

W. Zimmermann
Max-Planck-Institut
für Physik und Astrophysik
D-8000 München 4
Föhringer Ring 6
Federal Republic of Germany